STEPHEN JAY GOULD

STEPHEN JAY GOULD

Reflections on His View of Life

EDITED BY

Warren D. Allmon

Patricia H. Kelley

Robert M. Ross

OXFORD
UNIVERSITY PRESS

2009

OXFORD
UNIVERSITY PRESS

Oxford University Press, Inc., publishes works that further
Oxford University's objective of excellence
in research, scholarship, and education.

Oxford   New York
Auckland   Cape Town   Dar es Salaam   Hong Kong   Karachi
Kuala Lumpur   Madrid   Melbourne   Mexico City   Nairobi
New Delhi   Shanghai   Taipei   Toronto

With offices in
Argentina   Austria   Brazil   Chile   Czech Republic   France   Greece
Guatemala   Hungary   Italy   Japan   Poland   Portugal   Singapore
South Korea   Switzerland   Thailand   Turkey   Ukraine   Vietnam

Copyright © 2009 by Oxford University Press, Inc.

Published by Oxford University Press, Inc.
198 Madison Avenue, New York, NY 10016

www.oup.com

Oxford is a registered trademark of Oxford University Press

Library of Congress Cataloging-in-Publication Data
Stephen Jay Gould : reflections on his view of life /
edited by Warren D. Allmon, Patricia H. Kelley, and Robert M. Ross.
p. cm.
Includes bibliographical references and index.
ISBN 978-0-19-537320-2
1. Gould, Stephen Jay. 2. Biology. 3. Natural history.
I. Allmon, Warren D. II. Kelley, Patricia H. III. Ross, Robert M.
QH303.2.S74 2009
570—dc22
2008010562

1 3 5 7 9 8 6 4 2

Printed in the United States of America
on acid-free paper

Contents

Editors' Preface

A teacher... can never tell where his influence stops.

—Henry Adams (1907, 300), used by Steve Gould
as an epigraph in *The Panda's Thumb*

Although Steve Gould's death on May 20, 2002, provided the immediate impetus for this book, its original motivation came from a review of his book *Structure of Evolutionary Theory*, published just before his death. That review—by someone who in our view clearly had no idea what punctuated equilibrium or species selection were about—suggested to us that Steve's science was even more widely misunderstood than we had thought. We said to each other at the time that someone needed to "do something" about this situation.

Steve's death took most of his students and close colleagues by surprise, although a few of us were aware that he had been ill. For many of us, it left a great hole in our lives. After his death and the several memorial services that followed, the three of us were asked to organize a symposium in Steve's memory at the annual meeting of the Geological Society of America, which convened on November 2, 2003. We invited students and close colleagues of Steve to participate in this symposium, asking each to explore an aspect of his thought from his or her own relatively "intimate" perspective—that is, from the point of view of one who had known well, learned under, and/or worked with him for many years. Our logic was that such people would be more likely to have a clearer-than-average understanding of his thought and its significance. (Not all of the papers presented at that session are included in this book, and a few that were not presented have been added. Two were originally published elsewhere and are reprinted here.)

Steve Gould was a major and highly influential intellectual figure in science (particularly evolutionary paleobiology) and society over a span of about thirty years of his professional life. Indeed, some assessments during his lifetime deemed him the best-known scientist in the world; what other scientist, after all, merited a guest appearance on the television cartoon *The Simpsons*? Due to his prominence, a small Gould commentary industry had already become established prior to his death (e.g., Somit and Peterson 1992; Selzer 1993; Sterelny 2001). Furthermore, because Steve published two books (*Structure* [2002c] and *I Have Landed* [2001m], his tenth volume of essays from *Natural History* magazine*) in the months just before his death, a number of major review/essays on his life and work appeared around that time, supplemented after his death by various memorials, thereby expanding this industry considerably and laying a foundation for what may well be a significant Gouldiana literature in the future. An "essential" compilation of his writings has recently appeared (McGarr and Rose 2006), as well as an extract from *Structure* (Gould 2007), and at least one major biography is in preparation.

Despite such attention, the present volume is the first (and so far only) book to explore critically Steve Gould's numerous and varied scientific and intellectual contributions, what the connections among them are, and what their long-term impact may be on our understanding of the history of life. It is not a conventional memorial festschrift; such has been published elsewhere (Vrba and Eldredge 2005). It is also not (to use Dick Lewontin's phrase) a "compendium of encomia," nor (as Steve might have said, using one of his favorite words) an attempt at hagiography. Instead, we hope that this book is an informed yet honest assessment of Steve's contributions within the scientific, intellectual, and societal contexts of the late twentieth century. In some sense it is intended as a "reader's guide" to Gould.

---

*Throughout this volume, citations to publications by Gould himself refer to the cumulative bibliography at the end of the book.

Steve's work was widely quoted and criticized, but—at least in our experience—much less often read thoroughly and carefully and still less frequently fully understood. We would like to think of the essays here as written by "those who knew him best," but this would be presumptuous. We do think we knew him and his thoughts well, or at least a bit better than did most other scientists, including many of his critics. As the chapters of this volume demonstrate, however, familiarity does not necessarily breed agreement. In any case, we wanted to provide what we hope will be some perspective and clarity that we fear might be lost from the scientific community's understanding of Steve's contributions. We wanted to have our say, before the critics and "picklocks of biographers" (Benet 1930) have had their way with his legacy.

Most of the contributors to this volume were Steve's students, to whom he was first and foremost a teacher and mentor. He was not always warm or gentle, or even friendly, to his students, but he valued and inspired excellence, hard work, and accomplishment, and he stretched all of us farther than we thought we could go. He was indifferent to many of the things that excited us (as we were to many of the things that excited him). He was a difficult role model. He decided quickly whom he did and didn't favor, and you usually didn't get a second chance to make a first impression. He didn't always come to our talks at meetings or read our papers. But he worked hard to find us jobs, and he was always very generous to each of us—with his time (when we made appointments), his money, and especially with his mind. For some of us, he was among the most important influences in our entire lives. For all of us, our professional and personal lives are emptier now without him, and we are extraordinarily grateful to have known him well and to have been under his tutelage.

Warren D. Allmon
Patricia H. Kelley
Robert M. Ross

References

Adams, H. B. 1907. *The education of Henry Adams.* Washington, DC: Adams. Reprinted by the Library of America, New York, 1983.

Benet, S. V. 1930. The army of northern Virginia. www.civilwarpoetry. org/confederate/officers/generals.html (accessed 1/26/08).

McGarr, P., and S. Rose, eds. 2006. *The richness of life. The essential Stephen Jay Gould.* London: Jonathan Cape. (Also published by W. W. Norton, New York, 2007)

Selzer, J., ed. 1993. *Understanding scientific prose.* Madison: University of Wisconsin Press.

Somit, A., and S. A. Peterson, eds. 1992. *The dynamics of evolution.* Ithaca, NY: Cornell University Press.

Sterelny, K. 2001. *Dawkins vs. Gould. Survival of the fittest.* Cambridge: Icon Books.

Vrba, E. S., and N. Eldredge, eds. 2005. Macroevolution: Diversity, disparity, contingency. Essays in honor of Stephen Jay Gould. Supplement to *Paleobiology* 31 (2).

Contributors

Warren Allmon is Director of the Paleontological Research Institution and Hunter R. Rawlings III Professor of Paleontology in the Department of Earth and Atmospheric Sciences at Cornell University in Ithaca, New York. He received his AB from Dartmouth College in 1982 and his PhD from Harvard University under Steve Gould's supervision in 1988.

Richard Bambach is Professor of Geology, Emeritus at Virginia Tech and a Research Associate in the Department of Paleobiology at the National Museum of Natural History, Smithsonian Institution in Washington, DC. He received his BA from Johns Hopkins University in 1957 and his PhD from Yale University in 1969. He was a close friend and colleague of Steve Gould's for more than thirty years and collaborated with him on many projects.

Robert Dorit is Associate Professor in the Department of Biological Sciences at Smith College in Northampton, Massachusetts. He received his undergraduate degree from Stanford University in 1979 and his PhD in biology under Steve Gould's supervision in 1986.

Dana Geary is Professor in the Department of Geology and Geophysics at the University of Wisconsin in Madison. She received her BA from the University of California at San Diego, a masters degree from the University of Colorado, and in 1986 her PhD from Harvard University under Steve Gould's supervision.

Linda Ivany is Associate Professor in the Department of Earth Sciences at Syracuse University in Syracuse, New York. She received her undergraduate degree from Syracuse, a masters degree from the University of Florida, and her PhD from Harvard University under Steve Gould's supervision in 1997.

Patricia Kelley is Professor in the Department of Earth Sciences at the University of North Carolina at Wilmington. She received her BA from College of Wooster in 1975 and her PhD from Harvard University under Steve Gould's supervision in 1979.

David Kendrick is Associate Professor in the Department of Geology at Hobart and William Smith Colleges in Geneva, New York. He received his BA from Yale University in 1986 and his PhD from Harvard University under Steve Gould's supervision in 1997.

Philip Kitcher is John Dewey Professor in the Department of Philosophy at Columbia University in New York City. He earned his BA in Mathematics/History and Philosophy of Science from Christ's College, Cambridge, in 1969, and his PhD in History and Philosophy of Science from Princeton University in 1974. He was a close friend and colleague of Steve Gould's for more than twenty years and collaborated with him on many projects.

Richard Levins is John Rock Professor of Population Sciences at the Harvard School of Public Health in Boston, Massachusetts. He studied agriculture and mathematics at Cornell, then was a tropical farmer in Puerto Rico before earning his PhD at Columbia University in 1965.

Richard Lewontin is Alexander Agassiz Research Professor in the Museum of Comparative Zoology at Harvard University. He received his undergraduate degree from Harvard in 1951 and his PhD from Columbia University in 1954. He was a close friend and colleague of Steve Gould's for more than thirty years, and collaborated with him on many projects.

Bruce Lieberman is Professor in the Department of Geology at the University of Kansas in Lawrence. He received his BA from Harvard University in 1988, completing a senior thesis under Steve Gould's supervision and his PhD from Columbia University in 1994 under the supervision of Niles Eldredge.

Paul Morris is Biodiversity Informatics Manager at The Harvard University Herbaria and The Museum of Comparative Zoology. He received his BS from Colgate University in 1984 and his PhD from Harvard under Steve Gould's supervision in 1991.

Robert Ross is Associate Director for Outreach at the Paleontological Research Institution in Ithaca, New York. He received his BA from Case Western Reserve University in Cleveland, Ohio, in 1984 and his PhD from Harvard University under Steve Gould's supervision in 1990.

Jill Schneiderman is Professor in the Department of Earth Science and Geography at Vassar College in Poughkeepsie, New York. She received her BA from Yale University in 1981 and her PhD from Harvard University in 1988. Although her graduate study was in metamorphic petrology, she worked closely with Steve as a Teaching Fellow in Steve Gould's large undergraduate course "History of the Earth and of Life."

Roger Thomas is John Williamson Nevin Professor in the Department of Geology at Franklin and Marshall College in Lancaster, Pennsylvania. He received his undergraduate degree from Imperial College of the University of London in 1963 and his PhD from Harvard University in 1971. Although he came to Harvard to study under Bernhard Kummel, he was the first PhD student in paleontology to graduate with Steve Gould on his graduate committee.

Margaret (Peg) Yacobucci is Associate Professor in the Department of Geology at Bowling Green State University in Bowling Green, Ohio. She received her BA from the University of Chicago in 1991 and her PhD from Harvard University under Steve Gould's supervision in 1999.

STEPHEN JAY GOULD

∾ 1 ∾

The Structure of Gould

Happenstance, Humanism, History, and the Unity of His View of Life

Warren D. Allmon

I. Introduction

Once, in responding to critics who had attempted to link his views on another topic to punctuated equilibrium, Steve Gould wrote, "I do have other interests, after all" (1982f, 88; see also 2002c, 1005). This was of course very true. Steve read, thought, traveled, talked, and wrote across a wide expanse of time, space, and subjects. He sang Bach and Gilbert and Sullivan; loved architecture, baseball, and numerical coincidences; collected beautiful old books; met with the pope about nuclear war; corresponded with Jimmy Carter about God; once appeared on a TV talk show as an expert on conjoined twins; and published technical papers on allometry, snails, Irish Elks, eurypterids, pelycosaurian reptiles, clams, receptaculitids, the history of paleontology, and human cranial capacity. Despite this breadth, however, one of the central facts of his professional life was that essentially all of his interests were, proximately or ultimately, interconnected in a unusually coherent and explicitly stated intellectual view, not only of the history of Earth and its life but also of the philosophy of science and the nature of human thought.

Steve said as much. He described himself as an "urchin in the storm" for what he called his "personal, stubborn consistency of viewpoint" (1987f, 11) and said that he regarded "the subject of worldviews, or paradigms," as essential "for the unification of all creative human thought..." (1995k, 104). In *The Structure of Evolutionary Theory* (2002c, especially 24–48), he laid out the connections between the various parts of his views,[1] and this did not go completely unnoticed by reviewers and commentators. Philosopher Michael Ruse (who seemed to understand Steve more than most critics), has described (1992, 1999) the connections among the several aspects of Steve's view of life, and after Steve's death, a few reviewers and eulogizers commented on the linkages within his distinctive world view (e.g., Durant 2002; Stearns 2002; Bradley 2004; York and Clark 2005).

By and large, however, critics and commentators have not delved deeply into the fundamental logic and interconnectedness of Steve Gould's oeuvre. This oversight is unfortunate because it is, in my view, only by understanding the internal structure and logic of the full swath of Steve's thinking and writing (as I suggest below, they're more or less the same thing) that we can fairly judge their utility and value as contributions to evolutionary theory and paleobiology, clearly the areas on which he wished to make his most lasting mark. If his ideas are atomized into their component parts, they can be too quickly judged and too easily discounted, misunderstood, or unfairly criticized.[2] It is only by connecting the conceptual dots among the various components that the potential value of his ideas can be evaluated fairly.

It is ironic that it is difficult for us to understand Steve's view of life, for perhaps more than any other scientist, he left us a roadmap to his thought. "Many scientists," comments David Hull, "possibly most scientists, just do science without thinking too much about it" (1999, 1131). Steve was not among them. He laid out not just the nature of his own biases and influences, but the nature of the biases and influences that must encumber all science. He was a tireless advocate for the view that science is an inescapably human activity, based in empirical observations of the natural world but never separable from human biases and preconceptions. His "favorite line" (1992o; 1995k, 147) was from a letter Charles Darwin wrote to Henry Fawcett in 1861: "How odd it is that anyone should not see that all observation must be for or against some view if it is to be of any service!" and he was constantly mentioning the tension between the subjective and objective sides of science. Some examples:

Scientists often strive for special status by claiming a unique form of "objectivity" inherent in a supposedly universal procedure called the scientific method. We can attain this objectivity by clearly the mind of all preconception and then simply seeing, in a pure and unfettered way, what nature presents. This image may be beguiling, but the claim is chimerical, and ultimately haughty and divisive. For the myth of pure perception raises scientists to a pinnacle above all other struggling intellectuals, who must remain mired in constraints of culture and psyche. (1992 o; 1995k, 148)

Since all discovery emerges from an interaction of mind with nature, thoughtful scientists must scrutinize the many biases that record our socialization, our moment in political and geographic history, even the limitations (if we can hope to comprehend them from within) imposed by a mental machinery jury-rigged in the immensity of evolution. (1995q; 1995l, 345)

An old tradition in science proclaims that changes in theory must be driven by observation. Since most scientists believe this simplistic formula, they assume that their own shifts in interpretation only record their better understanding of novel facts. Scientists therefore tend to be unaware of their own mental impositions upon the world's messy and ambiguous factuality. Such mental manipulations arise from a variety of sources, including psychological predisposition and social context. (2001m, 360–61)

Our ways of learning about the world are strongly influenced by the social preconceptions and biased modes of thinking that each scientist must apply to any problem. The stereotype of a fully rational and objective 'scientific method,' with individual scientists as logical (and interchangeable) robots, is self-serving mythology.... This messy and personal side of science should not be disparaged, or covered up, by scientists for two major reasons. First, scientists should proudly show this human face to display their kinship with all other modes of creative human thought.... Second, while biases and preferences often impede understanding, these mental idiosyncrasies may also serve as powerful, if quirky and personal, guides to solutions." (1995k, 93–94)

When we recognize that we do not derive our concepts of history only from the factual signals that scientific research has extracted from nature, but also from internal limits upon the logical and cognitive modes of human thought, then we can appreciate the complex interaction of mind and nature...that all great theories must embody...[the idea] that mind and nature always interact to build our basic concepts of natural order—becomes especially relevant in our current scientific age, where prevailing beliefs about the sources of knowledge lead us to downplay the role of the mind's organizing potentials and limits, and therefore encourage us to regard our theories of nature as products of objective observations alone." (2001m, 280)

Impartiality [in science] (even if desirable) is unattainable by human beings with inevitable backgrounds, needs, beliefs, and desires. It is dangerous for a scholar even to imagine that he might attain complete neutrality, for then one stops being vigilant about personal preferences and their influences—and then one truly falls victim to the dictates of prejudice. Objectivity must be operationally defined as fair treatment of data, not absence of preference. (1996j, 36)

Yet, even though he emphasized the cultural embeddedness of science, Steve was not a relativist or strict constructivist. He praised "the adamantine beauty of genuine and gloriously complex factuality" (2001m, 207), and stated his firm belief that "we have truly discovered—as a fact of the external world, not a preference of our psyches—that the earth revolves around the sun and that evolution happens" (1995k, 93). "Human thought," he observed, "unlike the evolution of life, does include the prospect of meaningful progress as a predictable outcome, especially in science where increasingly better understanding of an external reality can impose a fundamental organizing vector upon a historical process otherwise awash in quirks of individual personalities, and changing fashions of cultural preferences" (2002c, 591). In many respects, he said, "I remain an old-fashioned, unreconstructed scientific realist" (2002c, 969).

Steve, in other words, told us where scientific ideas in general—and his ideas in particular—came from. He assumed, however, that we were the "educated readers" whom he constantly strived to reach, and expected us to work a little bit to locate and grasp this roadmap—amid the more than 800 items in his personal bibliography (see page 335 of this volume) and/or within the 1,464 pages of *Structure* (2002c)—and most of us simply do not take the time to do so. As several commentators and reviewers have remarked (e.g.,

Orr 2002a; Wake 2002; Quammen 2003; Ayala 2005), it is tragic and ironic that his magnum opus—in which he really does lay all of this out and connect the dots—is so large and so baroquely written that few are likely to ever read it in full. *Structure* will, writes Stephen Stearns, "be bought more often than read and used as a bookend more often than as a book. Much of it deserves attention, some of it is exciting, and some of it is beautiful, but the gems are hard to locate amidst the sesquipedalian verbiage" (2002, 2339).

In short, I fear that Steve's ideas risk being discarded piecemeal or ignored in toto because there are just too many of them, and it is this fear, more than anything, that provokes this essay and also the organizing of this book. In this chapter, I attempt to extract and explicitly lay out the major connections among the components of Steve Gould's worldview. My analysis follows his advice to subject scientific texts to the same "textual analysis" as is common in the humanities (2002c, 521). I try to use his own approaches of "mini-biography" and "intellectual paleontology of ideas" (2001m, 5), which he used on so many other scientists, to elucidate why he came to the conclusions he did. Steve repeatedly railed against the "whig interpretation" of history and the "old style of condescension for an intellectual childhood to compare with our stunning maturity" (1995x; 1998x, 84; see also 1985r, 1991t, 1995p), in which "we commit the greatest of all historical errors: arrogantly judging our forebears in the light of modern knowledge perforce unavailable to them" (1998m, 2000k, 18). "The proper criterion [for judging someone's work]," he said, "must be worthiness by honorable standards of one's own time." (1993l, 186), and it is this perspective I try to take here.

More generally, because Steve was so conscious of these influences, his work is a rare and valuable opportunity to explore the internal and external dynamics of one scientist's effort to construct a coherent and comprehensive conception of natural science. Even though he famously became interested in paleontology at age five (when his father took him to the American Museum of Natural History), he also brought to his mature science a full set of personal beliefs, interests, and biases. As one tries to follow the coherence of his views, we can use his massive literary output to try to investigate to what degree these views may have come about because of, or been strongly affected by, nonscientific ideas. As he wrote in

Structure, "we do need to know why an author proceeded as he did if we wish to achieve our best understanding of his accomplishments, including the general worth of his conclusions" (2002c, 34).

A crucial element in this analysis (and, as he would undoubtedly have said, of productive scientific ideas in general) is that Steve ran his ideas out to their furthest logical limit, even if abundant empirical support was lacking. He referred to this phenomenon (in discussing the work of others) as the "overextension of exciting ideas" (2001m, 303; also 1997m, 326), and "the ultimate fallacy of claiming too much" (2002c, 667). Maynard Smith (1995) complained that when punctuated equilibrium was first put forward, "it was presented as just what one would expect to see if the orthodox view, that species often arise by rapid evolution in small peripheral populations, is indeed accurate. If only they [Eldredge and Gould] had left the argument there!" That they did not, however, is hardly surprising. Most, if not all, exciting new scientific ideas—from bacterial theories of disease to extraterrestrial impacts as causes of mass extinction—are rapidly applied (by their original authors or others) beyond their immediate beginnings. Indeed broad application and explanation of diverse phenomena is one measure of how useful a scientific theory is. In general defense of such extension of the theory of punctuated equilibrium in particular, Steve wrote, "proponents of punctuated equilibrium would become dull specialists if they did not take an interest in the different mechanisms responsible for similarities in the general features of stability and change across nature's varied domains, for science has always sought unity in this form of abstraction" (2002c, 765–66).

Neither this chapter nor this volume can claim to be a thorough analysis of Steve's thought. A minor "Gould industry," devoted to assessing his intellectual legacy, has already begun (e.g., Brown 1999; Ruse 1999; Morris 2001; Sterelny 2001; Orr 2002a; Shermer 2002; Grantham 2004; McShea 2004; York and Clark 2005; Sepkoski 2005; Lewontin 2008) and will, one hopes, continue; there is a posthumous "greatest hits" volume (McGarr and Rose 2006), and at least one major biography is in preparation. It is the fundamental point of this chapter (and most of the other contributions to the present volume), however, that these and future analyses of whether he was *right* must start with whether he *made sense.* As he put it: "Brilliance, of course, only implies cogency, not correctness" (2002c, 585). My main concern here is not just whether Steve's views are true but that we understand them.

Here I argue that virtually everything that Steve ever wrote—which by his own account was a very large proportion of what he thought[3]—fits into a very clear intellectual framework set by a relatively small number of basic ideas, and that the connections between them—historic and intellectual—were and are very clear, and we can understand them better by exploring that framework explicitly.

II. Steve's *Weltanschauung* and its Discontents

A. *His view of life*

What was this coherent worldview? What was Steve Gould's "view of life"? To my knowledge, even in all of his voluminous writing, Steve never answered this in one succinct statement. But if he had, I think it might go something like this:

> Life and its history—indeed all of history—are highly and irreducibly complex, and dominated in most cases by unpredictable events. Stability results from structure, which results from this complexity; direction results largely from "random" events and unexpected outcomes, superimposed on—and usually dominant over—patterns created by deterministic processes; patterns of stability, complexity, and history create an inherently hierarchical structure that can only be understood hierarchically; change is often abrupt, disruptive, and unforeseeable in its consequences; progress and improvement in any kind of general sense do occur occasionally, but are not characteristic of most systems or intervals of history. Human evolution has proceeded along these lines as well; we are noteworthy for our consciousness, but are otherwise no different from any other species on Earth. Because our hubris has almost always incorrectly placed us outside and above the rest of nature, much of science consists of adjusting (usually diminishing) human status in the universe. Most of the various fascinating consequences of human consciousness are emergent properties of our brain's complexity; flexibility, contingency, and nondetermination are the hallmarks of our—and all other—evolutionary history. Human values are derived from this highly complex and contingent phenomenon of consciousness, and cannot be properly read, determined, or proscribed by or from

any external reality or influence. Science is the best method that humans have so far invented to gain understanding of the natural world but, like all human endeavors, it is subject to human foibles which need always to be vigorously identified and countered if science is to progress.

He did, however, write a number of paragraphs from time to time that summed up much of this comprehensive view. Some examples:

> In our Darwinian traditions, we focus too narrowly on the adaptive nature of organic form, and too little on the quirks and oddities encoded into every animal by history. We are so overwhelmed—as well we should be—by the intricacy of aerodynamic optimality of a bird's wing, of by the uncannily precise mimicry of a dead leaf by a butterfly. We do not ask often enough why natural selection had homed in upon this *particular* optimum—and not another among a set of unrealized alternatives. In other words, we are dazzled by good design and therefore stop our inquiry too soon when we have answered, "How does this feature work so well?"—when we should be asking the historian's questions: "Why *this* and not *that?*" or "Why *this* over here, and *that* in a related creature living elsewhere?"...History's quirkiness, by populating the earth with a *variety* of *unpredictable* but sensible and well-working anatomical designs, does constitute the main fascination of evolution as a subject. (1994q; 1995k, 370–71)
>
> The course of evolution is only the summation of fortuitous contingencies, not a pathway with predictable directions.... [We should grasp] evolution as a process causally driven by struggle among individuals for reproductive success, and not by any principle working bountifully for the good of species or any other "higher" entity in nature. We may then view life's history as an unpredictable set of largely fortuitous, and eminently interruptible, excursions down highly contingent pathways. (1995s; 1995k, 332–33)

Both natural and human history were present in virtually every element of his work. Both of these spheres, in Steve's view, shared similar properties. Although both are subject to physical laws, both are histories and therefore constrained within the realm of the physically possible by what has gone before and subject to contingencies, the unexpected "quirks" of happenstance. As discussed above, our struggles to understand both kinds of history are linked via the necessity of human foibles intervening in our comprehension; both are pursued by fallible and fascinating human beings.

B. At the center of the view: Punctuated equilibrium

Although it has scarcely been mentioned in reviews and commentary, I think that the semi-autobiographical section of *Structure* (2002c, 745–1024; esp. 774 ff, 972 ff) in which Steve describes the origins, logic, criticism, and history of punctuated equilibrium (hereafter, "PE") is among the book's most valuable highlights. Perhaps more than any other part of this frequently difficult-to-read book, it deserves almost all of its parentheticals, asides, and footnotes. As he obviously intended at least in part (e.g., 2002c, 973), I expect that it will be of great value to future historians of science, and it is the section that I imagine I will be assigning most often to future students, because it contains anything that they could ever conceivably want to know and can get nowhere else. (It has now very usefully been reprinted as a separate paperback volume; Gould 2007.) Most of all, the section makes clear the intellectual and "structural" core of *Structure*, and therefore of his view of life.

Steve was surprisingly inconsistent in acknowledging the central place of PE in his worldview. As illustrated by the quote in the first paragraph of this essay, he occasionally objected that he had "other interests." I am, however, much more persuaded by the realization he attributes to his friend Oliver Sacks who, he writes, "saw the theory of punctuated equilibrium itself…as my coordinating centerpiece, and I would not deny this statement." PE, Steve continues "stands for a larger and coherent set of mostly iconoclastic concerns… [it] led to the reformulation proposed herein for the first branch of essential Darwinian logic…. these aspects of punctuated equilibrium strongly contributed to my developing critiques of adaptationism…my sources extended outward into a diverse and quirky network of concerns that seemed, to me and at first, isolated and uncoordinated, and that only later congealed into a coherent critique" (2002c, 37, 39–40).

PE was and is widely misunderstood, at least in part because it is both a narrow idea and also a platform for a much larger set of ideas, both a theory about how speciation looks in the fossil record and also the basis for a much larger conception of how evolution works. Yet these dual roles are logically connected and can be understood if one tries to do so. As "a theory about the deployment of speciation events in macroevolutionary time," Steve said, "punctuated equilibrium explains how the sensible intermediacy of human timescales can yield a punctuational pattern in geological perspective—thus requiring the treatment of species as evolutionary individuals, and precluding the explanation of trends and other macroevolutionary patterns as extrapolations of anagenesis within populations" (2002c, 755–56).

Similarly, the origins of PE lie both in the details of paleontology and in the wider intellectual and scientific worldview. It was (and still is) based fundamentally on empirical observations about the fossil record, but it is also an obvious part of a much wider intellectual controversy over the nature of change. PE was both a reflection of these influences outside of paleontology, and the conduit for introducing and integrating them into what had been a relatively insular field. In their retrospective of PE on its twenty-first birthday, Steve and Niles Eldredge noted that PE arose within and was part of a distinctive cultural and intellectual milieu; modern science, they argued "has massively substituted notions of indeterminacy, historical contingency, chaos and punctuation for previous convictions about gradual, progressive, predictable determinism. These transitions have occurred in field after field. Punctuated equilibrium, in this light, is only paleontology's contribution to a *Zeitgeist*" (1993j, 227). This view was magnified in *Structure*.

> Punctuated equilibrium represents just one localized contribution, from one level of one discipline, to a much broader punctuational paradigm about the nature of change—a worldview that may...be judged as a distinctive and important movement within the intellectual history of the later 20th century....For the punctuational paradigm encompasses much more than a loose and purely descriptive claim about phenotypes of pulsed change, but also embodies a set of convictions about how the structures and processes of nature must be organized across all scales and causes to yield this commonality of observed results. (2002c, 970)

Yet despite all of this apparent clarity, it is my disturbingly consistent observation that many of my colleagues, including perhaps a majority of professors teaching paleontology, historical geology, and evolutionary biology, appear genuinely to misunderstand PE—where it came from, what it says, and what it implies. Steve used to say that there were two works that everyone talked about but no one read—the Bible and the *Origin of Species*. To this list we might justly add Eldredge and Gould (1972e).[4]

I do not wish to repeat the history or evidence for or arguments about PE here (see Geary, this volume). I would, however, make four points, which I think are important for a more general understanding of Steve's world view.

(1) PE came from a desire to unite paleontology with evolution. The origin of PE was closely tied to the aspirations of two young graduate students to prod paleontology out of the largely lethargic state in which they found it in the 1960s. Niles Eldredge and Steve Gould wanted to be paleontologists, but they also wanted to study the process of evolution. They were both clearly bothered by paleontology's poor reputation and frequently cited *Nature*'s summary: "Scientists in general might be excused for assuming that most geologists are paleontologists and most paleontologists have staked out a square mile as their life's work. A revamping of the geologist's image is badly needed" (Anonymous 1969). Yes, there were exceptions (Gould and Eldredge's advisor Norman Newell was a prominent one), but most invertebrate paleontologists were not well versed or even particularly interested in evolution at the time of the formulation of PE. (Some of the twentieth century's greatest invertebrate paleontologists never did write anything substantive on evolution.)

Recalling the origins of PE, Steve wrote that he and Eldredge "had been particularly frustrated . . . with the difficulty of locating gradualistic sequences for applying these [statistical] techniques, and therefore for documenting 'evolution' as paleontological tradition then defined the term and activity. When I received [Tom] Schopf's invitation to talk on models of speciation [at the 1971 national meeting of the Geological Society of America], I felt that Eldredge's 1971 publication had presented the only new and interesting ideas on paleontological implications of the subject—so I asked Schopf if we could present the paper jointly. I wrote most of our 1972 paper, and I did coin the term PE—but the basic structure of the theory belongs to Eldredge" (2002c, 775). (See also Schopf [1981] for further details on the strikingly serendipitous origin of the 1972 paper.)

There is also another important factor to consider in tracing the origins of PE. As noted by Stearns (2002) and Orr (2002a), it is revealing that Steve (2001c, 967) says that Thomas Kuhn's *Structure of Scientific Revolutions* (1962) was among the most important influences on the development of PE, not just because it substantively describes a punctuated tempo of change in scientific theories, but also because it methodologically lays out a roadmap for *revolution* in scientific theories. Two smart young paleontologists saw an opportunity to shake up their field, to transform it, to shift its center of gravity from "handmaiden for geology" to the "high table of evolutionary biology." Although PE clearly was originally based on empirical patterns from the fossil record—and, in its initial formulation, proposes nothing beyond application of a particular theory of speciation to paleontological data—Gould and Eldredge quickly realized that it was also a logical basis for liberating paleontology from biostratigraphy, for an independent status of macroevolution as a subfield of evolutionary biology based in part on the unique contribution of their chosen field, paleontology. This is heady stuff, and in this context it can hardly be surprising that Gould and Eldredge sought to run PE out to its maximal logical extent.

As Steve puts it, he and Eldredge set out "to apply microevolutionary ideas about speciation to the data of the fossil record and the scale of geological time... to show how standard microevolutionary views about speciation, then unfamiliar to the great majority of working paleontologists, might help out our profession to interpret the history of life more adequately" (2002c, 775, 777–78). The theory's emphasis on morphological stasis was an "empowering switch" that "enabled paleontologists to cherish their basic data as adequate and revealing, rather than pitifully fragmentary and inevitably obfuscating." Paleontology could therefore "emerge from the intellectual sloth of debarment from theoretical insight imposed by poor data—a self-generated torpor that had confined the field to a descriptive role in documenting the actual pathways of life's history. Paleontology could now take a deserved and active place among the evolutionary sciences" (2002c, 778).

Understanding that it was among Gould and Eldredge's goals to use PE to "revolutionize" paleontology (and evolutionary biology) helps to account for much of the criticism that PE and its subsequent elaboration received, and much of Gould and Eldredge's response. As Steve frequently complained, critics variously claimed that PE wasn't true, wasn't original, or wasn't interesting, much less that it was revolutionary. This certainly must have touched a nerve in the two young would-be fire brands.

(2) Steve caused a lot of his own problems. Much of the criticism that PE received was (and is) unjustified, but some resulted from confusion sown by Steve himself. This was not, as has been claimed, because he was ducking and dodging, changing his views to fit whatever would work. It was largely because, as mentioned above, he rapidly ran PE to (and perhaps beyond) its logical extremes, and also because he used hyperbole and incendiary language, even when he should have known better.

Steve (2002c, 981–84) attributed much of the negative reception of PE to the media coverage of the 1980 Chicago Macroevolution Conference. Some of the press, he argues, connected disagreements over mechanism at the meeting to then-resurgent creationism, and "kindled the understandable wrath of orthodox Darwinians and champions of the Modern Synthesis" (2002c, 983). Yet Steve himself was responsible for at least some of the negative reception and in *Structure* he (perhaps a bit reluctantly) admits this. When he lists his and Eldredge's "own faults and failures," he says:

> critics can identify three sources of potential confusion that might legitimately be laid at our doorstep, and might have been prevented had our crystal ball been clearer...I did use some prose flourishes that, in a context of considerable suspicion and growing jealously, probably fanned the flames of confusion. Although I never stated anything unclearly, and committed no logical errors that could legitimately have inspired a resulting misreading, I should have toned down my style in a few crucial places....We may have sown some confusion by using partially overlapping terminology for a specific theory (punctuated equilibrium), and for the larger generality (punctuational styles of change) in which that theory lies embedded. But this taxonomic usage does stress a legitimate commonality that we wished to emphasize. (2002c, 1010–11)

Two statements in particular, made in papers in 1977 and 1980 as Gould and Eldredge were beginning to explore the wider implications of PE in earnest, came back to haunt Steve; these two statements became lightning rods and "sound bites" for critics, many of whom had never read or understood their original context.

(a) The Synthesis is "effectively dead." In a paper celebrating the fifth year of the journal *Paleobiology*, infamously titled "Is a new and general theory of evolution emerging?" Steve suggested that the Neodarwinian synthesis, "as a general proposition, is effectively dead" (1980c, 120). This provoked enormous criticism and a series of spirited specific rebuttals (see 2001c, 1004 for references). In *Structure*, he admits that, perhaps, he should have been a bit more circumspect:

> Given the furor provoked, I would probably tone down—but not change in content—the quotation that has come to haunt me in continual miscitation and misunderstanding by critics: "I have been reluctant to admit it—since beguiling is often forever—but if Mayr's characterization of the synthetic theory[5] is accurate, then that theory, as a general proposition, is effectively dead, despite its persistence as textbook orthodoxy" (Gould 1980, 120). (I guess I should have written the blander and more conventional "due for a major reassessment" or "now subject to critical scrutiny and revision," rather than "effectively dead."...Yes, the rhetoric was too strong (if only because I should have anticipated the emotional reaction that would then preclude careful reading of what I actually said). (2002c, 1007)

He protests, however, that

> Critics generally complete their misunderstanding of my 1980 paper [1980c] by first imagining that I proclaimed the total overthrow of Darwinism, and then supposing that I intended punctuated equilibrium as both the agent of destruction and the replacement. But punctuated equilibrium does not occupy a major, or even a prominent, place in my 1980 paper....I did speak extensively—often quite critically—about the reviled work of Richard Goldschmidt, particularly about aspects of his thought that might merit a rehearing. This material has often been confused with punctuated equilibrium by people who miss the crucial issue of scaling, and therefore regard all statements about rapidity at any

level as necessarily unitary, and necessarily flowing from punctuated equilibrium. In fact...my interest in Goldschmidt resides in issues bearing little relationship with punctuated equilibrium, but invested instead in developmental questions that prompted my first book [1977e].... The two subjects, after all, are quite separate, and rooted in different scales of rapidity...I do strive to avoid the label of *homo unius libri.*" (2002c, 1005)

Steve responds to Dennett's (1997) harsh criticism in much the same way. Dennett (1997) quoted from the infamous 1980 paper to support his claim that Steve had advocated for a "non-Darwinian saltation" as the "first step in the establishment of a new species." The passage quoted by Dennett: "Speciation is not always an extension of gradual, adaptive allelic substitution to greater effect, but may represent, as Goldschmidt argued, a different style of genetic change—rapid reorganization of the genome, perhaps non-adaptive" (Gould, 1980c, 119). Steve responds to what he calls Dennett's "pitiful" case by saying that "this quotation doesn't even refer to PE, but comes from a section of my 1980 paper on the microevolutionary mechanics of speciation" (2002c, 1009).

Yet despite his admission that his earlier rhetoric might have been a bit excessive and even confusing at times, it is striking that Steve continued even as late as 2002 to make exactly the same kinds of extreme statements. For example, he says that critics misinterpreted PE as having "something to say about evolution in general... [It doesn't,] for punctuated equilibrium only confirms all the beliefs and predictions of the Modern Synthesis" (2002c, 1000–1001). In a very narrow sense, this is correct, but both Gould and Eldredge clearly did (and Eldredge continues to; see, e.g., Eldredge 1995) think they had "something to say about evolution in general" and clearly implied that they thought what they "had to say" would, at least in part, transcend the Synthesis. Steve similarly claims that he never "made the Goldschmidtian link" (2002c: 1007), yet he so strongly implied it (in several places: 1977s; 1980v; 1982h) that only the most careful reader would have (at least initially) grasped his distinction.[6]

(b) **Marxism at his daddy's knee.** In their first major foray into exploring the wider implications of PE, Gould and Eldredge (1977; 1977c) included discussion of the cultural embeddedness of theory, contrasting the Victorian setting of Darwin's gradualism with other possible cultural settings of punctuational styles of change. They concluded with what became one of the most-repeated Gouldisms: "It may also not be irrelevant to our personal preferences that one of us learned his Marxism, literally at his daddy's knee." (Gould and Eldredge 1977, 146). This statement too, was subject to wide citation and criticism.

In *Structure*, Steve reflected on the decades of opprobrium this line engendered by reviewing in some detail his and Eldredge's structuring of the passage:

> I do not see how any careful reader could have missed the narrowly focused intent of the last section in our 1977 paper, a discussion of the central and unexceptionable principle, embraced by all professional historians of science, that theories must reflect a surrounding social and cultural context. We began the section by trying to identify the cultural roots of gradualism in larger beliefs of Victorian society…We couldn't then assert, with any pretense to fairness or openness to self-scrutiny, that gradualism represents cultural context, while our punctuational preferences only record unvarnished empirical truth….We therefore began by writing [p. 145] that "alternative conceptions of change have respectable pedigrees in philosophy." We then discussed the most obvious candidate in the history of Western thought: the Hegelian dialectic and its redefinition by Marx and Engels as a theory of revolutionary social change in human history….But the argument required one further step for full disclosure. We needed to say something about why we, rather than other paleontologists at other times, had developed the concept of punctuated equilibrium. We raised this point as sociological commentary about the *origin* of ideas, not as a scientific argument for the *validity* or the same ideas….So I mentioned a personal factor that probably predisposed me to openness towards, or at least an explicit awareness of, a punctuational alternative to conventional gradualistic models of change: "It may also not be irrelevant to our personal preferences that one of us learned his Marxism, literally at his daddy's knee."…I have often seen this statement quoted, always completely out of context, as supposed proof that I advanced punctuated equilibrium in order to foster a personal political agenda. I resent this absurd misreading. I spoke only about a fact of my intellectual ontogeny; I said nothing about my political beliefs (very different from my father's, by the way, and a private matter that I do not choose to discuss in this forum).

I included the line within a discussion of personal and cultural reasons that might predispose certain scientists towards consideration of punctuational models. . . . In the next paragraph, I stated my own personal conclusions about the general validity of punctuational change—but critics never quote these words, and only cite my father's postcranial anatomy out of context instead: "We emphatically do not assert the 'truth' of this alternative metaphysic of punctuational change. Any attempt to support the exclusive validity of such a monistic, a priori, grandiose notion would verge on the nonsensical." (2002c, 1018)

Fair enough, but I am also reminded that Steve often said that one should look at the core of an argument, not the fine points, to get at what they really think. For example, he responded to critics of PE who suggested that the theory contained nothing new by complaining about "the frequent grousing of strict Darwinians who often say something like: 'but we know all this, and I said so right here in the footnote to page 582 of my 1967 paper...'" (2002c, 1023). "General tenor," he said, "not occasional commentary, must be the criterion for judging a scientist's basic conceptions." (Gould and Eldredge 1993, 444). By this standard, it is I think safe to say that Steve wanted to push the comparison of PE to other punctuational ideas to the full extent possible, and he paid the price in criticism for occasionally going too far.

Steve's responses to criticism of these and other similarly inflammatory passages in his writing legitimately raise the question of how he could *not* have seen how potentially confusing such statements were. There are several possible explanations: (1) He did realize how provocative such statements would be and genuinely didn't care, and in fact intentionally intended to stimulate controversy. He did, after all, write that "iconoclasm always attracts me" (2001m, 369); (2) He made such statements unconsciously, later really did realize that he had made more than just a stylistic mistake, and "backpedaled hard" (Dennett 1995, 283–84); (3) He couldn't imagine that his readers wouldn't read carefully enough to understand the distinctions so clear in his own mind (and actually mostly there in what he wrote). I personally think it was some combination of the first and third of these. As I have already mentioned, Steve thought it was completely conventional and legitimate to rush to the logical boundaries with a new idea, test the theoretical limits, and then pull back where needed; he said as much (albeit sometimes in fine print) and simply assumed everyone would understand.

(3) The logic of PE (and its implications) is clear. Anyone putting ideas into any public forum opens the doors for potential criticism. The more ideas you put out there—and the more iconoclastic and provocative they are—the more you risk being criticized. In his writing and conversation, Steve grouped criticism of him into three categories. The first was correction of empirical or objective points, which he said he not only accepted but loved. "The factual correction of error," he wrote, "may be the most sublime event in intellectual life, the ultimate sign of our necessary obedience to a larger reality and our inability to construct the world according to our desires" (1993l, 452).

The second was simple, personal nastiness, in the form of willful misrepresentation and snide remarks, which he said was deeply hurtful to him. (One of my clearest memories as a graduate student is that he advised a group of us one day, as we were discussing a paper highly critical of him: "when you go out into the world, don't engage in this kind of *ad hominem* attack.") The unfairness of much of this criticism has been cited by others (e.g., Ruse 2000; Wagner 2002). Steve attributed much of this kind of commentary to "little more than complex fallout from professional jealousy, often unrecognized and therefore especially potent" from "our most negatively inclined colleagues" (2002c, 1000).

The third was criticism that resulted from (conscious or unconscious) misunderstanding of what he had tried to say. Although some of the most severe and high-profile criticism focused around Steve's critique of adaptationism, particularly in sociobiology and evolutionary psychology (e.g., Davis 1984; Dennett 1997; Pinker 1997; Wright 1999), these issues are in my view epiphenomenal on the core of Steve's view, which is the theory of PE.

In my two decades of teaching, reading the technical literature, going to scientific meetings, and encountering professional colleagues, no single phenomenon has impressed, puzzled, and frustrated me more (aside from creationism) than the misrepresentation and misunderstanding of PE and its larger evolutionary implications. Not being a professional logician limits my ability to level a coherent technical critique of the responses I have encountered. I can only say (in a statement that sounds so naive that I can hardly write it) that I simply cannot understand how something that appears to be so entirely logical to me can appear so otherwise to others.

This point is central to the argument and analysis of this essay. Putting empirics aside, the logical necessity of many if not most of the immediate implications of PE is compelling, if not indisputable, and this has been pointed out repeatedly by others (e.g., Hull 1980; Sober 1984; Lloyd 1988; Eldredge 1989; Vrba 1989; Lieberman 1995). The argument in its simplest form is as follows: If all or even most species in a clade are in stasis, then most evolutionary change in morphology is not occurring within species, and therefore must be occurring between species. If this is the case, trends must largely be the result of sorting among species, rather than extension and extrapolation of within-species anagenesis. This requires at least a modestly hierarchical view of the evolutionary process and an emphasis on speciation that the Modern Synthesis did not have. I don't see how it can be otherwise. PE, wrote Steve in summarizing this logic, "supplies the central argument for viewing species as effective Darwinian individuals at a relative frequency high enough to be regarded as general—thereby validating the level of species as a domain of evolutionary causality, and establishing the effectiveness and independence of macroevolution. Punctuated equilibrium makes its major contribution to evolutionary theory, not by revising microevolutionary mechanics, but by individuating species (and thereby establishing the basis for an independent theoretical domain of macroevolution... [This shift] ineluctably places much greater emphasis upon chance and contingency, rather than predictability by extrapolation." (2002c, 781–83) PE, Steve once said with succinctness, "leads to hierarchy, not saltationism" (1986a, 62). This argument says nothing about the validity of species selection, which I discuss further below.

Certainly the most quoted criticism of this argument—and of Gould as a scientist—came from John Maynard Smith (1995):

Gould occupies a rather curious position, particularly on his side of the Atlantic. Because of the excellence of his essays, he has come to be seen by non-biologists as the preeminent evolutionary theorist. In contrast, the evolutionary biologists with whom I have discussed his work tend to see him as a man whose ideas are so confused as to be hardly worth bothering with, but as one who should not be publicly criticized because he is at least on our side against the creationists. All this would not matter, were it not that he is giving non-biologists a largely false picture of the state of evolutionary theory.

Ironically, Maynard Smith (1984) is also the originator of the much-repeated line that paleontology is once again at the "high table" of evolution, largely as a result of PE and its derivatives (see, e.g., Eldredge 1995; Ruse and Sepkoski 2008).

Other examples of the same genre of criticism appear in comments by other distinguished authors. Dan McShay (2004), for example, admits that he has "never been able to understand why species selection requires punctuated equilibrium," but worries "that this is my own obtuseness, because Steve and others seem so sure of the connection" (2004, 48). In his overview of Steve's career, Allen Orr (2002a, 137) complains that "it's hard to see what species selection has to do with punctuated equilibrium anyway." In a still more cluelessly critical vein, Mark Ridley writes in his review of *Structure*:

> According to Gould, the theory of punctuated equilibrium implies that species are individuals, not classes. But I do not see the logical connection. Evolution in general, not punctuated equilibrium in particular, is the reason species do not form classes. If anything, the relative constancy of species after their sudden origin would make them more like a class...again, I do not see that species selection follows from either punctuated equilibrium or the individuality of species....Gould argues that punctuated equilibrium means that species are individuals and that the individuality of species enables species selection to operate. I have no problem with the three factual claims—of punctuated equilibrium, of the individuality of species, and of species selection. But I do not agree that the three are linked causally or conceptually. If they are not, Gould's system does not work. (2002, 11)

After years of attempting to rebut some of these critiques, Steve offered an analysis fully in line with his long-standing view of how science works. If smart people don't "get it," he said, then that is a sure sign that "it" is outside their conceptual worldview:

I have long faced a paradox in trying to understand why many intelligent critics seem unable to understand or acknowledge our reiter-
ated insistence that the radical claim of punctuated equilibrium lies
not in any proposal for revised microevolutionary mechanisms... but
rather at the level of macroevolution.... When smart people don't
"get it," one must conclude that the argument lies outside whatever
"conceptual space" they maintain for assessing novel ideas in a given
area. Many evolutionists, particularly those committed to the strict
Darwinism of unifocal causation at Darwin's own organismic level,
or below at the genic level, have never considered the hierarchical
model, and apparently maintain no conceptual space for the notion
of effective selection at higher levels. (2002c, 1013)

I think Steve was correct in this critique. After all, he himself
refers to hierarchy as the most difficult intellectual conundrum
he ever confronted (2002c, 598). Yet I think his analysis is incomplete. It is true that most of his critics did not understand (or even
try to understand) hierarchy, but they also did not grasp the other
ramets of his worldview, and how they cohered into an overarching
conception of nature.

III. Humanism

A. A humanistic naturalist

Although he did not say so frequently in his early work, by the end
of his career Steve often identified himself with the humanities and
the humanist perspective. For example, he wrote that he was "a
naturalist by profession, and a humanist at heart" (2001m, 396); "I
love, best of all," he said, "the sensitive and intelligent conjunction
of art and nature—not the domination of one by the other" (1998x,
2). "If any overarching theme pervades this body of writing," he
said about his *Natural History* essays, "I suppose that a groping effort
toward the formulation of a humanistic natural history must unite
the disparate" (1998x: 4). "I do love nature," he wrote, perhaps
somewhat defensively, "as fiercely as anyone who has ever taken up
a pen in her service. But I am even more fascinated by the complex
level of analysis just above and beyond... that is, the history of
how humans have learned to study and understand nature. I am
primarily a 'humanistic naturalist' in this crucial sense.... That is,
I am enthused by nature's constitution, but even more fascinated

by trying to grasp how an odd and excessively fragile instrument—
the human mind—comes to know this world outside, and how the
contingent history of the human body, personality, and society
impacts the pathways to this knowledge" (1998x, 5).

Harvey Blume (2002) noted that "Gould's science and literary
style owed more to art and artists than to algorithms." This humanistic interest had deep roots in his life. Steve's childhood was
clearly one in which books and culture mattered a great deal. He
says that he "shared the enormous benefit of a respect for learning
that pervades Jewish culture, even at the poorest economic levels"
(1999n, 8). These humanistic interests led him to become a double
major at Antioch College in geology and philosophy, which in turn
led him to the examination of two ideas that were to have major
implications for his later work—uniformitarianism and form. As
an undergraduate he wrote a paper on "Hume and uniformitarianism." "This work led me" he wrote, "to a more general analysis of
the potential validity of catastrophic claims, and particularly to an
understanding of how assumptions of gradualism had so stymied
and constrained our comprehension of the earth's much richer
history" (2002c, 44–45). Such thinking also clearly contributed to
his predilection for "punctuated" patterns of change.

Although Steve's love of formalism and structuralism clearly
had a basis in his empirical work, I think this interest was fundamentally based in his humanistic leanings. He loved D'Arcy
Thompson's book *Growth and Form* (1942) and wrote his senior
thesis on Thompson's theory of morphology (eventually published
as 1971b). To the end of his life he remained proud of his first
review article (1966c) on this subject, "written and published while
I was still a graduate student" (2002c, 42). As a direct result, Steve
then took up allometry for some of his first empirical and theoretical studies, fascinated by the problems of correlations of growth
and the resulting structural constraints (e.g., 1968b; 1969g; 1971c
and e). In 1970, he published a paper on form (1970c) that took a
strongly adaptationist approach, much to his later dismay (2002c,
41). Yet most of his work on form focused on what he would come
to call the formalist or structuralist perspective, or "laws of form"—
the notion that growth itself, like history, was a powerful channeler
of the potential directions of evolutionary change. Ultimately, this
interest in form was to lead to what may well end up being one of

his most lasting and influential scientific contributions, *Ontogeny and Phylogeny* (1977e), as well as the famous Spandrels paper with Richard Lewontin (1979k). "I read the great European structuralist literatures in writing my book on *Ontogeny and Phylogeny*," he said; "I don't see how anyone could read, from Goethe and Geoffroy down through Severtzov, Remane and Riedl, without developing some appreciation for the plausibility, or at least the sheer intellectual power, of morphological explanations outside the domain of Darwinian functionalism" (2002c, 43).

This interest in structure and form of course vastly transcended its humanistic origins in Steve's thought to become a central feature of his view of the evolutionary process. Rules of structure, he wrote, "deeper than natural selection itself, guarantee that complex features must bristle with multiple possibilities—and evolution wins its required flexibility thanks to messiness, redundancy, and lack of perfect fit" (1993l, 120). Indeed, the very possibility of future evolution—what he called "evolvability"—depended in large part on the nonselective side consequences of these structural rules. (See Thomas, this volume, for further discussion.)

Beyond these specific scientific themes, Steve's interest in humanism infused everything he did with a panoramic view that virtually required him to connect science with art, literature, and history. His insistence on examining the history and social setting of ideas was not merely an antiquarian exercise but rather central to his view of how humans think. "I cannot imagine a better test case for extracting the universals of human creativity," he said, "than the study of deep similarities in intellectual procedure between the arts and the sciences" (1999c; 2001m, 51). Writing about Vladimir Nabokov, Steve said that the lepidopterist-novelist "sought to...illustrate the inevitably paired components of any integrated view that could merit the label of our oldest and fondest dream of fulfillment—the biblical idea of 'wisdom'" (1999c; 2001m, 51–52).

Steve's humanistic interests connected directly to his view of how science works. Because science is a human activity, examination of the human origin of ideas, particularly the personal background of the thinkers who developed them, was essential to a more adequate understanding of the ideas themselves. Some examples:

Theory-free science makes about as much sense as value-free politics. Both terms are oxymoronic. All thinking about the natural
world must be informed by theory, whether or not we articulate our
preferred structure of explanation to ourselves....Moreover, theory
is always, and must be, colored by social and psychological biases of
surrounding culture; we have no access to utterly objective observation or universally unambiguous logic. (1993p; 1995l, 419–20)

Scientific progress depends more upon replacing theories than
adding observations (and waiting until they coalesce into a proper
explanation), and if all theories are bolstered by cultural biases,
then any process of replacement requires an unmasking of previous
structures. (1993p; 1995l, 420)

Creative science is always a mixture of facts and ideas. Great
thinkers are not those who can free their minds from cultural
baggage and think or observe objectively (for such a thing is impossible), but people who use their milieu creatively rather than as a
constraint....Such a conception of science not only validates the
study of history and the role of intellect—both subtly downgraded
if objective observation is the source of all good science. It also
puts science into culture and subverts the argument—advanced by
creationists and other modern Yahoos, but sometimes consciously
abetted by scientists—that science seeks to impose a new moral
order from without. (1981e; 1987f, 103)

Also in line with his humanism was Steve's strident advocacy of
interdisciplinariness, and his complaints about "the increasingly
rigid and self-policed boundaries" (2001m, 29) between academic
disciplines. (I was deeply disappointed when I left graduate school
for my first job and discovered that the rest of academia did not
share this commitment. Even now, when interdisciplinariness is
more on the lips of administrators than ever, in practice it faces
substantial obstacles in academic culture, and remains largely unen-
couraged and unrewarded.) He wrote numerous essays, as well as
four books, on the connection of the arts and humanities with
science (e.g., 1986m; 1992l; 1993h; 2000l). The master naturalist
and traveler Alexander von Humboldt (1769–1859), said Steve,

rightly emphasized the interaction of art and science in any
deep appreciation of nature...this vision may now be even more
important and relevant today....For never before have we been
surrounded with such confusion, such a drive to narrow special-

ization, and such indifference to the striving for connection and integration that defines the best in the humanist tradition. Artists dare not hold science in contempt, and scientists work in a moral and aesthetic desert...without art. Yet integration becomes more difficult to achieve than ever before, as jargons divide us and anti-intellectual movements sap our strength. (2001m, 108)

It is ironic, given his profound humanistic interest, that Steve found such personal and intellectual delight in the dethroning and diminution of human status, of smashing the idols of our hubris, of passionately arguing that the world was not only not made for humans, but that it did not care for us at all. One of his favorite and most-repeated quotes was from Freud (1935) (Steve usually abbreviated it [e.g., 1995k, 325; 1996d, 17; 2001m, 217]; it is given here in full):

Humanity has in the course of time had to endure from the hands of science two great outrages upon its naive self-love. The first was when it realized that our earth was not the center of the universe, but only a tiny speck in a world-system of a magnitude hardly conceivable; this is associated in our minds with the name of Copernicus, although Alexandrian doctrines taught something very similar. The second was when biological research robbed man of his peculiar privilege of having been specially created, and relegated him to a descent from the animal world, implying an ineradicable animal nature in him: this transvaluation has been accomplished in our own time upon the instigation of Charles Darwin, Wallace, and their predecessors, and not without the most violent opposition from their contemporaries. But man's craving for grandiosity is now suffering the third and most bitter blow from present-day psycho-logical research which is endeavoring to prove to the ego of each one of us that he is not even master in his own house, but that he must remain content with the veriest scraps of information about what is going on unconsciously in his own mind. We psycho-analysts were neither the first nor the only ones to propose to mankind that they should look inward; but it appears to be our lot to advocate it most insistently and to support it by empirical evidence which touches every man closely.[7]

This statement, Steve said, "suggests a criterion for judging the completion of scientific revolutions—namely, pedestal-smashing itself. Revolutions are not consummated when people accept the physical reconstruction of the universe thus implied, but when they grasp the meaning of this reconstruction for the demotion of human status in the cosmos" (1995s; 1995k, 325). This "pedestal-smashing" was an indelible and enduring element of Steve's thought and approach to intellectual life. The more disappointing to cherished human hopes an idea was, the more he liked it. In fact, he thought that an idea was truer *because* it was against our comfortable beliefs: "Most satisfying tales," he said, "are false" (1996o, 318). He did not find this attitude depressing in the least: "The deflation of hubris is blessedly positive, not cynically disabling" (2001m, 227). "The debunking of canonical legends...serves a vital scholarly purpose at the highest level of identifying and correcting some of the most serious pitfalls in human reasoning....we like to explain pattern in terms of directionality, and causation in terms of valor. The two central and essential components of any narrative—pattern and cause—therefore fall under the biasing rubric of our mental preferences" (2000o; 2001m, 55–56).

Nature, to Steve, was one of innate unpredictability and twists and turns, not just regularity dictated by physical law (which of course he accepted). The idea that order could be created by "blind" natural selection delighted him. "How delicious," he gushed, "to contemplate that these 'benevolent' results [good organic design and harmonious ecosystems] arise only as side consequences of a mechanism operating 'below' divine superintendence, and pursuing no 'goal' but the selfish propagation of individuals—that is, organisms struggling for personal reproductive success, and nothing else" (1992s; 1995k, 341).

Although many other taxonomies would be just as fruitful, I think of Steve's humanism as falling into four broad categories: writing, human equality, religion, and the role of a public intellectual.

B. The republic of letters: Essays, books, and the status of scientific writing

Many commentators have said that Steve was as much (or more) a writer as a scientist; many critics of his science have agreed. Indeed, most of his obituaries and memorials cited his popular writing as his signal achievement. If there was ever a scientist who demonstrated the truth of Canadian physician William Osler's observation that "In science, the credit goes to the man who convinces the world, not to the man to whom the idea first occurs" (Bean and Bean 1961, 112), it was Steve Gould. But Steve's life in letters was much more than his popularity as an essayist and best-selling "popular writer." His published legacy leaves us with a number of important unanswered questions about the nature of scientific writing. Some of these issues have already been subjected to textual analyses (e.g., Lyne and Howe 1986; Selzer 1993). Here I comment on three aspects of his writing: the connection between his popular and technical publications; the form of the essay; and his love affair with language and literature in general.

(1) **Popular vs. technical.** Steve said that he saw no distinction between his technical and popular writing, and intended his "popular" essays "for professionals and lay readers alike—an old tradition, by the way, in scientific writing from Galileo to Darwin, though effectively lost today." (1995k, xiv). He refused "to treat these essays as lesser, derivative, or dumbed-down versions of technical or scholarly writing for professional audiences." Rather, he said, he insisted on "viewing them as no different in conceptual depth (however distinct in language) from other genres of original research" (2001m, 6). Even one of Steve's harshest critics praised this feature of his writing: "he follows the admirable policy of writing at the same time for amateurs and professionals. I envy his ability to do this" (Maynard Smith, 1992).

There were strong similarities between the two kinds of work. For example, Steve noted that he had frequently presented in his "popular" essays "genuine discoveries, or at least distinctive interpretations, that would conventionally make their first appearance in a technical journal for professionals....I have frequently placed into these essays original findings that I regard as more important, or even more complex, than several items that I have initially published in conventional scholarly journals" (2001m, 6–7). Also, in both popular and technical work he was "most moved by general themes," but found them "vacuous unless rooted in some interesting particular" (1987f, 10).

Yet the intellectual connection that Steve perceived between his popular and scholarly writing has a more general and provocative meaning and implication that all scientists should consider further. Steve's published work spans a continuum from Op-Ed pieces that were simply fun or completely nonscientific to peer-reviewed taxonomic monographs and dense philosophical analyses of hierarchy. He perceived no bright line between popular and nonpopular work, and his erudition and prolixity allowed him (much to the dismay of some critics) to get away with it. Most other scientists probably couldn't (or wouldn't) try to copy this model. Yet in an age of proliferating blogs, self-published books, and online databases, not to mention exploding volumes of knowledge and discovery (reviewed and nonreviewed) in all fields, the nature of "published" work is rapidly changing. As a confirmed and proud luddite, Steve mostly did not work in such a world, but he stretched the bounds of scientific literature in his own distinctive way. It is worth considering whether his hybridized style of nonpeer-reviewed but still scholarly publication (exemplified most notably by his *Natural History* magazine essays) may be a viable genre for future "scientific" work. Steve clearly wanted his colleagues to cite these pieces as primary literature. (I have done so several times, because they contain scientifically valuable insights, ideas, opinions, and the occasional genuinely new empirical discovery that are unavailable elsewhere; e.g., Allmon 2007.) Steve's "popular" essays were, however, only very infrequently cited by scientists as primary sources in the technical evolutionary literature (Ruse 1999), much to Steve's displeasure:

> I confess that I have often been frustrated by the disinclinations, and sometimes the downright refusals, of some (in my judgement) overly parochial scholars who will not cite my essays (while they happily quote my technical articles) because the content did not see its first published light of day in a traditional, peer-reviewed publication for credentialed scholars. (2001m, 6–7)

(2) **The essay as scientific literature.** Steve reveled in the essay as a literary form, repeatedly pointing out its venerable origins in the work of Michel de Montaigne (1533–92), whose *Essays* (1580) "defined as crucial to the genre...ordinary things (with

deeper messages)" (1995k, ix). He noted that the word "essay" is derived from a French word meaning literally "try" or "attempt" (2001m, 9). Each of his three hundred essays for *Natural History*, for example, were based on "a gem of a detail [which] always sought to ground a generality" (1995k, xi). And he had a rich storehouse of such gems. "I cannot forget or expunge any item that enters my head,"[8] he said, "and I can always find legitimate and unforced connections among disparate details. In this sense, I am an essay machine; cite me a generality, and I will give you six tidbits of genuine illustration" (1995k, xi–xii). Maynard Smith (1992) summarized nicely what I have frequently thought after reading a Gould essay: "they often tell me something that I ought to have known but didn't." Steve said he kept up his remarkable streak of monthly essays without a break in large part because he learned from them; they were voyages of personal discovery (as when he rediscovered a volume of Edmund Burke on his shelf as part of researching an essay on women natural history writers: "If I didn't write these monthly essays, Burke would probably have stayed on my shelf until the day I died" [1995k, 197]).

Yet Steve's essays were not in the "conventional" mode of the "natural history essay," that is more-or-less straightforward descriptive celebrations of the beauty and wonder of nature. His "personal theory about popular writing in science," he said, divided natural history essays into two modes: "Galilean [in recognition of Galileo's writing his major works in the vernacular Italian rather than the elitist Latin], for intellectual essays about nature's puzzles, and Franciscan [after St. Francis of Assisi], for lyrical pieces about nature's beauty." "I am, he said, "an unrepentent Galilean. I work in a tradition extending from the master himself, to Thomas Henry Huxley in the last century, down to J. B. S Haldane and Peter Medawar in our own. I greatly admire Franciscan lyricism, but I don't know how to write in that mode" (1994t; 1995k, 10).

This preference was not just an issue of literary style. It was connected to the fact that he "always found the theory of how evolution works more fascinating than the realized pageant of its paleontological results" (2002c, 38), and for humanistic and intellectual issues over what he called the " 'wonderment of oddity' or 'strange ways of the beaver' tradition" of essay writing (1995u; 1998x, 394). "Sorry to be so disparaging," he added parenthetically after this revealing statement. "The stories are terrific. I just often yearn for more intellectual generality and less florid writing." "I would be an embarrassing flop in the Franciscan trade," he wrote elsewhere. "Poetic writing is the most dangerous of all genres because failures are so conspicuous, usually as the most ludicrous form of purple prose.... Cobblers should stick to their last and rationalists to their measured style" (1991a, 12–13). This style was also connected to his general lack of interest in ecology (see Allmon et al., this volume) and his long-standing critique of adaptationism and emphasis on contingency in evolution. "Nature writing in the lyrical mode," he said, "often exalts the apparent perfection and optimality of organic design. Yet...such a position plunges nature into a disabling paradox, historically speaking. If such perfection existed as a norm, you might revel and exult all the more, but for the tiny problem that nature wouldn't be here (at least in the form of complex organisms) if such optimality usually graced the products of evolution...optimality provokes wonder but provides no seeds for substantial change...Creativity in this sense demands slop and redundancy" (1990m; 1993l, 97–98).

Steve's writing style—in both his popular and technical writing—was an object of much praise and envy. Dust-jacket blurbs of his essay volumes lauded his "elegant prose," "wit and style," and "characteristically energetic, down-to-earth lucidity." Reviews cited glowingly the essays' "provocative and delightfully discursive" style (Wilford 1991) and Steve's "ability to astonish and amuse us" (Lehmann-Haupt 1980). Although he disagreed stridently with Steve's conclusions, even Richard Dawkins said he wrote well (e.g., 1990). John Updike (1985) observed that, although as Steve's career progressed he was "writing more lengthily but, my faint impression is, more felicitously." The essays were anthologized for undergraduate English classes in several colleges and universities (Palevitz 2002).

He was proud of being able to take technical scientific writing in directions it did not usually go: "for some perverse reason that I have never understood, editors of scientific journals have adopted several conventions that stifle good prose, albeit unintentionally—particularly the unrelenting passive voice required in descriptive sections, and often used throughout" (2000z, xii). In reviewing one of Steve's volumes of essays, Slobodkin (1988, 503) noted one of the characteristics that distinguished his writing: "in most scientific prose the author strives for clarity in the dual sense of expository simplicity and in making oneself transparent so that the empirical world is visible through the text but the peculiarities of the author are invisible.... The uniqueness of Steve is that he dances between us and his subject."

Yet clearly not everyone liked Steve's style. Although one reviewer praised *The Flamingo's Smile* by saying that the standard Gould essay was "so clear that any educated person can read it and understand" (Glass 1987, 426), another expressed about the same volume a view that perhaps best epitomizes Steve's later writing and how it was received by many scientists: "Graceful these essays are not—there are too many digressions and flat-footed reiterations, too little concern for pace and rhythm and economy and polish. For all the precision of his thought and research, his syntax and language are sometimes confoundingly imprecise" (Quammen 1985).

By the end of his career, Steve's style had unarguably become more elaborate, reaching an apotheosis in his final works. *Structure*, in particular, was criticized first and foremost for its size: an "elephantine opus" (Quammen 2003, 74), and "heavy enough for a stewardess to have insisted that I store it in an overhead compartment for takeoff and landing lest it endanger the passengers" (Stearns 2002, 2339). Reviewers complained about its "almost pathological logorrhea" (Ridley 2002) and "remarkably undisciplined prose" (Orr 2002a, 133), and for undergoing almost no editing or peer review (Monastersky 2002; Ayala 2005, 113). The writing, said a reviewer, "is sometimes so verbose, convoluted, and digressive that sentences have to reread in order to understand their content" (Zimmerman 2003, 454). "Such billowing clouds of verbal flatulence," opined an even less kind commentator, "herald a new phenomenon—the literate bioterrorist—or maybe a biologically literate deconstructionist, more interested in generating complex clauses than in communicating anything." The book, he continued, "is too verbose, too densely written, too bombastic and self-referential...and too long." and "stands as a monument to good, professional editing, which it didn't receive. Gould—who famously refused to allow any modification of his unique prose—got his way at the end, and his book is the worse for it" (Barash 2002, 284).

In February 1993 I was presented for the first time with the harrowing and thrilling opportunity of formally reviewing one of Steve's papers (eventually published as 1994f), and I too noted the inflated prose with mild disapproval: "Although the discursive style is fun to read and informative," I wrote in my review, "I note the severe space limitations and resulting publication delays that currently plague this journal, and regretfully suggest that the text can be shortened by perhaps 20–30% without serious damage to its scientific content." (Most of the text was, in the end, published as originally written.) Scott Wing, former editor of the journal *Paleobiology*, said that he accepted writing from Steve that he "wouldn't tolerate from others." Steve's prose, said Wing, was like Russian dolls, "with parenthetical remarks within parenthetical remarks within parenthetical remarks" (Monastersky 2002, A18).

(3) Literature for literature's sake. When I was a graduate student (1982–88), Steve was just beginning to collect rare books seriously, and he kept many of them in his office where they were available to us students. Eventually, as his collection grew, there was a lock on the cabinet, and finally the rare books disappeared entirely when he moved to New York City to his famed loft lined with bookshelves (see, e.g., Stephens 1997). Most media accounts of his book collecting made him sound like just another eccentric Harvard bibliophile and did not communicate the core of his interest in antiquarian books. He did love them for themselves, but he also used them as primary sources for his research. Books, he wrote, reiterating a point I frequently heard him make in conversation, "are the wellspring and focus of our lives as scholars" (1987f, 10). Many of his essays were based on old books he purchased, in which he discovered marginalia or other ephemera, or from which he made new observations that spurred an insight (e.g., 1990o, 1993v, 1995w, 1997r, 1998s, t, 1999y, 2000p, 2000u).

He was enormously proud that his essays were based mostly on primary sources (in the original languages). This was not just an antiquarian concern; his insistence on tracking down original sources in the history of science resulted in what he viewed as several significant discoveries (e.g., 1988j, 1993s). Beyond this, he defended his obsession on the grounds of general scholarly integrity:

Very few people, including authors willing to commit to paper, ever really read primary sources—certainly not in necessary depth and completion, and often not at all...yet another guarantee of authorial passivity before secondary sources, rather than active dialogue, or communion by study, with the great thinkers of our past. I stress this point primarily for a practical, even an ethical, reason....When writers close themselves off to the documents of scholarship, and rely only on seeing or asking, they become conduits and sieves rather than thinkers. When, on the other hand, you study the great works of predecessors engaged in the same struggle, you enter a dialogue with human history and the rich variety of our intellectual traditions. You insert yourself, and your own organizing powers, into this history—and you become an active agent, not merely a "reporter." Then, and only then, can you become an original contributor, even a discoverer, and not only a mouthpiece. (1998x, 6)

Steve's considerable ability with languages was a point of great personal pride; "at a time when so few Americans can deal in anything but English...," he said, "I can read the languages in which the main documents of evolutionary theory are written" (Monastersky 2002, A17; see also 2001c, 36).

C. They were despised and rejected: The fact of human equality

Steve clearly had a soft spot for the underdog, probably because he saw himself as one. He occasionally alluded in his essays to a childhood that included substantial abuse from his peers. For his childhood interest in dinosaurs, for example, he said he "was viewed as a nerd and misfit on that ultimate field of vocational decision—the school playground at recess. I was called 'Fossil Face'; the only other like-minded kid in the school [Richard Milner] became 'Dino'...The names weren't funny, and they hurt" (1995k: 222). Richard Milner recalls first meeting Steve when they were both twelve years old in the sixth grade in Queens. Milner described Gould as "a short, chubby, bright-eyed boy with a broad grin" and confirms that Steve hated his nickname, but said he accepted it "with good humor" (Milner 2002, 30). Another childhood acquaintance recalls Steve as a "chubby and somewhat awkward 14-year-old" (Mackler 2002). When I was a teaching assistant for him, I recall Steve objecting to our changing the grades of students who personally complained to us because he said he had always been too shy to do so and thought there were many like him.

Beyond his individual experience, as a Jew (albeit a secular one), Steve viewed himself as a member of a sometimes disparaged and maligned group, and the history of discrimination, anti-semitism, and immigration quotas was therefore very personal to him. His own ancestors had arrived from Hungary, Poland, and Russia during the first decade of the twentieth century, a fact that he frequently referred to in his writing. Referring to Henry H. Goddard (1866–1957), who argued for restrictions on immigration to the United States in the early twentieth century, Steve dedicated *Mismeasure of Man* (1981l, 1996j) "To the memory of Grammy and Papa Joe [his maternal grandparents], who came, struggled, and prospered, Mr. Goddard notwithstanding." He said that he wrote the book for reasons that "mixed the personal with the professional. I confess, first of all, to strong feelings on this particular issue. I grew up in a family with a tradition of participation in campaigns for social justice, and I was active, as a student, in the civil rights movement at a time of great excitement and success in the early 1960s" (1996j, 36).

Mismeasure was largely a critical success (but it also provoked enormous negative reaction; see, e.g., Jensen 1982; Carroll 1995; Rushton 1996, 1997) and Steve brought out a second edition with a new foreword (1996j), largely to respond to *The Bell Curve* (Herrnstein and Murray 1994). In a preview of the new edition, Steve made it clear what his ultimate fear about *The Bell Curve* was. The book, he said, presented an "apocalyptic vision of a society with a growing underclass permanently mired in the inevitable sloth of their low IQs. They will take over our city centers, keep having illegitimate babies (for many are too stupid to practice birth control), commit more crimes and ultimately require a kind of custodial state, more to keep them in check (and out of our high IQ neighborhoods)" (1994j).

The theme of human equality ran through many of his essays on human evolution, in which he pointed to "human equality as a contingent fact of history" (1984aa, 1997o) and the unreality of human races (1974r). These essays were also strong arguments for the a "bush" versus a "ladder" view of human evolution (see, e.g., 1976m, 1986a, 1987o, 1987p), and so a punctuational over a gradualistic view of evolutionary change, as well as the powerful role of contingency in evolution in general. This entire line of thought was also closely connected to his critique of adaptationism, biological determinism, sociobiology, and evolutionary psychology in human biology which was also in turn connected to his critique of gene selectionism, and thereby to his thinking on hierarchical theories of evolution.

Steve was much struck by the implications the "great chain of being" (Lovejoy 1936) for both nonhumans and humans; ranking of nonhuman nature, he argued, led inevitably to ranking of humans (e.g., 1981l, 1983x, 1983y). He loved the English writer Alexander Pope (1688–1744) but shivered at one of his passages in *Essay on Man* (1734),[9] and its echoing by others, such as a now-forgotten female popularizer of conchology, Mary Roberts: "To this splendid superstructure [wrote Roberts in 1834], nothing can be added; neither can any thing be taken from it, without producing a chasm in creation, which, however imperceptible to us, would materially affect the general harmony of nature. All things were made by Him, and without him cannot any thing subsist; besides, it seems as if he designed to teach us by the admirable arrangement of his creatures, that the different gradations in society are designed by his providence, and appointed for our good" (1993r; 1995k, 196).

Summarizing his views on what he saw as the long and sorrowful legacy of human discrimination, he wrote:

In many years of pondering over fallacious theories of biological determinism, and noting their extraordinary persistence and tendency to reemerge after presumed extirpation, I have been struck by a property that I call "surrogacy." Specific arguments raise a definite charge against a particular group—that Jews stink, that Irishmen drink, that women love mink, that Africans can't think—but each specific claim acts as a surrogate for any other. The general form of argument remains perennially the same, always permeated by identical fallacies over the centuries. Scratch the argument that women, by their biological nature, cannot be effective heads of state and you will uncover the same structure of false inference underlying someone else's claim that African Americans will never form a high percentage of the pool of Ph.D. candidates. (2001m, 352)

Steve wrote with particular passion when discussing anti-Semitism. At the end of an essay on an early interpretation of fossils that also included strong anti-Semitic statements, Steve wrote a compelling epitome of how his views of the relationship between science and nonscience connected to his views of human values:

> The improvement of knowledge cannot guarantee a corresponding growth of moral understanding and compassion—but we can never achieve a maximal spread of potential benevolence...without nurturing such knowledge. Thus the reinterpretation of jew stones as [fossil] sea urchin spines...can be correlated with a growing understanding that Jews, and all human groups, share an overwhelmingly common human nature beneath any superficiality of different skin colors or cultural traditions. And yet this advancing human knowledge cannot be directed toward its great capacity for benevolent use, and may actually (and perversely) promote increasing harm in misapplication, if we do not straighten out our moral compasses and beat all those swords....into plowshares, or whatever corresponding item of the new technology might best speed the gospel of peace and prosperity through better knowledge allied with wise application rooted in basic moral decency. (2001m, 174)

D. The fullness of life: The roles and status of religion and science

Steve described himself variously as "a humanist and non-theist" (1995k, 40); "a Jewish agnostic" (1998x, 270); "a paleontologist by training, and with abiding respect for religious traditions" (2001m, 214); "not, personally, a believer or a religious man in any sense of institutional commitment or practice" (1998x, 281); and "an agnostic in the wise sense of T. H. Huxley, who coined the word in identifying such open-minded skepticism as the only rational position because, truly, one cannot know" (1999n, 8–9). That is, he clearly did not believe in a personal God or deity, but he was also closely tied to his own Jewish heritage. He relates that he "had no formal religious education," not even a bar mitzvah, because his "parents had rebelled against a previously unquestioned family background." They "retained pride in Jewish history and heritage, while abandoning all theology and religious belief." "In my current judgment," he adds parenthetically, "they rebelled too far" (1999n, 8). [10] From an early age, Steve valued the cultural role of religion, but not its revealed or supernatural part. It is not possible, in my view, to understand Steve's views of religion and science unless one grasps his views on the source of human values and ethics, which in turn come out of this firmly Jewish and humanistic tradition.

At first glance, Steve appears to have shared completely the atheistic and materialist views of critics of religion such as Richard Dawkins and Daniel Dennett. (For example, he frequently said in lectures about the possibility of extraterrestrial life that astrobiology was similar to theology in that it was "a discipline with no subject matter.") He also repeatedly and stridently denied that science or nature (or, by implication, God) could be the source of human values or ethics, because almost *any* message can be (and has been) so derived: "answers to questions about ethical meaning cannot come from science" (1992w; 1995k, 75). "Nature simply is what she is," without any inherent moral or ethical message or signal for human life (2001m, 108–9). This philosophy was straight out of both Darwin and Enlightenment humanism: "When we stop demanding more than nature can logically provide...we liberate ourselves to look within" (2001m, 217–18).

> Our failure to discern a universal good does not record any lack of insight or ingenuity, but merely demonstrates that nature contains no moral messages framed in human terms. Morality is a subject for philosophers, theologians, students of the humanities, indeed for all thinking people. The answers will not be read passively from nature; they do not, and cannot, arise from the data of science. The factual state of the world does not teach us how we, with our powers for good and evil, should alter or preserve it in the most ethical manner....the answer to the ancient dilemma of why such cruelty (in our terms) exists in nature can only be that there isn't any answer—and that framing the question 'in our terms' is thoroughly inappropriate in a natural world neither made for us nor ruled by us. It just plain happens....If nature is nonmoral, then evolution cannot teach any ethical theory at all. (1982m; 1983d, 42–44)
>
> Once we recognize that the specification of morals and the search for a meaning in our lives cannot be resolved by scientific data in any case, then Darwin's variational mechanism will no longer seem threatening, and may even become liberating as a rationale for abandoning a chimerical search for the purpose of our lives, and the source of our ethical values, in the external workings of nature. (2001m, 248)

Yet Steve did not engage in the strident criticism of religion for which Dawkins and Dennett are well known (e.g., Dawkins 2006; Dennett 2006). Instead, he put forth what, on the surface, appeared to be a very different view, which he called "non-over-lapping magisteria," or NOMA (1997n, 1999n). This view held that science and religion occupy separate but equal realms of human endeavor, or magisteria, and neither could or should make claims on the other's legitimate domain of influence. "No scientific theory, including evolution," he argued, "can pose any threat to religion, for these two great tools of human understanding operate in complementary (not contrary) fashion in their totally separate realms: science as an inquiry about the factual state of the natural world, religion as a search for spiritual meaning and ethical values" (2001m, 214).

Steve clearly saw this as a very important social issue: "People of goodwill wish to see science and religion at peace, working together to enrich our practical and ethical lives" (1999n, 4). "[T]he myth of a war between science and religion remains all too current, and continues to impede a proper bonding and concilia-tion between these two utterly different and powerfully important institutions of human life. How can a war exist between two vital subjects with such different appropriate turfs—science as an enter-prise dedicated to discovering and explaining the factual basis of the empirical world, and religion as an examination of ethics and values?" (1994o; 1995k, 48–49). We need science to do what it does, he argued, but "We will also need—and just as much—the moral guidance and ennobling capacities of religion, the humani-ties, and the arts, for otherwise the dark side of our capacities will win, and humanity may perish in war and recrimination on a blighted planet" (2001m, 269).

> Science can supply information as input to a moral decision, but the ethical realm of "oughts" cannot be logically specified by the factual "is" of the natural world—the only aspect of reality that science can adjudicate....I win my right to engage moral issues by my member-ship in *Homo sapiens*—a right vested in absolutely every human being who has ever graced this earth, and a responsibility for all who are able. If we ever grasped this deepest sense of a truly universal community—the equal worth of all as members of a single entity,

the species *Homo sapiens*, whatever our individual misfortunes or disabilities—then Isaiah's vision could be realized, and our human wolves would dwell in peace with lambs, for "they shall not hurt nor destroy in all my holy mountain." We are freighted by heritage, both biological and cultural, granting us capacity both for infinite sweetness and unspeakable evil. What is morality but the struggle to harness the first and suppress the second? (1995k, 318)

NOMA, however, did not fare well among theologians or philosophers (see, e.g., Polkinghorn 1998; Haught 2000, 2003; Ruse 2000). The basic reason lay in Steve's definition of religion. In order to get religion to not conflict with science, said critics, Steve had to define religion in a way that excluded much of what religious people value, namely a caring God with supernatural powers. To make NOMA work, said theologian John Haught, for example, Steve had to "first reduce 'religion' to ethics" (2000, 25). Haught later elaborated on this critique: Steve could only reconcile science and religion, he said:

> by understanding religion in a way that most religious people themselves cannot countenance. Contrary to the nearly universal religious sense that religion puts us in touch with the true depths of the real, Steve denied by implication that religion can ever give us anything like reliable knowledge of *what is.* That is the job of science alone. As far as Steve was concerned, our religious ideas have nothing to do with objective reality. Scientific skeptics may appreciate religious literature, including the Bible, for its literary and poetic excellence. But they must remember that only science is equipped to give us factual knowledge. Doubters may enjoy passages of Scripture that move them aesthetically, or they may salvage from religious literature the moral insights of visionaries and prophets.... Still, Steve could not espouse the idea that religion in any sense gives us truth. No less than Dennett and Dawkins, when all is said and done, he too held that only science can be trusted to put us in touch with what is. At best, religion paints a coat of "value" over the otherwise valueless "facts" disclosed by science. Religion can enshroud reality with "meaning," but for Steve this meaning is not intrinsic to the universe "out there." It is our own creation. (Haught 2003, 6–7)

Some of these critics accused Steve, in his role as evolutionist laureate in the battle with creationism, of articulating NOMA in part to make evolution more palatable to what he knew was a largely religious American general public. Perhaps this was indeed part of his motivation. Even if it was, however, this charge largely misses the main source of the view that NOMA represents: human values were, for Steve, no less real and "out there" than rocks or snails, but they could not be reduced to or directly determined *by genes*. They were for him, like so many other aspects of human consciousness, emergent (and contingent) epiphenomena of the incredible complexity of the human brain. Just because values and ethics are "our own creation," this did not for Steve make them less real.

Ultimately, and ironically, NOMA failed because it was an attempt to do what Steve consistently criticized in others: make reality match our hopes. His family background and intellectual leanings made him a nonbeliever, but his cultural heritage imbued him with a deep and heartfelt appreciation of the value of non-revealed aspects of religion. His abiding humanism—perhaps combined with some (subliminal?) strategic spinning—compelled him to seek and find a personal reconciliation of science and religion, but the religion that he thought could coexist in such equality with science is a religion that few believers would accept (see Allmon 2009, for further discussion).

E. Intellectual adventures within ourselves: The role of the (public) intellectual

An important aspect of Steve's humanism was his self-conscious status as a scholar and (eventually a very public) intellectual (Lewontin 2008). His writing, particularly his popular essays, is filled with digressions and discursions about this topic. This self-appointed status made paleontologists both proud and embarrassed. Proud because, in many respects, Steve's conspicuous intellect brought out the best in us, encouraging us to be deeper scholars and to think about things in different ways. Sometimes, however, we were a bit reluctant to claim him because, although he was ours, he was sometimes, well, just a little much. Decades before his appearance on *The Simpsons*, for example, he was producing mixtures of admiration and dismay at our own professional meetings for his intellectual and rhetorical pyrotechnics. Two episodes in this category stand out in my memory. In 1985, in summing up a professional short course on mollusks before a standing-room-only crowd at a major geology meeting, he discoursed at length on hyaena penises (see 1985d). In 1989, after listening to him give a major talk on the reinterpretation of the Burgess Shale (which was then so obsessing him) to a packed hall at an international meeting of evolutionary biologists, a senior colleague turned to me and said, with a combination of affection and bewilderment, "Steve is a caricature of himself."

More substantively, Steve's prolific exploration of the qualities of scholarship both in others and in himself provides a fascinating (and inspiring) case study of both the opportunities and pitfalls of such a broad and anastomosing view of the world. In an essay on Goethe as scientist, Steve quoted the German polymath in a passage that is remarkably applicable to himself. Quoting his own translation of Goethe (1831), he wrote that "a man of lively intellect feels that he exists not for the public's sake, but for his own...every energetic man of talent has something universal in him, causing him to cast about here and there and to select his field of activity according to his own desire" (1993l, 155). "The truly awesome intellectuals in our history," he wrote in another essay, "have not merely made discoveries; they have woven variegated, but firm, tapestries of comprehensive coverage. The tapestries have various fates: Most burn or unravel in the footsteps of time and the fires of later discovery. But their glory lies in their integrity as unified structures of great complexity and broad implication" (1993l, 125). "Good scholars," he said, "struggle to understand the world in an integral way (pedants bite off tiny bits and worry them to death). These visions of reality...demand our respect, for they are an intellectual's only birthright. They are often entirely wrong and always flawed in serious ways, but they must be understood honorably and not subjected to mayhem by the excision of patches" (1993l, 136).

Being such a consciously public intellectual, for Steve, also came with solemn duties, and he tried to imbue in his students a strong sense of scholarly obligation. It was incumbent on each of us, he said, to be a generalist without being a dilettante; to connect one idea with another in a world filled with dissociated information and academic over-specialization; to understand that the history of ideas matters as much as the ideas themselves;

and that you can and should be a teacher and a researcher and a communicator to the public, and in fact to be less is not to meet your obligations as a scholar. Our internal intellectual adventures were to be shared with others, and this simply came with the territory:

> Our greatest intellectual adventures often occur within ourselves— not in the restless search for new facts and new objects on the earth or in the stars, but from a need to expunge old prejudices and build new conceptual structures. No hunt can promise a sweeter reward, a more admirable goal, than the excitement of thoroughly revised understanding—the inward journey that thrills real scholars and scares the bejesus out of the rest of us. (2001m, 355)

Taking stands on important issues was also part of being an intellectual for Steve, and he took public positions and campaigned actively on at least four such issues: human equality, creationism, textbooks, and natural history museums. I have already discussed his stands on human equality. Steve's public crusading against creationism was even more famous (e.g., 1981f, 1982k, 1987i, 1987t). Philosopher Michael Ruse frequently described his experience with Steve during the 1981 Little Rock creationism trial: "For me these recollections epitomize what Stephen Jay Gould was all about: First, that he was there at all—many other prominent figures, beginning with Carl Sagan, had been too busy to take time out to go down to the South and fight the creationists. But Steve felt it was his public duty, and he never gave it another thought" (Ruse 2003). Historian of creationism Barbara Forrest similarly said of Steve that it was remarkable that a "person as important in science as he was thought it was worthwhile to get involved" in fighting the creationists. "He lent his reputation to get the attention in the media," she continued. "He did what I wish more scientists would do" (Palevitz 2002).

Steve waged a similar, if less spectacular campaign against copying in textbooks (1988j, 1990m; see also O'Keefe 2002, xv). He called biology textbooks "the most impenetrable and permanent of all quasi-scientific literatures" (2001m, 310). Copying by textbook authors was not only damaging because it led to persistence of errors. It also clearly offended him and his pride in using primary sources. (He co-authored a textbook himself [1981c], but I heard him say several times that he didn't like the experience and would never write another.)

One campaign even closer to my own heart was Steve's championing of natural history museums. He admitted to "ambivalence...about the *Jurassic Park* phenomenon, and about dinomania in general" because it threatens to corrupt natural history museums with the promise of greater popularity and accompanying financial stability. "As a symbol of our dilemma," he observed:

> consider the plight of natural history museums in the light of commercial dinomania. In the past decade, nearly every major or minor natural history museum has succumbed (not always unwisely) to two great commercial temptations: to sell a plethora of scientifically worthless and often frivolous, or even degrading, dinosaur products by the bushel in their gift shops; and to mount, at high and separate admission charges, special exhibits of colorful robotic dinosaurs that move and growl but (so far as I have ever been able to judge) teach nothing of scientific value about these animals.[11] If you ask my colleagues in museum administration why they have permitted such incursions into their precious and limited spaces, they will reply that these robotic displays bring large crowds into the museum, mostly of people who otherwise never come. These folks can then be led or cajoled into viewing the regular exhibits, and the museum's primary mission of science education receives a giant boost. I cannot fault the logic of this argument, but I fear that my colleagues are expressing a wish or a hope, not an actual result, and not even an outcome actively pursued in most museums. (1995k, 235)

Steve consistently made an eloquent plea for natural history museums to do what they do best: to present and interpret authentic objects of nature. "It is our job," he said, speaking of natural historians and museum people:

> to stay whole, not to be swallowed in compromise, not to execute a pact of silence, or endorsement, for proffered payoff. The issue is more structural than ethical: we are small, though our ideas are powerful. If we merge without maintaining our distinctness, we are lost.... Our task is hopeless if museums, in following their essences and respecting authenticity, condemn themselves to marginality,

insolvency, and empty corridors. But fortunately, this need not and should not be our fate. We have an absolutely wonderful product to flog—real objects of nature.... Luckily—and I do not pretend to understand why—authenticity stirs the human soul. The appeal is cerebral and entirely conceptual, not at all visual. Casts and replicas are now sufficiently indistinguishable from the originals that no one but the most seasoned expert can possibly tell the difference.

Our success," he concluded, in words that warm the heart of every natural history museum director, "cannot be guaranteed, but we do have one powerful advantage, if we cleave to our essence as guardians of authenticity" (1995k, 234, 236–67).

IV. History

A. Why study history?

My profession, Steve said, referring to paleontology, "embodies one theme even more inclusive than evolution—the nature and meaning of history" (1985z, 18). History, he said, "must not be dismissed as a humanistic frill upon the adamantine solidity of 'real' science, but must be embraced as the coordinating context for any broad view of the logic and reasoning behind a subject so close to the bone of human concern as the science of life's nature and structure" (2002c, 46).

Whether it was from the human side, or from that accidental encounter with the *T. rex* when he was five, Steve was clearly interested in history from a very early point in his life. This led him to what he called his "first two scientific commitments"—paleontology and evolution. He was, however, not just taken by history as narrative but as a fundamental process, and not just by the history of life but also the history of human thought about that life. All of these together comprised a single but multistranded web of connections throughout his thinking.

Steve's interest in history was at least threefold. First was the history of science. More than almost anyone else in paleontology and evolutionary biology, he was fascinated by the history of these fields, and this interest was clearly assuming a larger proportion of his attention at the end of his life. While many practicing scientists turn to the history of their field late in their careers, Steve viewed the history of the discipline as an essential part of being an active practitioner within it, and he imbued his students with this view as well. I have frequently been struck since leaving graduate school by what an unusual view this is. Many, if not most, biologists and geologists know scarcely more about the history of their field than is contained in the obligatory first chapters of textbooks, and largely view the history of the field as a quaint antiquarian exercise. Steve, in contrast, saw it as central to good scholarship. Analysis of superseded world views, he argued, helps us to grasp the significance of the theories and ideas we now put forth. Examining the history of science, he said, allows us to see that smart people have struggled with issues that we might now think are solved. Historiography is thus an essential part of doing science today: "To unravel the archaeology of human knowledge, we must treat former systems of belief as valuable intellectual 'fossils,' offering insight about the human past, and providing precious access to a wider range of human theorizing only partly realized today" (2001m, 168).

Second, Steve was a relentless advocate for the intellectual value of the historical—as distinguished from the experimental—sciences. He argued that practitioners of fields such as paleontology, historical geology, evolutionary biology, and cosmology should never see themselves as pursuing less rigorous questions than students of more ahistorical fields such as physics or chemistry (e.g., 1986a, 1989d, 1994g, 1999b, 2001b).

> Historical science is not worse, more restricted, or less capable of achieving firm conclusions because experiment, prediction, and subsumption under invariant laws of nature do not represent its usual working methods. The sciences of history use a different mode of explanation, rooted in the comparative and observational richness of our data. We cannot see a past event directly, but science is usually based on inference, not unvarnished observation (you don't see electrons, gravity, or black holes either). (1989d, 279)
>
> The firm requirement for all science . . . lies in secure testability, not direct observation. . . . History's richness drives us to different methods of testing, but testability is our criterion as well. . . . We search for repeated pattern, shown by evidence so abundant and so diverse that no other coordinating interpretation could stand, even though any item, taken separately, would not provide conclusive proof. (1989d, 282)

The common epithet linking historical explanation with stamp collecting represents the classic arrogance of a field [physics] that does not understand the historian's attention to comparison among detailed particulars, all different.... The historical scientist focuses on detailed particulars... because their coordination and comparison permits us, by consilience of induction, to explain the past with as much confidence (if the evidence is good) as Luie Alvarez could ever muster for his asteroid by chemical measurement,... We shall never be able to appreciate the full range and meaning of science until we shatter the stereotype of ordering [different scientific fields] by status and understand the different forms of historical explanation as activities equal in merit to anything done by physics or chemistry. (1989d, 281)

The "lesser" status of historical science may be rejected on two grounds. First, it is not true that standard techniques of controlled experimentation, predictability, and repeatability cannot be applied to complex histories.... Nature... presents us with experiments aplenty, imperfectly controlled compared with the best laboratory standards, but having other virtues (temporal extent, for example) no t attainable with human designs. Second... [h]istory... is knowable in principle... testable, and different. We do not attempt to predict the future.... But we can postdict about the past—and do so all the time in historical science's most common use of repeatability.... Finally, history's richness drives us to different methods of testing, but testing (via postdiction) is our method as well. [Following Darwin, we look for a "concilience of inductions":]... types of evidence so numerous and so diverse that no other coordinating interpretation could stand—even though any item, taken separately, could not provide conclusive proof—must be the criterion for evolutionary inferrence. (1986a, 64–65)

This notion of the separate but equal status of historical science was put into practice in Steve's very successful course on the history of life at Harvard, which he taught for more than twenty years (see 1984g, and Ross, this volume).

Third, he was a tireless advocate for the importance of history itself as an essential element of the evolutionary process. History in his view was less the stately unfolding of a preordained or predictable course of events than a mostly unpredictable series of events that constrain (both positively and negatively) subsequent conditions and potential. All evolutionary biologists are taught that evolution is Markovian, with each step depending on the previous one, but Steve internalized and then promulgated this notion to an extraordinary degree. History was for him virtually a thing, a force, like gravity. "History matters," he was fond of saying. By this he meant that history as sequence of events bestows on its products an inescapable (but largely unpredictable) legacy. It was this flow and power of history as a process that perhaps led him to focus more on how evolution works than on the specific organisms it produces (the "theory rather than the pageant" [2002c, 38]).

This interest in the importance of history in evolution was closely tied to Steve's critique of adaptationism and to his emphasis on imperfection and exaptation as sources of raw material for "evolvability." If natural selection was all-powerful, he argued, it would build whatever phenotype was required in an optimal way for local circumstances, and history would *not* matter. This, he held, was exactly what the extreme Darwinian selectionist position posited: "The most common denial of history made by self-styled Darwinian evolutionists resides in claims for optimality—conventionally for the mechanics of morphology, more recently for behavior and ecology" (1986a, 66). His interest in history was also the explanation for his admiration of French paleontologist Louis Dollo (1857–1931), famous for "Dollo's Law" of irreversibility in evolution. "Irreversibility" Steve said, was a profound "signature of history" (1993l, 92). He called Dollo one of his intellectual heroes, and maintained an active interest in his ideas throughout his career (e.g., 1970e, 1994e).

B. Ladders and bushes: The critique of progress

The "most fundamental question in palaeontology," Steve said, is "does the history of life have an intrinsic direction (toward greater morphological complexity, increased diversity, etc.)?" (1976c, 231; see also 1977b); that is, is there *progress?* It is hard to pick just one theme that Steve thought was more important than any other, but if one must choose, it would have to be the issue of progress in evolution. His view was unmistakable: "Progress is a noxious, culturally embedded, untestable, nonoperational, intractable idea that must be replaced if we wish to understand the patterns of history" (1988g, 319; see also 1996d).

Steve's critique of progress united many strands of his thought. The morphological stasis of PE implied a more limited role for conventional natural selection than the Modern Synthesis had suggested; combined with his interest in structuralism, as already discussed, this led to the critique of adaptationism (Gould and Lewontin 1979; 1979k), which implied that progress, in the sense of general improvement, was even less likely than Darwin thought. PE also implied that evolutionary trends are driven less by selection-driven anagenesis and more by sorting among species, leading to a view of evolution as more of a directionless "bush" than a unidirectional "ladder." Progress, furthermore, was (and still is) a deeply held Western cultural value, a source of personal and national purpose, meaning, and comfort. As such, it was in the crosshairs of Steve's intense distaste for any view that smacked of seeing in nature what makes us feel good: "ladders are culturally comforting fictions, and copious branching is the true stuff of evolution" (1993l, 67).

> If the purely adaptationist vision were valid, we might gain the comfort of seeing ourselves, and all other creatures, as quintessentially "right," at least for our local environments of natural selection. But evolution is the science of history and its influence. We come to our local environments with the baggage of eons; we are not machines newly constructed for our current realities. (1993l, 369)

The theme and phrasing of "ladders vs. bushes" were common in Steve's writing. "Many of my essays," he said, "stress this theme of mentally liberating bushes versus constraining ladders because I believe that no other misconception so skews public understanding of evolution" (2001m, 324). "Humans are not the end result of predictable evolutionary progress, but rather a fortuitous cosmic afterthought, a tiny little twig on the enormously arborescent bush of life, which, if replanted from seed, would almost surely not grow this twig again, or perhaps any twig with any property that we would care to call consciousness" (1995s; 1995k, 327).

Progress is also a problem for analyses in the history of science, Steve argued, since it implies "whig" and "presentist" views of the past. "Models of inevitable progress," he said, "whether for the

panorama of life or the history of ideas, are the enemy of sympathetic understanding, for they excoriate the past merely for being
old (and therefore both primitive and benighted)" (1993l, 186).

C. History and hierarchy

By his own account, Steve realized in 1972 that PE implied a hierarchical view of evolution, but he and Eldredge didn't quite know
what to make of it. As they worked through the implications of
PE and ran them out to their logical conclusions, hierarchy came
to dominate both of their thinking (e.g., 1982f, 1982g; Eldredge
1989, 1995, 1999). As was the case with so many other strands of
his thought discussed here, Steve's passion for hierarchy clearly
had both empirical and theoretical roots. Stasis obviously implies
it, but hierarchy also likely appealed to Steve theoretically because
it was yet another way in which history could really substantively
matter in evolution. If all evolution is reducible to natural selection acting on individuals to optimize them for their present environment, then history is little more than a parade of perfection
and strict reductive determinism. If, however, a variety of discrete
processes act at different hierarchical levels (above and below the
level of the individual), which are themselves produced by the
historical sequences of evolutionary change, then history—and
historical science—are essential elements of a full understanding
of evolution. Hierarchy also appeared to offer the best opportunity for an independent macroevolutionary theory, based in
paleontology, thereby fulfilling the ambitions of those two young
Columbia graduate students.

Steve also found particularly fertile fodder for uniting these
disparate strands in looking below the level of the individual, at
the meaning of the growing tide of information from molecular
genetics (see Dorit, this volume):

> The collapse of the doctrine of one gene for one protein, and one
> direction of causal flow from basic codes to elaborate totality, marks
> the failure of reductionism for the complex system that we call
> biology—and for two major reasons. First, the key ingredient for
> evolving greater complexity is not more genes, but more combina-
> tions and interactions generated by fewer units of code—and many
> of these interactions (as emergent properties, to use the technical

jargon) must be explained at the level of their appearance, for they cannot be predicted from the separate underlying parts alone. So organisms must be explained as organisms, and not as a summation of genes. Second, the unique contingencies of history, not the laws of physics, set many properties of complex biological systems. Our thirty thousand genes make up only one percent or so of our total genome. The rest... originated more as accidents of history than as predictable necessities of physical laws. Moreover, these noncoding regions, disrespectfully called "junk DNA," also build a pool of potential for future use that, more than any other factor, may establish any lineage's capacity for further evolutionary increase in complexity. (2001m, 227)

D. *Replaying the tape: The role of contingency*

Durant (2002, 391) commented that Steve's first bout with cancer (1982) was "surely enough to persuade anyone of the importance of contingency in life," but Steve's interest in contingency clearly goes back much farther. At some point early in his career, Steve relates that his "general love of history in the broadest sense spilled over into my empirical work as I began to explore the role of history's greatest theoretical theme in my empirical work as well—contingency," which he defined as "the tendency of complex systems with substantial stochastic components, and intricate nonlinear interactions among components, to be unpredictable in principle from full knowledge of antecedent conditions, but fully explainable after time's actual unfoldings" (2002c, 47). Steve credited his graduate advisor Norman Newell's interest in sudden and catastrophic causes of mass extinction during the 1960s with stimulating his enthusiasm for the unpredictable effects of abrupt change (1998f).

An extremely important contributor to Steve's embrace of contingency was certainly what he frequently called the "MBL studies" (because much of the work was done at the Marine Biological Lab at Woods Hole, Massachusetts; see also Bambach, this volume). In the early 1970s a group that included David Raup, Thomas Schopf, Daniel Simberloff, Jack Sepkoski, and Steve worked on trying to specify how ordered phyletic patterns, as Steve wrote, "heretofore confidently attributed to selection for little reason beyond the visual appearance of order itself, could plausibly be generated within purely random systems" (2002c, 27). These studies obviously affected Steve profoundly, leaving him "humbled by the insight that our brains seek pattern, while our cultures favor particular kinds of stories for explaining these patterns—thus imposing a powerful bias for ascribing conventional deterministic causes, particularly adaptationist scenarios in our Darwinian traditions, to patterns well within the range of expected outcomes in purely stochastic systems" (2002c, 43).

Contingency, said Steve, "embraces one of the deepest and grandest issues that we can fruitfully engage in science—the nature and status of history in comparison with the more conventional style of explanation by predictable and repeated occurrence under timeless and invariable laws of nature" (2001b, 195). It also became for him the epitome of the general effect of history on evolution. Around it he was eventually to integrate his critiques of progress, adaptationism, gradualism, predictability, and biological determinism, as well as his interests in evo-devo, hierarchy, constraint, unpredictability, and the dashing of the fondest of conventional human hopes. He acknowledged that all evolutionists accepted some role for chance; the difference was (as he often said) in relative frequency: "I envision," he said, "that almost every interesting event of life's history falls into the realm of contingency" (1989d, 290). "[M]any aspects of even the broadest patterning of life's history," he maintained, such as "why and when do multicellular organisms arise, why and when do mammals eventually inherit the environments of large terrestrial vertebrates from dinosaurs—fall largely (or at least importantly) into the domain of contingent explanation" (2001b, 197). He could wax especially lyrical about this perspective: "Contingency is rich and fascinating; it embodies an exquisite tension between the power of individuals to modify history and the intelligible limits set by laws of nature. The details of individual and species' lives are not mere frills, without power to shape the large-scale course of events, but particulars that can alter entire futures, profoundly and forever" (1993l, 77). "We tend to look at history," he said in a 1988 interview:

> as though it were a series of predictable optimal states, and that's where most of our problems come from. The real message of history is that you have this kind of massive contingency where everything

that exists now is totally unpredictable....I think that's the most important lesson in history, and I think it would help us understand why we live in a world where a lot of things don't make sense...that troubles us deeply, because our cultural biases lead us to think that things don't make sense. Maybe if we understood how history really works, we would realize from these massive ill-fittings that a lot of things really don't make sense. You don't have to try to explain everything that's troubling as though it really was good when it arose as a Darwinian adaptation. It's an adaptationist assumption that if we do anything, it must have its evolutionary source in something that was once right or appropriate. But it doesn't have to. (Batten 1988)

Contingency also became a focal point for the integration (by Steve and others) into paleontology one of the major events in late twentieth century geology—the increasing acceptance of nongradual, nonuniformitarian change (e.g., Kauffman 1987; Albritton 1989; Ager 1993). Steve was both observer and participant in the growing development and popularity of these ideas (e.g., 1965b, 1967d, 1975t, 1984h); he both reflected and helped to create the *Zeitgeist*. "This issue of uniformitarian vs. catastrophic change," he wrote later, "stands as one of the grand questions of science, for the debate pervades so many disciplines and bears so strongly upon some of the most profound puzzles of our lives" such as the nature of causality and the nature of change (1995k, 164). Although many developments contributed to this intellectual sea-change, the single most important event was surely the Alvarez extraterrestrial impact theory for the end-Cretaceous extinction (Alvarez et al. 1980), and Steve was one of the first paleontologists to embrace this idea (e.g., 1984c, 1984x, 1985c, 1985j, 1986a, 1987x, 1989g).

It is difficult now to recall or understand what a boiling cauldron of scientific activity the early and mid-1980s were for much of paleontology (see, e.g., Glen 1994). The Alvarez hypothesis was greeted with great skepticism by most of the paleontological community and was hotly debated, in print and at professional meetings. It stimulated a burst of empirical work on the Cretaceous-Tertiary as well as other mass extinction events. Right on its heels was the 26 million year periodicity hypothesis (Raup and Sepkoski 1984), which in turn generated more controversy and a huge array of other hypotheses about possible extraterrestrial causes of mass extinction (see Raup 1986). The early 1980s was also of course the time when punctuated equilibrium and its implications were being debated—in the wake of the 1980 Chicago Macroevolution meeting, Stanley's book on the same topic (1979), and Gould's highly provocative papers (1980b, 1980c, 1982f, 1982g). It was in this context that Steve assembled what can only be referred to as his own grand synthesis—of punctuation, mass extinction, contingency, and hierarchy.

The idea apparently began to develop in his mind at the 1983 annual meeting of the Geological Society of America meeting in Indianapolis, when he heard Adolph Seilacher present his "vendobiont" theory for the Ediacara fossils (Seilacher 1984, 1989). On November 9, just a few days after attending the meeting, Steve wrote an excited and revealing letter to Luis and Walter Alvarez, David Jablonski, David Raup, Seilacher, and Jack Sepkoski:

Dear Luis, Walter, Dave, Dave, Dolf, and Jack,

It all came together for me in Indianapolis, as this rather hastily written Natural History column (to appear next February [1984r]) will testify—and I want to thank all you gentlemen for the insights.

I used to think (long ago and with my strong internalist, Platonist, D'Arcy Thompsonian biases) that mass extinction was just a whiz-bang phenomenology with no lasting importance (besides delaying things for a while each time) for patterns in the history of life. Then, I think, I just ignored it for a while, since I was so caught up in punctuated equilibrium as an unorthodox theory for pattern in normal times. Now you have all helped me to realize that it truly is a separate process, and a cardinal shaping force for patterns in life's history (Dave Raup, at least, will remember my old argument that "vectors" in life's history (or their non-existence) has always been the fundamental question of paleontology.). And we do not have a general theory for it as yet. All this taken together must constitute the chief excitement of paleobiology for the near future at least. I have hardly begun to consider all the implications, but I do think that we finally have the basis to grope for a general theory of pattern by considering unorthodox processes both for normal times (if punctuated equilibrium has any lasting meaning it will be here) and for mass extinctions. Their interaction must be

the dominant generator of pattern. I'll bet that most of microevolutionary thought for Darwinian transformation of local populations
won't be outstandingly relevant. What do you think?
  Sincerely,
  [signed] Steve

The eventual result of this epiphany was "The paradox of the
first tier: An agenda for paleobiology" published in the tenth
anniversary issue of *Paleobiology* (1985f). This paper was, in my
view, the most bold, coherent, logical, elegant, extreme, and over-
reaching technical paper Steve ever wrote.[12] It pulled together
a huge array of ideas and hypotheses, some well founded and
others at the time only tenuous, into a single overarching hierarchical view of evolution, from the ecology of natural selection
in local populations, to the effects of periodic mass extinctions
separated by tens to hundreds of million years. Closely behind
this impressive hierarchical edifice came Steve's provocative,
controversial, and influential explication of the reinterpretation of the Burgess Shale (1985x, 1986q, 1989d, 1990q, 1991i,
1992k, 1992m, 1993k), which claimed that contingency had
played a, if not *the*, dominant role in sorting out the survivors of
the Cambrian explosion from the less-fortunate "weird wonders"
so beautifully preserved in this extraordinary fossil deposit.
Altogether, this work solidified "a worldview that celebrates
quick and unpredictable changes in a fossil record featuring
lineages construed as largely independent historical entities."
Steve added tellingly that he found "such a world stunning and
fascinating in its chaotic complexity and historical genesis;" he
said he would "happily trade the comforts of the older view for
the joys of contemplating and struggling with such multifarious
intrigue" (1995k, 103).
  Steve also frequently wrote about the role of contingency in intellectual history, for example for typewriter keyboard arrangements
(1987k) or scientific illustrations (1999x), and he noted approv-
ingly (2001b) the increasing interest in contingency as an important factor among scholars of human history (e.g., McPherson
1988). The recently successful "what if" volumes of popular history
(e.g., Cowley 1999) similarly make use of a perspective sometimes
called the "counterfactual" to explore the possibly large implica-

tions of small events that might have happened differently in the past. Such arguments are not advocating total nondeterminism or "randomness" as causes of history; they are simply making the case that Steve made repeatedly: that unpredictable, unique, historical events, *by their very nature*, will exert a much stronger effect on the ultimate course of future history than most Western historians (and scientists) had previously acknowledged, and that historians (and paleontologists and evolutionary biologists) ignore such events at their peril.

E. The critique of determinism

In the minds of many biologists, Steve Gould was mostly seen, and is now mostly remembered, for his strident criticism of sociobiology, and its descendants evolutionary psychology and human evolutionary ecology (e.g., 1969c, 1974f, 1974s, 1974t, 1976n, 1978g, 1978h, 1979v, 1980x, 1983e, 1983m, 1984m, 1984w, 1994j, 1995n; see also Kitcher, this volume). I am neither qualified nor inclined to analyze the substantive details of either side of what was frequently a nasty debate. Here I wish only to point out that Steve's critique of sociobiology was, like almost every other facet of his intellectual life, closely and logically connected to multiple other interests and themes. As I have already discussed, he clearly had personal political and social views that were at odds with those of some advocates of sociobiology, and the effects of this disagreement cannot be discounted. He also, however, disagreed with sociobiology's focus on adaptive explanations for aspects of human behavior for the same reason he critiqued all such applications of "the adaptationist program"—because in his view they took too little account of the nonadaptive, historically contingent features of organisms that he thought were the crucial stuff of much evolutionary change. He disagreed with the application of hereditarian interpretations of discrete measures of human intelligence (such as IQ) for the same reason, because he thought that many features of organisms, including much of what we call human consciousness and intelligence are emergent characteristics of the highly complex human brain that evolution, probably mainly by natural selection, had built for other reasons.

V. Ever since Steve: Assessing a legacy

Steve Gould thought and wrote more than most practicing scientists about what controls the ultimate historical fates of a particular scientist's intellectual legacy. In one of his own favorite essays, for example, he describes the career and afterlife of paleontologist Nathaniel Southgate Shaler, who in the late nineteenth century was "by far, Harvard's most popular professor," and who, thirty years after his death, "at the Harvard tercentenary of 1936...was named twelfth among the fifty people most important to the history of Harvard" (1988q; 1991a, 313). Yet today, he is virtually unknown. "Why has he faded," Steve asks, "and what does his eclipse teach us about the power and permanence of human thought?" (1988q; 1991a, 318). Steve provided no unambiguous answer, except to note that, unlike Shaler, who more or less stuck to his mentor Louis Agassiz's views of divine direction in the history of life, his friend and fellow Harvard professor William James—"one of America's great gifts to the history of human thought"—"questioned Agassiz from day one...probed and wondered, reached and struggled every day of his life" (1988q; 1991a, 318–19). Steve's clear implication is that iconoclasts will ultimately prevail.

As is true elsewhere in this essay, I cannot make a thorough analysis or a confident judgment. I will only note three aspects of analysis of the Gouldian legacy that future historians (and scientists) might want to keep in mind.

Changes

As Stearns (2002, 2339) nicely puts it, Steve "deserves quite a bit more credit than his severest critics would grant (zero)" but less than Steve himself would award himself "(a great deal indeed)." It is a fact of the current state of evolutionary biology that Steve left a significant legacy of substantively changed views. These changes (for which he can take at least partial credit) include at least the following:

(1) **Stasis.** Although it remains difficult to put a firm number on its frequency, it is clear that morphological stasis is widespread in the fossil record, at least in many groups of benthic marine macroinvertebrates, and perhaps in many other groups as well, and may well be predominant in many clades under most circumstances.

This was not predicted by the Modern Synthesis and was almost wholly unknown or unappreciated prior to 1972. It is simply not, in my experience, true that, as Orr put it recently: "By the nineties, most evolutionary biologists had simply stopped paying attention to punctuated equilibrium....Punctuated equilibrium was down, if not out" (2002a, 136). On the contrary, it has, at least in part, become integrated into the evolutionary canon (e.g., Price 1996, 367–74; Freeman and Herron 1998, 475–81; Futuyma 1998, 689–94; Stearns and Hoekstra 2005, 433–34).

(2) **Evo-devo.** *Ontogeny and Phylogeny* was prescient and influential in its emphasis on the developmental basis for evolutionary change. Stearns (2002) notes that structuralism was largely lacking from the Modern Synthesis and says that Steve was correct to emphasize its importance. In doing so, says Stearns, Steve "did play an important role in preparing the anglophone community to receive the results [of molecular developmental genetics; "hoxology" as Steve called it], to know why they were important, and to place them in the context of historically significant questions." But, argues Stearns, the continental Europeans didn't need any such preparation, and therefore, if Steve "had never existed, I suspect that the field of evo-devo would have been in approximately the state today that it actually finds itself in," although the history of evo-devo that we experienced was "more interesting and colorful...because of him even if we could have gotten there without him" (2002, 2343).

(3) **The Softening of the Synthesis.** Adaptation and natural selection are still at the core of modern evolutionary biology. As Orr recently begins a technical paper: "Evolutionary biologists are nearly unanimous in thinking that adaptation by natural selection explains most phenotypic evolution within species as well as most morphological, physiological, and behavioral differences between species" (2002b, 1317). It is also, however, widely (if not always loudly) acknowledged that a substantially greater diversity of views about evolutionary processes is acceptable today compared to a generation ago, and this is in part clearly due to Steve's influence. Orr himself acknowledges that "Gould's attacks on adaptationism may have been extreme, but fanciful Just So stories are now, thankfully, rarer" (Orr 2002a, 138). Similarly, as Stearns (2002) eloquently puts it:

[t]he complacency and rigidity of evolutionary biology in the 1960s were real. The consistency of evolutionary phenomena with population genetics was incorrectly extended to a general belief that population genetics was sufficient to account for evolution. This gave population genetics a privileged position as the standard against which evolutionary thought should be measured, and it created an atmosphere in which important evolutionary phenomena not directly tied to genetic mechanisms were often defined away or ignored, to the great frustration of those interested in them.... Steve's greatest contribution was his effectiveness in shattering the complacency of the field and broadening the range of respected discourse.... He was a real leader in opening our minds to important things that had been missed, and he did our field a great service in reminding the public that there is more to biology than molecular biology and that there are interesting unanswered problems whose solutions will not require DNA sequences. (Stearns 2002, 2345)

Species selection

While the abundant evidence for stasis provides ample empirical confirmation of at least a core of PE, the continuing paucity of evidence for species selection, after more than a quarter century of searching, appears to me to be a serious problem, one that Steve did not adequately acknowledge and in fact in *Structure* rhetorically obfuscates. (I do not intend to discuss the details of species selection here, but only to comment on the style of Steve's argument.) The logic of species sorting as a result of PE is clear, as is the meaning of emergent characters (and/or emergent fitness; see 2001c, 658–59). Yet, as Steve says, "accepting a common logic but challenging the empirical importance of legitimate phenomena [is] a good substrate for productive debate in science" (2001c, 646). As several commentators (e.g., Erwin 2004) and Steve himself have noted, species selection simply does not have many empirical examples. Steve, however, thought that this argument is "unfair," and noted (correctly) that "a few excellent (and elegant) cases have been well documented, so this process cannot rank as a distant plausibility waiting for an improbable verification, as some critics have charged" (2001c, 709). The fact, he argues, that "well-documented cases of species selection do not permeate the literature" is because "[w]e have barely begun to acknowledge (much less to define or operationalize) this process, and we have still not entirely agreed upon the criteria for recognition" (2001c, 710).

Perhaps this is true, but in fact there are by my count *exactly three* well-documented and widely accepted examples of species selection, and all are more than twenty years old (Vrba 1980; Hansen 1980, 1982; Jablonski 1987). Many other hierarchy-imbued Gould students have been out there working with our respective empirical baliwicks and have yet to identify even additional possible examples. It is true that we did not (until *Structure*) have an explicit cookbook of criteria for searching for such examples, and this may have had some dampening effect, but I (as a fairly sympathetic observer) still find the paucity of evidence to be at least strongly suggestive evidence of paucity.

What is most interesting in my present context is how Steve treats this situation in his lengthy discussion of species selection in *Structure*. Ironically, he gives us a roadmap for analyzing such a situation in his own critique of Dawkins's gene selection: "When the logic of an argument requires that the empirical world operate in a certain manner, and nature then refuses to cooperate, unwavering supporters often try to maintain their advocacy by employing the tactic of conjectural 'as if,'" or ceteris paribus (2001c, 628). This is exactly what Steve does for species selection, relying on a highly detailed (and, as far as I am concerned, completely reasonable) theoretic analysis of the logic of higher order selection to say that it simply *should* be out there. When it comes to discussing the empirical record, however, he frankly stretches our credulity when he uses rhetoric more suitable for a much larger dataset: "our best examples of species selection," he says, "work through differential rates of speciation rather than varying propensities for extinction" (2002c, 649–50), making it sound like there are enough to really make such a distinction. Similarly, in referring to the widely cited example of different larval strategies in Cenozoic volutid gastropods (Hansen 1980, 1982), Steve calls it "a classic example, much discussed in the literature," when in fact it is arguably no more than the best of a tiny number of examples.

It is a further irony that he makes these rhetorical special pleas, because elsewhere he made just the reverse argument when discussing the occurrence of directional trends in the fossil record. A "case or two in the fossil record does not establish a pattern," he says. "Directional trends produced by wedging do occur, but they scarcely cry for recognition from every quarry and hillslope. The overwhelming majority of paleontological trends tell no obvious story of conquest in competition" (1993l, 304). Similarly, Steve wrote that we must "treasure our exceptions...But we must also be aware that single cases are fragile, and that sturdy facts are pervasive patterns in nature, not individual peculiarities. Most 'classic stories' in science are wrong" (1981u, 384). If we were to apply the same logic to species selection it would not come out looking good. My own view in the end is perhaps most similar to that of Flannery: "While I suspect that the concept of species selection is destined not to survive, at least in its present form, this is such a strongly contested field of biology that I would certainly not lay money—even at short odds—against Steve's eventual triumph" (2002, 53).

Stimulation

Even for many areas in which Steve's substantive conclusions have not stood up well in light of subsequent data or theory, many evolutionary biologists acknowledge that his ideas were enormously productive in stimulating research. Steve frequently argued that it was OK to be wrong for the right reason or right for the wrong reasons (e.g., 1996p; 1998x, 155; 1997m; 1998x, 323). He noted that errors could be useful "prods" to clarification and discovery, and quoted with approval the economist Vilfredo Pareto who said: "Give me a fruitful error any time, full of seeds, bursting with its own corrections. You can keep your sterile truth for yourself" (2002c, 614). (My own favorite version of the same view—which I once again learned as an undergraduate but came to appreciate only under Steve's influence—is from Mexican muralist Jose Clemente Orozco and is inscribed on the wall of Dartmouth's Baker Library: "Errors and exaggerations do not matter. What matters is boldness in thinking...in having the temerity to proclaim what one believes to be true without fear of the consequences. If one were to await the possession of the absolute truth, one must be either a fool or a mute.")

Other commentators have praised Steve for the fertility (if not the correctness) of his views:

There's no question he's been one of the most influential and visible paleontologists, and indeed evolutionary biologists, in the last 50 years...Steve has provided an overarching vision and this astonishing ability to move among disciplines and integrate these ideas into producing a coherent picture. (David Jablonski, quoted in Monastersky 2002, A17)

Most researchers...recognize that the concept [PE] has been invaluable in encouraging paleontologists to examine the fossil record with a rigor and attention to detail that previously was largely lacking. (Flannery 2002, 52)

Key parts of punctuated equilibrium may be wrong, but paleontological data are, largely due to Steve, richer than ever. Species selection may not make sudden sense of the fossil record, but a reinvigorated paleontology sits at evolutionary biology's high table. (Orr 2002a, 138)

I think the Modern Synthetic dogma is wrong. Steve did play some role in making us question the dogma. (H. Allen Orr, quoted in Monastersky 2002, A18)

VI. An End of a Beginning of an Appreciation and Farewell

In these days of seeking "balance" between "work and life" or "career and family," I just as often hear that many people willingly choose one over the other. Those who choose work sometimes say that it is what feeds them and makes them feel alive. Those who choose to devote more time to family sometimes say that their accomplishment and investment are in a secure and fulfilling marriage, and/or successful, healthy children. The academic equivalent of this balancing exercise is the struggle in which most faculty engage, to both advise students and pursue their own work. It was not until I had graduate students of my own that I came to appreciate the "great asymmetry" (cf. Gould 1998a) of this struggle: a student will (usually) have only one major doctoral advisor during their career, while one advisor will (usually) have many doctoral students during theirs. The significance and attention given by the advisor to the student is therefore almost always less than that given by the student to the advisor. Analogies between advisor-student and parent-child are, I realize, tenuous and probably dangerous as well. Yet I cannot help but reflect that, during a career that spanned less than forty years,

Steve Gould accomplished more scholarly productivity than most people could do in four lifetimes, and at the same time "raised" and sent out into the world at least thirty doctoral students, the majority of whom are still academically active and productive today. He did not choose; he found balance; he did it all.

In an interview not long before his death, Steve Gould was asked about his long-term wishes for *Structure* (2002c). He replied that "the biggest hope that any author would have if he put so much of a lifetime into something of this size is that it would be seen as a way station in the development of evolutionary theory that was useful and helped to focus things. Directed some energy. Got some things right, formulated something in a comprehensive and useful way" (Monastersky 2002, A17). Despite its flaws, *Structure* certainly does all of this. More important, Steve's career output does all this and much more. Thus, if we are to judge an academician's life's works as both the knowledge increased and the intellectual offspring produced, Steve Gould will share the legacy he predicted for one of his own heroes, Lavoisier: "His works, of course, will live—and he needs no more" (1998s; 2000k, 113).

References

Allmon, W. D. 2007. The evolution of accuracy in natural history illustration: Reversal of printed illustrations of snails and crabs in pre-Linnaean works suggests indifference to morphological detail. *Archives of Natural History* 34 (1): 174–91.

———. 2009. The "God spectrum" and the uneven search for a consistent view of the natural world. In J. S. Schneiderman and W. D. Allmon, eds., *For the rock record: Geologists confront intelligent design creationism.* Berkeley: University of California Press (in press).

Ayala, F. J. 2005. The structure of evolutionary theory: on Stephen Jay Gould's monumental masterpiece. *Theology and Science* 3 (1): 97–117.

Barasch, D. P. 2002. Grappling with the ghost of Gould. Review of S. J. Gould, *The structure of evolutionary theory. Human Nature Review* 2:283–92.

Batten, Mary. 1988. Charting life's unpredictable pathways: Stephen Jay Gould. *Calypso Log* 14 (4): 14–17.

Bean, R. B., and W. B. Bean. 1961. Sir William Osler's aphorisms from his bedside teachings and writings. Springfield, IL: Charles C. Thomas.

Blume, H. 2002. The origin of specious, and why reductionists are winning the Darwin wars. *American Prospect* 13 (17): 41–44.

Bradley, C. 2004. Stephen Jay Gould: an appreciation. *New Politics* 9 (4): 110–21.

Brown, A. 1999. *The Darwin wars: How stupid genes became selfish gods*. New York: Simon and Schuster.

Carroll, J. B. 1995. Reflections on Stephen Jay Gould's *The mismeasure of man* (1981). *Intelligence* 21: 121–34.

Cowley, R., ed. 1999. *What if? Eminent historians imagine what might have been*. New York: Putnam Adult.

Dawkins, R. 1976. *The selfish gene*. Oxford: Oxford University Press.

———. 1990. Review of S. J. Gould, *Wonderful life*. *Sunday Telegraph* (London), Feb. 25.

———. 2006. *The God delusion*. Boston: Houghton Mifflin.

Dennett, D. C. 1995. *Darwin's dangerous idea. Evolution and the meanings of life*. New York: Simon and Schuster.

———. 1997. Darwinian fundamentalism: An exchange. *New York Review of Books*, Aug. 14, 64–65.

———. 2006. *Breaking the spell. Religion as a natural phenomenon*. New York: Viking.

Durant, J. 2002. In memory of Stephen Jay Gould. *Public Understanding of Science* 11: 389–92.

Eldredge, N. 1989. *Macroevolutionary dynamics. Species, niches, and adaptive peaks*. New York: McGraw-Hill.

———. 1995. *Reinventing Darwin. The great debate at the high table of evolutionary theory*. New York: John Wiley & Sons.

———. 1999. *The pattern of evolution*. New York: W. H. Freeman.

Erwin, D. H. 2004. One very long argument. Review of S. J. Gould, *The structure of evolutionary theory*. *Biology and Philosophy* 19: 17–28.

Freeman, S., and J. C. Herron. 1998. Evolutionary analysis. Upper Saddle River, NJ: Prentice-Hall.

Freud, S. 1935. *A general introduction to psycho-analysis*. New York: Liveright Publishing.

Futuyma, D. J. 1998. Evolutionary biology. 3rd ed. Sunderland, MA: Sinauer Associates.

Glass, B. 1987. Review of S. J. Gould, *The flamingo's smile*. *Quarterly Review of Biology* 62 (4): 425–26.

Haught, J. F. 2000. *God after Darwin. A theology of evolution*. Boulder, CO: Westview Press.

———. 2003. *Deeper than Darwin. The prospect for religion in the age of evolution*. Boulder, CO: Westview Press.

Herrnstein, R., and C. Murray. 1994. *The bell curve: Intelligence and class structure in American life*. New York: Free Press.

Hull, D. 1999. Uncle Sam wants you. Review of M. Ruse, *Mystery of mysteries. Is evolution a social construction? Science* 284: 1131–33.

———. 2002. A career in the glare of public attention. *Bioscience* 52 (9): 837–42.

Jablonski, D. 2002. A more modern synthesis. Review of S. J. Gould, *The structure of evolutionary theory*. *American Scientist* (July–August): 368–72.

Jenkins, A. R. 1982. The debunking of scientific fossils and straw persons. *Contemporary Education Review* 1 (2): 121–35.

Kuhn, T. 1962. *The structure of scientific revolutions.* Chicago: University of Chicago Press.

Lessem, D. 1986. Stephen Jay Gould's view of life. *Boston Globe Magazine,* Nov. 16, 16, 17, 89–101.

✳ Lewontin, R. C. 2008. The triumph of Stephen Jay Gould. Review of *The richness of life: The essential Stephen Jay Gould.* S. Rose, ed., New York: W. W. Norton, 2007; and *Punctuated equilibrium,* by S. J. Gould, Cambridge, MA: Harvard University Press, 2007. *New York Review of Books* 55 (2): 39–41.

Lovejoy, A. O. 1936. *The great chain of being: a study of the history of an idea.* Cambridge, MA: Harvard University Press.

Lyne, J., and H. F. Howe. 1986. "Punctuated equilibria": rhetorical dynamics of a scientific controversy. *Quarterly Journal of Speech* 72: 132–47.

Mackler, J. 2002. Stephen Jay Gould, a man for all seasons. *Socialist Action,* July, www.socialistaction.org/news/200207/gould.html.

Maynard Smith, J. 1984. Palaeontology at the high table. *Nature* 330: 516.

———. 1992. Taking a chance on evolution. Review of S. J. Gould, *Wonderful Life* and *Bully for Brontosaurus* and E. Mayr, *Toward a new philosophy of biology. New York Review of Books* 39 (9): 234–36.

———. 1993. The theory of evolution. Canto edition. (Reprint of the 3rd ed. [1975], with a new introduction.) Cambridge: Cambridge University Press.

———. 1995. Genes, memes, and minds. Review of D. C. Dennett, *Darwin's dangerous idea: evolution and the meanings of life. New York Review of Books* 42 (19): 46–48.

Mayr, E. 1963. *Animal species and evolution.* Cambridge, MA: Belknap Press of Harvard University Press.

McGarr, P., and S. Rose, eds. 2006. *The richness of life. The essential Stephen Jay Gould.* London: Jonathan Cape.

McPherson, J. 1988. *Battle cry of freedom.* New York: Oxford University Press.

McShea, D. W. 2004. A revised Darwinism. *Biology and Philosophy* 19: 45–53.

Milner, R. 2002. Farewell, Fossilface. A memoir of Stephen Jay Gould (1941–2002). *Skeptic* 9 (4): 30–35.

Monastersky, R. 2002. Revising the book of life. *Chronicle of Higher Education,* March 15, A14–A18.

Morris, R. 2001. *The evolutionists. The struggle for Darwin's soul.* New York: W. H. Freeman.

O'Keefe, D. J. 2002. *Persuasion. Theory and research.* 2nd ed. Thousand Oaks, CA: Sage Publications.

Orr, H. A. 2002a. The descent of Gould. How a paleontologist sought to revolutionize evolution. *New Yorker,* Sept. 30, 132–38.

———. 2002b. The population genetics of adaptation: the adaptation of DNA sequences. *Evolution,* 56 (7): 1317–30.

Palevitz, B. A. 2002. Love him or hate him, Stephen Jay Gould made a difference. *Scientist* 16 (12): 12.

Polkinghorn, J. 1998. *Belief in God in an age of science.* New Haven: Yale University Press.

Price, P. W. 1996. Biological evolution. Fort Worth, TX: Saunders College Publishing.

Quammen, D. 1985. Evolution and the .400 hitter. Review of S. J. Gould, *The flamingo's smile. New York Times Book Review,* Sept. 22, 24.

———. 2003. The man who knew too much. Review of S. J. Gould, *The structure of evolutionary theory. Harper's Magazine,* June, 73–80.

Raup, D. M. 1986. *The Nemesis affair. A story of the death of dinosaurs and the ways of science.* New York: W. W. Norton.

Ridley, M. 2002. The evolution revolution. Review of S. J. Gould, *The structure of evolutionary theory. New York Times Book Review,* March 17, 11.

Ruse, M. 1999. *Mystery of mysteries. Is evolution a social construction?* Cambridge, MA: Harvard University Press.

———. 2000. Being mean to Steve. Metaviews 004.2000.01.12. www.meta nexus.net

———. 2003. The mismeasure of science. Review of S. J. Gould, *The hedgehog, the fox, and the magister's pox: Mending the gap between science and the humanities. Natural History* 112(6): 52–55, 58.

Ruse, M., and D. Sepkoski, eds. 2008. *Paleontology at the high table.* Chicago: University of Chicago Press (in press).

Rushton, J. P. 1996. Race, intelligence, and the brain: The errors and omissions of the revised edition of S. J. Gould's *The Mismeasure of Man. Personality and Individual Difference* 23: 169–80.

———. 1997. The mismeasures of Gould. *National Review,* Sept. 15, 30–34.

Schopf, T. J. M. 1981. Punctuated equilibrium and evolutionary stasis. *Paleobiology* 7 (2): 156–66.

Seilacher, A. (1984). Late Precambrian and Early Cambrian metazoa: Preservational or real extinctions? In H. D. Holland and A. F. Trendall, *Patterns of change in earth evolution,* 159–68. Berlin: Springer Verlag,

———. 1989. Vendozoa: organismic construction in the Proterozoic biosphere: *Lethaia* 22: 229–39

Selzer, J., ed. 1993. *Understanding scientific prose.* Madison: University of Wisconsin Press.

Sepkoski, D. 2005. Stephen Jay Gould, Jack Sepkoski, and the "quantitative revolution" in American paleobiology. *Journal of the History of Biology* 38 (2): 209–37.

Shermer, M. B. 2002. This view of science: Stephen Jay Gould as historian of science and science historian, popular scientist and science popularizer. *Social Studies of Science* 32 (4): 489–524.

Slobodkin, L. 1988. Review of S. J. Gould, *An urchin in the storm*. *American Scientist* 76: 503–4.

Stanley, S. M. 1979. *Macroevolution. Pattern and process*. San Francisco: W. H. Freeman.

Stearns, S. C. 2002. Less would have been more. Review of S. J. Gould, *The structure of evolutionary theory*. *Evolution* 56 (11): 2339–45.

Stearns, S. C., and R. F. Hoekstra. 2005. Evolution. 2nd ed. New York: Oxford University Press.

Stephens, S. 1997. Making it new in Soho. Stephen Jay Gould and Rhonda Roland Shearer recast a downtown loft. *Architectural Digest* 54 (2) (February): 108–15.

Sterelny, K. 2001. *Dawkins vs. Gould. Survival of the fittest*. Cambridge: Icon Books.

Thompson, D. W. 1942. *On growth and form*. Cambridge: Cambridge University Press.

Updike, J. 1985. Evolution be praised. Review of S. J. Gould, *The flamingo's smile*. *New Yorker*, Dec. 30, 76–79.

Wagner, P. J. 2002. Excursions in macroevolution. Review of J. S. Levinton, *Genetics, paleontology, and macroevolution*, 2nd ed., and J. B. C. Jackson, S. Lidgard, and F. K. McKinney, eds., *Evolutionary patterns: Growth, form, and tempo in the fossil record*. *Evolution* 56 (9): 1876–79.

Wake, D. 2002. A few words about evolution. Review of S. J. Gould, *The structure of evolutionary theory*. *Nature* 416: 787–88.

Wright, R. 1999. The accidental creationist. Why Stephen Jay Gould is bad for evolution. *New Yorker*, Dec. 13, 56–65.

York, R., and B. Clark. 2005. Review essay: The science and humanism of Stephen Jay Gould. *Critical Sociology* 31 (1–2): 281–95.

Zimmerman, W. F. 2003. Stephen Jay Gould's final view of evolution. Review of S. J. Gould, *The structure of evolutionary theory*. *Quarterly Review of Biology* 78 (4): 454–59.

2

Diversity in the Fossil Record and Stephen Jay Gould's Evolving View of the History of Life

Richard K. Bambach

I. Introduction

Steve Gould was the most publicly visible paleontologist of the last third of the twentieth century. A brilliant intellect coupled with his forceful and aggressive writing style, iconoclastic views, and passionate advocacy for paleontology as a discipline made Steve exceptionally influential at a time when paleontology was changing from "the handmaiden to stratigraphy" into its modern status as "paleobiology."

The development of Steve's career essentially paralleled the sequence of levels in his mature view of evolutionary theory. In early years Steve studied microevolutionary issues relating to form, development, and adaptation in species. He then moved on to focus on macroevolutionary issues such as speciation, trends, and patterns of diversity within clades. Finally, he dealt with the holistic aspects of disparity and contingency in the history of life, culminating in his synthesis of the structure of evolutionary theory.

Steve was interested in ideas, not the details of description. Although he was associated with the initial development of methodologies for analyzing data on diversity, he focused on what the patterns might mean, not what they actually were or what might explain specific events of diversity change. For him, the importance of diversity change was to demonstrate that macroevolution was independent from microevolution—and that distinction was the basis for his conviction that there is a hierarchical structure to evolutionary theory.

Steve Gould's initial research program on morphology and the ontogeny of morphology culminated in his great book *Ontogeny and Phylogeny*, published in 1977. However, Steve shifted away from concentrating on species interactions with the environment, the main topic of his publications from 1969 and 1970, as he came to view punctuated equilibrium as the main pattern of evolutionary change (1972e, 1977c). From the mid-1970s through the early 1980s his work stressed the implications of evolutionary pattern for evolutionary theory. He often emphasized the macroevolutionary relationship of phylogeny and clade diversity—the level at which species sorting related to punctuated equilibrium resided and where rigorous analyses using rule-based modeling gave the hope of understanding patterns in the history of life. Hierarchy became the major theme of his mature view of evolutionary theory as he developed his concept of the third tier of diversity change (the effect of mass extinctions in knocking apart the accumulated achievements at lower tiers). Impressed in his later career with the "paradox of the first tier" (1985f), Steve argued forcefully that progress did not characterize evolution, that the history of life was contingent, not predictable, and that most major events of evolutionary interest had been concentrated early in the history of life. His late works create the impression that to him the origins of anatomical disparity in the Cambrian explosion and the origins of metabolisms in the prokaryotes even further back in time were the items of fundamental importance—the diversification of life in the Phanerozoic was just contingent events and the expansion of variance (1996d).

Throughout his career, Steve did observational ("lab bench") science, too. But, oddly, here his primary focus was on Bermudan and Caribbean land snails, a minor element of the history of life. Snails only a few tens of thousands of years old were never going to make the headlines given to dinosaurs, hominid fossils, or the earliest records of microbial life. But Steve was faithful to *Poecilozonites* and *Cerion* (the latter even as a vanity license plate), not because they were intrinsically important or central to the profession of paleontology, but because he found them fruitful exemplars for his more general interests. *Cerion* appears in a dozen titles among his own papers cited in his summary opus *The Structure of Evolutionary Theory*. Those papers range in date over a quarter of a century, from 1971 to 1997.

In this chapter I examine the statements Steve made related to diversity and topics that impinge on diversity. Tracing Steve's use of examples of diversity change and the way he cited diversity lets us see the shifts in perspective that marked different phases of his work. Appropriate short quotes reveal Steve's thinking in his own words and let us see the evolution of his ideas. In what is to follow we also see glimpses of how his rhetoric called attention to his goals. He was always suggesting agendas for us—and arguing in an almost political sense that one needed to reject tradition and adopt new perspectives to gain deeper understanding. That aggressive approach made him controversial. It attracted attention and got his ideas widely debated but weakened his appeal, even though many of his viewpoints have now grown to be accepted practice.

II. 1969–1972: "Prehistory"—the origins of major intellectual themes touching on diversity

Steve Gould's studies of *Poecilozonites* (1969i, 1969j) reveal his initial steps toward two of the major intellectual highlights in his career: punctuated equilibrium and the potential for paleontology to be a "nomothetic" (rule-based) discipline. Although the empirical pattern of punctuated equilibrium is of only peripheral interest for the study of diversity, branching of lineages ("more making" in Niles Eldredge's terminology [1995, 181]), central to the concept of punctuated equilibrium, is a necessity for increasing taxonomic diversity. Steve's evolutionary interests and his use of multivariate quantitative analytical techniques also led to his association with Dave Raup, Tom Schopf, and Dan Simberloff (nicknamed the MBL group because they began meeting together at the Marine Biology Laboratory at Woods Hole). Their collaborative work produced the papers of Steve's that are most relevant to diversity analysis. And from that work his view of paleontology as a potentially "rule-based" discipline emerged.

An evolutionary microcosm (1969i)

The monograph titled "An evolutionary microcosm: Pleistocene and Recent history of the land snail *P. (Poecilozonites)* in Bermuda" (Gould 1969g) was developed from his dissertation work, but with considerable additional data. It is worth considering at length because it is the foundation of much of Steve's career and contains the seeds of many ideas Steve developed more fully in succeeding years.

In the introduction he argued, as he often did in later years, that extrapolation could not extend without limit, and, therefore, was not the correct way to explain things at different scales: "I doubt that principles derived from studies of living populations carried out in the course of a man's lifetime can provide a completely satisfactory model for processes occurring during the millennia that elapse in the history of nearly every significant phylogenetic event." (1969g, 410). Note that here he left the door open to "gradualistic" phenomena. In fact, despite considerable later varied debate engendered by confusion about changes in rate of evolution associated with speciation (2002c, 796–98 and 972–1024, especially 1010–14), the punctuations in punctuated equilibrium were always those observed on geological time scales and not different than rates in ecological time well known to biologists. In his 1969 monograph he also observed that the unique data of paleontology, as it follows the history of life through time, could contribute to refining evolutionary theory: "An incorporation of insights gained from the study of vast time spans might increase the generality of evolutionary theory in much the same way that a consideration of high velocities modified Newtonian physics" (1969g, 410).

The pattern of stasis of form in a lineage, with most morphological change occurring at the time of branching in cladogenesis (speciation), was not discussed as a generality in the *Poeciolozonites* monograph, but both stasis and punctuation were described in the work. For one example, *P. nelsoni*, Steve observed that, instead of a "zigzag evolution" of morphology shifting as carbonate dunes and red soils moved back and forth at particular localities, two subspecies—one adapted to dune environments and the other to red soil conditions—migrated with their environments as the environments shifted, each maintaining its genetic and morphologic character: "even if an occasional local population survives in the "alien" environment, it will not alter in form to identity with the usual subspecies of that environment, but will retain enough of its features to be recognizable.... such local populations have twice been found in Bermuda" (1969g, 469). In a second example, the *P. bermudensis zonatus* stock, two subspecies that maintained distinct morphology over a 300,000-year interval (even while adjusting somewhat to changing climate fluctuations) also produced four incipient speciation events, one in one lineage, three in the other. Relative stasis for both subspecies lineages plus the four branching events are described and illustrated (1969g, figs. 20, 21, 22). The *P. bermudensis zonatus* story differed from the traditional interpretation: "the proposition that two taxa, one a paedomorphic derivative of the other... is the simplest phyletic interpretation... I found a more complex story, involving an unusually literal interpretation of paleontological data. Each of the four discontinuous occurrences represents an independent episode of paedomorphosis" (1969g, 473).

Steve did not emphasize the rapidity of evolution of the paedomorphs; no argument was made that these incipient speciation events were "punctuations." However, the occurrence of four branching events in just one portion of the Pleistocene, each of which produced a demonstrably different morphology than the persisting parental lineage, set the stage for Steve's later views of rapid morphological change at branching events. Steve also made a particularly important point by demonstrating that the four branching events had occurred as allopatric events: "as would be expected under current notions of geographic speciation, the origin of a paedomorphic offshoot invariably occurs at the periphery of the known range of its parental form" (1969g, 478), and he cited chapter 16 of Ernst Mayr's *Animal Species and Evolution* (Mayr 1963) as the source for the idea of speciation in peripheral isolates. Then Steve presented data that indicated each paedomorph had first appeared in a restricted peripheral locale and concluded, "since the four paedomorphic offshoots occupy geographic subdivisions of the species range and differ genetically from the parental stock, separate taxonomic status is warranted" (1969g, 479). In this example of iterative evolution Steve had documented both stasis and allopatric evolutionary change, features that Niles Eldredge would emphasize in his pioneering 1971 paper in *Evolution* and Eldredge and Gould (1972e) would generalize as punctuated equilibrium.

Steve observed that phyletic change (anagenesis) also had occurred within the two relatively stable subspecies in the *Poecilozonites bermudensis zonatus* stock, and he commented that: "illustrated here are the two major evolutionary events of phylogeny: speciation or the multiplication of lineages and phyletic evolution or the transformation of lineages" (1969g, 469–70).

Despite his later reputation as opposing "the adaptationist programme," Steve claimed here that the phyletic changes he found in the *P. bermudensis zonatus* stock were adaptive, as he also suggested for the paedomorphic branching events in the stock, and he even used ecological argument as one reason why he felt these events differed from the pattern of stasis he had observed in the *P. nelsoni* stock. For *P. nelsoni* the "apparent temporal fluctuations in morphology were interpreted as artifacts of an imperfect record. Two subspecies lived side by side…and their alternating superposition in a single section reflects the migration of their preferred environments" (1969g, 487). The two forms were two persistent lineages, each adapted to the setting in which their fossils were found and not zigzag variants of one form, because their differences "are large and numerous (involving color—usually considered a better indicator of taxonomic distinction than form—as well as shape)" (1969g, 487). For the *P. b. zonatus* stock, however, the two persistent geographic subspecies could be distinguished throughout the stratigraphic sequence, despite the within-lineage (phyletic) changes each displayed as the environment changed. He illustrated this in four figures, two showing form persistence (1969g, figs. 21, 22) and two showing change within each subspecies (1969g, figs. 23, 24). There were no changes in shape or differences in color, just in size and shell thickness. The changes in the *P. b. zonatus* stock through time were, in fact, zigzag shifts in single lineages because the "differences in *P. b. zonatus* are few and small, involving only those features which adapt the snail to its new environment (…) and only those magnitudes small enough to be encompassed by the phenotypic plasticity of a subspecies" (1969g, 488). He concluded that, "the major temporal variations of morphology in the *P. b. zonatus* stock are adaptive in nature"

(1969g, 491). For the branching events he had also concluded that, "paedomorphosis served as one pathway to the attainment of a thinner shell, which would have been adaptive in the limited calcium environment of red soils" (1969g, 482–83), the only setting in which the paedomorphs appeared. Because the characteristics of size and shell thickness found in red soil environmental settings in Bermuda had developed in several different ways, Gould recognized that they were beneficial in that environment.

Although it seems curious to see adaptive change so strongly advocated in a Steve paper, given his later reputation for questioning the universality of adaptation (1979k), Steve never denied that selection to develop utility occurred. In fact, he wrote some of his best-known essays, such as "The Panda's Thumb" (1978q), on the topic. To be sure, he took pains to point out morphology that was not adaptive (e.g., 1974d, 1984x), and he later argued that the term "adaptation" should be restricted only to features evolved *de novo* for their current function (with the term "exaptation" being used for structures co-opted from a prior use (or non-use) to serve a new and different function) (1982e), but he never denied selection-driven change, only that utility should be demonstrated before adaptive claims are made.

In the conclusions of the *Poecilozonites* monograph, Steve noted the problems extrapolation can cause and the potential of careful phylogenetic analysis in paleontology. "In more general terms, micro-evolutionary studies focus on the production of diversity in response to isolation and environmental differences. The extrapolation of these emphases leads to a 'species divergence model,' which views the evolution of higher taxa as a simple extension of microcosmic processes of speciation—i.e., the higher taxon is viewed as a larger branch on the traditional tree, the boughs and branches of which continually diverge. This extrapolation does not give sufficient emphasis to the massive parallelism and trends toward increased mechanical efficiency that proceed in a relatively constant physical environment. These are major determinants of patterns in transspecific evolution but have little relevance to phenomena of infraspecific variation" (1969g, 497). While a bit peculiar for Steve in some respects (we catch Steve stating there are "trends toward increased mechanical efficiency"!), this statement is an early expression of Steve's career-long advocacy of a hierarchical nature to evolutionary theory. It is also a distant source for his future thinking about the nature of trends and the distinction between disparity and diversity, ideas that dominate his later work.

On careful analysis of phylogeny, Steve says, "paleontology, when it deals with the documentation of phylogeny, operates in the realm of history....But history becomes scientific when inductive generalizations are derived from series of events by the extraction of repetitive aspects from their integral uniqueness. By establishing a role for induction in history, repetitive occurrence leads to the formulation of laws: this is the major contribution of parallelism, convergence, and iteration to the explanation of evolutionary events—for explanation involves subsumption of observed conditions under general laws" (1969g, 497). Here Steve begins to present his argument that paleontology can be a nomothetic (rule-based) discipline, although this early statement differs from the emphasis on contingency that characterizes his work in the 1990s.

A unique venture into paleoecology (1969h)

The only paper of Steve Gould's that I would categorize as purely paleoecological is his 1969 paper in the *Proceedings of the North American Paleontological Convention*. At the convention it was presented in a symposium on the evolution of communities, a particularly hot topic at that time. It deals with the associations of microgastropod species in Bermuda that were present in the same suite of collections that Steve used in his *Poecilozonites* monograph. As with the *Poecilozonites* monograph, little relates to taxonomic diversity over time. But just as the *Poecilozonites* study revealed the origins of many concepts that characterize Steve's career, the North American Paleontological Convention paper is where the importance of analytical paleontology, the backbone of most serious work on the analysis of diversity over time, is made explicit.

The paper deals with the multivariate analyses used to group 48 samples of microgastropods into four major associations and identifies the ecological significance of the assemblages as they related to changing environmental conditions during the Pleistocene in Bermuda. Although analytical techniques in paleontology had been actively discussed for over a decade by John Imbrie, David Raup, Ralph Gordon Johnson and others, the combination of cluster analysis, discriminant function analysis, and factor analysis in Steve's paper was new to community analysis in paleontology. In discovering that four associations typified particular combinations of environmental conditions, Steve pointed out that the work "required a factor analytic assist to fathom the more arcane associations. Both the composition of Q-mode eigenvectors and the clustering of species in R-mode analysis suggest that four biologically-important assemblages be recognized among the 12 microgastropod species" (1969h, 494). This language was novel for the descriptive paleoecology of the time.

The most interesting aspect of the paper today is Steve's expressed hope for such methodology. It is a manifesto for analytical paleontology and what Steve would eventually call a nomothetic evolutionary discipline: "I would only say that I see in these methods the promise that we can found a science of paleoecology. If science, in order to be called such, must explain according to general principles, then most of what has passed under the name of paleoecology—the correlation of organic form with sediment type and the descriptive reconstruction of paleoenvironments—is not ecology. Ecology is preeminently the causal study of organic systems in all their complexity and interaction. Until we can detect repeatable patterns amidst this complexity and relate them to general statements, we have little more than a descriptive account of some awesome phenomena" (1969h, 493). Aside from the oddity of seeing Steve focusing on paleoecology rather than evolution, his advocacy for developing theory (general statements) applicable to paleontology still calls us to our duty.

Punctuated equilibrium (1972e)

Unquestionably the best known paper in Steve Gould's oeuvre is, "Punctuated Equilibria: An Alternative to Phyletic Gradualism," written in collaboration with Niles Eldredge (1972e). Punctuated equilibrium (called "equilibria" in the papers co-authored by Niles and Steve, but termed "equilibrium" in all of Gould's later papers) and its pervasive influence in evolutionary thinking is the subject of many studies. I don't have the space, or assigned responsibility, to discuss this theme (see chapters by Geary and Lieberman, this volume), but, as with the earlier works already cited, some aspects are fundamental in connecting Steve's particular interests to topics related to diversity.

The paper is a true collaboration. Niles Eldredge and Steve Gould had become acquainted in the mid 1960s, while Steve was a graduate student and Niles an undergraduate at Columbia University. Eldredge's 1971 paper in *Evolution* pointing out that gradual evolutionary change in fossil lineages is seldom seen and suggesting allopatric speciation as the reason for the common sudden appearance of new species in the fossil record is an obvious precursor to Eldredge and Gould 1972. Niles was the first to generalize about both stasis and sudden appearances. He was, however, already familiar with the conclusions of Steve's *Poecilozonites* work, which had documented both stasis in the face of environmental change and allopatric branching of lineages, and he cites it in the *Evolution* paper. Niles' own trilobite data, however, provided a second example, suggesting the potential of generality not claimed in Steve's snail study. Steve gave Niles principal credit for realizing allopatric speciation implied sharp breaks: "eventually we (primarily Niles) recognized that the standard theory of speciation—Mayr's allopatric or peripatric scheme (1954, 1963)—would not, in fact, yield insensibly graded fossil sequences when extrapolated into geological time, but would produce just what we see: geologically unresolvable appearance followed by stasis" (1989e, 118). Steve had made his set of initial observations; Niles saw the potential for generalization because he independently had a nearly parallel set of observations, but on a different scale, and the two collaborated in true collegial fashion on the punctuated equilibrium paper because each had come to the same interacting conclusions in their individual studies. The paper was prepared for the pioneering symposium volume, *Models in Paleontology*, edited by Tom Schopf, a book that serves as a founding document for the trend toward analytical paleontology and paleobiological interpretation that has characterized paleontology from the 1970s onward (see Allmon, this volume).

The point that theory pervades everything scientists do so that no work can be purely "objective" was a favorite of Steve's. Concerning the idea of inducing theory from data, Eldredge and

Gould note, "the inductivist view forces us into a vicious circle. A theory often compels us to see the world in its light and support. Yet, we think we see objectively and therefore interpret each new datum as an independent confirmation of our theory. Although our theory may be wrong, we cannot confute it. To extract ourselves from this dilemma, we must bring in a more adequate theory; it will not arise from facts collected in the old way" (1972e, 86). In the particular context of the paper on punctuated equilibrium, they use this idea to argue that "gradualistic" views of species trans-formation, coupled to the failure to differentiate between phyletic (within lineage) and branching patterns of evolutionary change, have held back paleontology: "The issue is central to the study of speciation in paleontology. We believe that an inadequate picture has been guiding our thoughts on speciation for 100 years. We hold that its influence has been all the more tenacious because paleontologists, in claiming that they see objectively, have not recognized its guiding sway. We contend that a notion developed elsewhere, the theory of allopatric speciation, supplies a more satis-factory picture for the ordering of paleontological data" (1972e, 86). They also connected branching, a necessary consequence of allopatric speciation, to the topic of diversity: "new species can arise in only two ways: by the transformation of an entire popu-lation from one state to another (phyletic evolution) or by the splitting of a lineage (speciation). The second process must occur: otherwise there could be no increase in numbers of taxa and life would cease as lineages became extinct" (1972e, 87).

They illustrate their ideas with their own studies of *Poecilozonites* (Gould 1969g) and phacopid trilobites from the Devonian in eastern North America (Eldredge 1971, 1972) and conclude, "our two examples, so widely separated in scale, age, and subject, have much in common as exemplars of allopatric processes.... Both are characterized by *rapid* evolutionary events punctuating a history of stasis. These are among the expected consequences if most fossil species arose by allopatric speciation in small, peripherally isolated populations. This alternative picture merely represents the appli-cation to the fossil record of the dominant theory of speciation in modern evolutionary thought. We believe that the consequences of this theory are more nearly demonstrated than those of phyletic gradualism by the fossil record of the vast majority of Metazoa"

(1972e, 108). They also argued that, "the theory of allopatric speciation implies that a lineage's history includes long periods of morphologic stability, punctuated here and there by rapid events of speciation in isolated subpopulations" (1972e, 109–10).

In this initial paper Eldredge and Gould were remarkably politic in their advocacy of punctuated equilibrium in contrast to phyletic gradualism: "The idea of *punctuated equilibria* is just as much a preconceived picture as that of phyletic gradualism. We readily admit our bias towards it and urge readers, in the ensuing discussion, to remember that our interpretations are as colored by our preconceptions as are the claims of champions of phyletic gradualism by theirs" (1972e, 98). Curiously, although they cited some other examples besides their own, Eldredge and Gould did not cite the major paleontological support for their view: standard biostratigrapic practice. Correlation in paleontology has always depended on zones, because the morphology of most species remains sufficiently constant that their occurrences over time cannot be subdivided into a succession of finer scale temporal intervals. From the inception of the concept of the stage by d'Orbigny in 1842 and the zone by Oppel in the 1850s, fine scale stratigraphic subdivision has always been by intervals or units, not by continuous gradation. If gradualism over longer intervals had been common, paleontologists would have been able to follow temporal change as a continuum, rather than as a sequence of zonal intervals. I have always felt this established the generality of the punctuated pattern beyond doubt.

The germ of what would become a major theme late in Steve's career, disparity contrasted to diversity, also lurks in the observation contrasting gradualistic extrapolation with punctuated change made late in the punctuated equilibrium paper: "The presence of 21 classes [of echinoderms] by the Ordovician, coupled with their presumed monophyletic descent, requires extrapolation to a common ancestor uncomfortably far back in the Precambrian if Ordovician diversity is the apex of a gradual unfolding.... We expect that successively higher ranks of the taxonomic hierarchy will contain more and more taxa: a class with one genus is anomalous and we are led either to desperate hopes for synonomy or, once again, to our old assumption—that we possess a fragmentary record of a truly diverse group.... With the picture of punctuated equilibria, however, classes of small membership are welcome and echinoderm evolution becomes more intriguing than bothersome. Since speciation is rapid and episodic, repeated splitting during short intervals is likely when opportunities for full speciation following isolation are good" (1972e, 110).

Eldredge and Gould also discussed the possibilities of "differential success of species exhibiting morphological change in a particular direction" as a way to produce apparent trends, another topic that would later be important to Steve. This method of making trends relates to diversity, because diversity change within clades is another aspect of differential success of species. Differential success of species leading to trends would be named species selection by Stanley (1975).

III. 1973–1980: Focus on nomothetic paleontology

In the mid to late 1970s Steve Gould was at the heart of the shifting emphasis in paleontology that is best summarized by the growth in the use of the term "paleobiology" as a synonym for paleontology. The pattern described as punctuated equilibrium was widely studied, because it could be directly tested with data from the fossil record. Following the "Models in Paleobiology" symposium held at the Geological Society of America meeting in the fall of 1971, David Raup, Tom Schopf, Steve Gould, and Dan Simberloff—all of whom became major figures—met together several times at the Marine Biology Laboratory at Woods Hole, Massachusetts (hence, the MBL group name). They produced a series of papers that are among those most relevant in all of Steve's output to the topic of diversity. Their use of computer modeling made testing hypotheses against null models a common practice, and Steve capitalized on their work to advocate that paleontology develop a nomothetic, as well as idiographic, base. He featured diversity patterns as examples demonstrating his general concept.

Stochastic models (Raup et al. 1973 [Gould 1973j])
The first paper authored by the MBL group (Raup et al. 1973) was Steve's first explicit "nomothetic" paper. Steve showed his penchant for the gaudy turn of phrase in the subject heading used for the introduction, "Prospectus," as well as in the application of

the term "nomothetic" to the work: "This is the first in a projected series of papers that might bear the general title 'nomothetic pale-ontology.' 'Nomothetic' is a term used by psychologists, historians, and philosophers to designate an approach to historical science favoring the study of 'cases and events as universals, with a view to formulating general laws.' The conventional approach, on the other hand, is termed 'idiographic': 'the study of cases or events as individual units, with a view to understanding each one separately' (Random House Dictionary)" (Raup et al. 1973, 526).

This first project by the MBL group used a computer model of phylogeny with just a few basic parameters: "In summary, the input constants required by the program are (1) initial probability of extinction, (2) initial probability of branching, (3) equilibrium diversity, (4) the damping constant governing variation in diver-sity, and (5) the minimum size of a clade" (Raup et al. 1973, 532). The phylogenetic histories of clades were related to their patterns of diversity change through time: originations adding diversity, extinctions cutting diversity back. Computer simulations were run using various probabilities of origination and extinction. The resulting diversity patterns were viewed as clade histories, and the results of the simulation were compared to the patterns of real clades of reptiles: "The results of the program and the comparison with reptiles demonstrate that an exceedingly simple stochastic model can produce branching and diversity patterns very like those described in the real world" (Raup et al. 1973, 539). However, about half the reptilian clades and lineages had been affected by the end-Cretaceous extinction, and the authors noted that, "given the probabilities used in the simulation, it is extremely unlikely that such a coincidence of extinctions [among clades] would occur" (Raup et al. 1973, 538). Thus, this work showed that it was possible to achieve somewhat realistic results with rather simple modeling and also to differentiate unusual events from otherwise stochastic patterns.

Towards a nomothetic paleontology (Raup and Gould 1974 [Gould 1974i])

Raup and Gould produced a follow-up paper examining the evolution of morphology, a favorite subject of Steve's (1974i). "Nomothetic" now made it into the title ("Stochastic Simulation and

Evolution of Morphology—Towards a Nomothetic Paleontology"),
and Jack Sepkoski made his first appearance in the paleontological
literature in the acknowledgements ("We are especially grateful to
J. J. Sepkoski for taking the time to write COLINK (for comparing
phenetic and cladistic phylogenies) and for general comments
and criticisms" [1974i, 322]). Jack was finishing graduate study
at Harvard and preparing to join the faculty with Raup at the
University of Rochester.

The project modeled the evolution of morphology, with lineages
in randomly generated clades evolving in the style of punctuated equilibrium. Ten hypothetical characters were permitted
to change only at branch points, and new character states only
appeared in the new branches. The frequency distribution of
various combinations of character states could be determined,
and the pathway of character changes leading to any new character could be traced. Interestingly, each "clade" had a characteristic range of morphology, even though the process of generating
clades and altered characters was random, not selection driven.
Each character had its own "history," and the paths leading to
some new character states looked like selection generated trends,
even though they were all random walks. This is characteristic of
a Markovian system because morphology could shift only one step
at a time, and change always shifted from a morphology already
present; transitions could not be from or to any morphology.
This result impressed Raup and Gould with the power of random
processes to produce apparent order: "Our demonstration that
random processes can produce most of the patterns generally
associated with directional causes constitutes a challenge to the
formalist position....Adaptation is better demonstrated by studies
of the mechanics of form in relation to environment than by
evidences of directional change through time....The fact that
examples of steadily changing characters do not occur with much
higher frequency in the fossil record than in the computer simulations suggests that, over long stretches of time, undirected selection may be the rule rather than the exception in nature" (1974i,
321). This result, along with Steve's studies of allometry, clearly lie
behind his motivation for participation with Richard Lewontin five
years later in the "Spandrels of San Marco" paper (1979k), their
classic critique of the adaptationist programme.

Paleobiology *Volume 1, Issue 1 (Schopf et al. 1975 [Gould 1975c])*

The MBL group had a paper in the first issue of *Paleobiology* (and Steve was involved in two other items in that issue as well [1975b,d], one of which began on page one). The result of the MBL study (Schopf et al. 1975) suggested that rates of evolution (and, consequently, rates of diversity change) as seen in the fossil record might be correlated with morphological complexity rather than rates of genetic change: "In both the real world and the computer simulation, the bias of differential morphologic complexity may account for the observation that 'only complicated animals evolve.' Most paleontologic studies of the 'rate of evolution' may tell us more about morphologic complexity than about evolutionary rates of genomes" (Schopf et al. 1975, 63). That supposition still needs to be investigated further. Much of Steve's work over the years set agendas for us, and we haven't finished with them yet.

Real and random clades (Gould et al. 1977a)

Steve was the senior author on the last MBL group paper (1977a), the most relevant of the series to the analysis of diversity patterns. Five parameters for "clade shape," the shapes of spindle diagrams of the diversity history of clades best known from the spindle diagrams of family diversity within classes published by Sepkoski (1981, fig. 1), are defined: (1) size—the sum of the number of taxa in each time interval during which a clade existed, (2) duration—the number of time intervals during which a clade existed, (3) maximum diversity—the largest number of member taxa in a clade at any time, (4) center of gravity—the temporal point that divides the history of a clade into two equal size portions based on cumulative diversity (the temporal point at which half the total size of the clade has been achieved), and (5) uniformity—a measure of the stability of diversity in a clade through time, calculated as the proportion of the area of a rectangle circumscribed about the space occupied by the clade spindle form. The distribution of those parameters in real and randomly generated clades was analyzed. Stochastic simulated clades were generated using the simulation program developed first in Raup et al. (1973). Because the number of originations and extinctions must be equal for an extinct clade, the probabilities of branching and extinction of lineages in the simulations were set to be equal and the simulations were run at a wide range of probability values. For comparison, the clade shapes for "real" clades were determined for 144 sets of orders within classes, 206 sets of families within orders, and 1442 sets of genera within families, using data from a variety of literature sources.

The stochastic simulations illustrated how clade shape varied as probability of branching and extinction varied from low to high. Longer durations (but generally lower maximum diversities) were associated with low probabilities; shorter durations (but generally higher maximum diversities) were associated with higher probabilities of branching and extinction. This parallels the later observation of Norman Gilinsky (1994) that decreased volatility of diversity dynamics characterizes longer-lived still living taxa compared to shorter-lived extinct taxa. This led Gilinsky to conclude that large-scale "taxon sorting" is responsible for the general decrease in total extinction rate observed through the Phanerozoic. Also, because the increased volatility at higher probabilities of branching and extinction increases the likelihood of rapid truncation of clades (and also decreases average uniformity), average center of gravity increased from just under to just over half way through the history of simulated clades as probability of extinction (and branching) were increased in the simulations.

Gould et al. (1977a) found that scaling of simulated clades and real clades were sufficiently different that direct comparison was difficult for dimensional values, such as size. However, it was possible to compare extinct with living clades for both real and simulated sets of clades ("living" determined for the simulations by clades not yet terminated ["extinct"] at the end of the computer runs). "How different, then, is the real world from the stochastic system? How, in other words, is the real world 'taxonbound' and 'timebound'—i.e., in need of specific, causal explanations involving uniquenesses of time and taxon at various stages of earth history. The answer would seem to be 'not very'—the outstanding feature of real and random clades is their basic similarity" (1977a, 32). Although living clades were generally larger than extinct clades, Gould et al. noted that, for their average stochastic products, "the resulting differences in size between A [alive] and E [extinct] clades fully match their discrepancy between real A and E clades" (Gould et al. 1977a, 35). Uniformity was complex to evaluate, but Gould et al. decided that, "for both amphibians and mammals, low values of UNI [uniformity] are probably a result of real, biological interaction with other groups of vertebrates. The real world may be, in this respect, 'timebounded.' Some times really are 'good' for certain groups (in these cases because a successful competitor had either not yet evolved, or recently become extinct)" (1977a, 37). They also suggested an explanation for a consistent difference in center of gravity between early-originating and later-originating clades observed in both simulations and real examples: "Moreover, we have already noted...that mean CG [center of gravity] for simulated clades originating in this early pre-equilibrial phase is less than 0.5, while equilibrial clades never fall below 0.5. The explanation for Cambro-Ordovician invertebrates and Paleocene mammals must be the same: origination rates exceeded extinction rates as the world filled up. Our DE [damped-equilibrium] model seems to simulate the real world and we can understand its behavior thereby" (1977a, 39). Steve would return to this particular data in a more complex analysis in 1987 (Gould et al. 1987b).

Stanley et al. (1981) demonstrated that the scaling used in the MBL program (1973j, 1977a) was actually not realistic in that the fluctuations of diversity in the stochastic simulations by the MBL group could not be replicated when rates were scaled to represent species origination and extinction at realistic standing species diversities. However, the parameters of clade shape developed in Gould et al. (1977a) remain useful for evaluating the diversity histories of clades at higher taxonomic levels. For example, Gilinsky and Bambach (1986) found most clades are not significantly different in overall diversity history from the mean of many randomly generated clades with the same range of diversity dynamics (a conclusion corroborated analytically by Gilinsky and Good 1989). However, for most real clades, the center of gravity of the clade occurs somewhat earlier in time than the bootstrap simulated mean for that clade. Also, the real maximum diversity generally slightly exceeds the bootstrap simulated mean for each clade. The root cause of this general tendency for a small excess of diversity in the earlier part of clade histories is probably the asymmetry between origination and extinction demonstrated by Gilinsky and Bambach (1987), with high origination concentrated early in clade history. This provides the change in origination compared to extinction that Stanley et al. (1981) argue should occur to produce observed clade dynamics.

Punctuated equilibrium revisited: (1977c)

In the second issue of the third volume of *Paleobiology*, Steve was back with Niles Eldredge for an expanded review of the concept of punctuated equilibrium. Much of the last ten pages of the paper (about one-third of the text) discussed large-scale aspects of "punc-eq" that apply to diversity issues and general theoretical considerations. This includes their acceptance of Stanley's suggestion of "species selection" (Stanley 1975), on which Gould and Eldredge base some of their expanded theoretical arguments.

Since diversity patterns are the patterns of the deployment of species (or higher taxa) in geological time, the central issue that links punctuated equilibrium with aspects of diversity is summed up in the statement that, "punctuated equilibria is a model for discontinuous tempos of change at one biological level only: the process of speciation and the deployment of species in geological time" (1977c, 145). For example, Gould and Eldredge claim that because morphologic change occurs predominantly in branching events (speciation), trends are produced by differential success among species, i.e., increased diversification of species in one direction versus others, as they had noted in Eldredge and Gould (1972e). However, they expand the theoretical issue to include the patterns of success of clades, using diversity as an indicator of success: "Consider successful clades that are both diverse and long-lived....But just as life history parameters of maturation time and reproductive effort have been used to explain "success" in ecological time, so must the macroevolutionary analog of speciation rate be included in our study of successful clades" (1977c, 143). Aspects that would increase diversity ("increaser clades"), such as consistently high speciation rate and "preemption of adaptive zone by abundant speciation," or features that would permit a clade to maintain diversity because of resistance to extinction ("survivor clades"), such as large populations, large environmental range, or "triumph over other species in direct competition," are described and they identify all as "strategies for success" (1977c, 144). They conclude, "we believe that the need to translate micro to macroevolution through the level of speciation guarantees that paleobiology shall not be a derivative field, but shall provide essential theory to any complete science of evolution" (1977c, 145). In this last statement we see the great ambition Steve always held for paleontology, which he was instrumental in bringing to fruition. Steve revisited the topic of trends regularly in his future work and the differentiation of levels of evolution (micro, macroevolution) led to his developing interest in hierarchy in evolutionary theory.

Eternal metaphors (1977b)

Steve's expanding focus on large-scale concepts is exemplified by one of his most interesting general reviews, "Eternal Metaphors of Palaeontology" (1977b): "In this paper, I wish to propose that (1) the basic questions palaeontologists have asked about the history of life are three in number; (2) the formulation of these questions preceded evolutionary thought and found no resolution within the Darwinian paradigm; (3) the major contemporary issues in palaeobiology represent the latest reclothing of these ancient questions" (1977b, 1). The three questions Gould posited are, "(1) Does the history of life have definite directions; does time have an arrow specified by some vectorial property of the organic world (increasing complexity of structure, or numbers of species, e. g.).... (2) What is the motor of organic change? More specifically, how are life and the earth related? Does the external environment and its alterations set the course of change, or does change arise from some independent and internal dynamic within organisms themselves?... (3) What is the tempo of organic change? Does it proceed gradually in a continuous and stately fashion, or is it episodic?" (1977b, 1–3). Diversity plays a part in each of these three questions. It is explicit in number one, the causes of diversity change are the subjects of question two, and rates of diversity change are considered under question three.

Steve then developed a "taxonomy" of viewpoints paleontologists hold on these issues, using a bifurcating sequence of categories (steady state vs. directional, environmentalist vs. internalist, punctuational vs. gradualist). Thus he could contrast antithetical positions, his favorite method of argument. Finally, he reviewed pre- and post-evolutionary views of the field in a very illuminating fashion. When he arrived at "contemporary palaeobiology" he noted he "will not analyze my contemporaries in detail; still, two recent examples should exemplify my contention" (1977b, 17). The two examples were Jim Valentine's studies on diversity changes and their possible association with plate tectonic movements as discussed in his 1973 book—Gould even used Valentine's family diversity curve as an illustration (1977b, fig. 3)—and the 1973 MBL group study of modeling phylogeny and diversity (Raup et al. 1973), plus several papers that considered various aspects of that study. Steve presented diversity studies to illustrate both directional and steady-state attitudes about the eternal metaphors among current workers.

In the concluding section of the paper Steve set up several contrasting views of rate and scale in evolution and said, "I believe that the concept of independent levels provides a resolution" (1977b, 22). He then noted: "Sewell Wright (1967, 120) proposed a profound analogy: just as mutations are random with respect to the direction of selection in ecological time, so might speciation itself be random with respect to the direction of trends in evolutionary time" (1977b, 22). He used this statement to reframe his now well-established view of the importance of speciation in generating trends ("trends represent the differential success of subsets from a random spectrum of speciations" [1977b, 22]). After discussing species selection and naming Wright's rule ("If a key aspect of the phenomenon must have a name, I would prefer 'Wright's rule of differential success,' or—of [sic] this be judged too cumbersome—just 'Wright's rule'" [1977b, 24]), he concluded, "in any case, Stanley (1975) correctly argues that Wright's rule 'decouples' micro-evolution from macro-evolution, thus affirming the independence of palaeontology as an evolutionary subdiscipline.... There can scarcely be a more important task for palaeontolgists than defining the ways in which macro-evolution depends upon processes not observed in ecological time" (1977b, 24). Steve's hierarchical view of evolutionary theory was taking concrete form, and his ambition for paleontology was again boldly stated.

A nomothetic, evolutionary discipline (1980b)

The mature statement of Steve's dream of paleobiology as a "rule-based" discipline came in 1980 with his paper, "The promise of paleobiology as a nomothetic evolutionary discipline" (1980b). Here he proclaims his view that paleontology has important things to say about evolution and that it should have high scientific status on its own merits. The brief for the value of paleontology is not concealed; early on Steve admits, "this paper is not a review article; it is a partisan statement" (1980b, 96). He makes his most impassioned claim for the value of paleontology stating, "a general theory of paleontology can only emerge from its status as guardian of the record for vast times and effects. If every evolutionary principle can be seen in a *Drosophila* bottle or in the small and immediate adjustment of local populations in the *Biston betularia* model, then paleontology may have nothing to offer biology beyond exciting documentation. But if evolution works on a hierarchy of levels (as it does), and if emerging theories of macroevolution have an independent status within evolutionary theory (as they do), then paleontology may become an equal partner among the evolutionary disciplines" (1980b, 98). As we have seen already, diversity was central to his view of macroevolution as a field of its own. Gould also emphasizes that paleontology has begun to adopt a valuable theory and testing methodology, "Simpson's style of science has finally taken root in paleontology. *Models in Paleobiology* (Schopf 1972) was purposely constructed as an exemplification of it. This journal [*Paleobiology*] is its conscious embodiment" (1980b, 99).

Steve argues that to be fully effective paleontology must continue to respect its rich data base and wonderful idiographic material, even as it develops theory: "The asymmetry of fact and theory dictates that science without the second may be dull; but without the first, it is garbage. Facts needn't cover all areas like a seamless blanket, but their general absence guarantees sterility" (1980b, 100). However, as always, he is concerned about purely inductivist attitudes: "Much paleobiological work continues in the "empirical law" tradition—it accumulates cases in the hope that some useful generality will emerge. I strongly suspect that such work, although intrinsically valuable for its elucidation of cases, will furnish no new or expansive generalities" (1980b, 101). He cites three examples: community reconstruction, mechanistic functional morphology, and biostatistics, noting that each is unproductive when used simply to corroborate pattern already known from the Recent. "Yet each of these three areas has also generated fruitful models and hypotheses," he concludes, when scholars have followed C. S. Peirce's idea of "abduction"—"literally by the creative grabbing and amalgamation of disparate concepts into bold ideas that could be formulated for testing" (1980b, 102).

Diversity becomes important in the later parts of the paper. In preparation for this paper, Steve informally polled twenty colleagues to discover what developments in the discipline paleontologists considered successful or disappointing over the previous twenty years ("since the Darwinian centennial in 1959" [1980b, 108]). The only topic to be mentioned by more than three-fourths of those polled (sixteen of twenty responses) was stochastic modeling of diversity. "Studies of the tempo and mode of evolution based upon hypotheses of punctuated equilibrium and species selection" came in second at thirteen responses and no other topic received more than eleven citations. Stochastic modeling of diversity received a "split vote" on the successful/disappointing scale with eight positive and eight negative responses. Steve noted, "no other subject engendered more passion, garnered more votes, or elicited more varied description—from revolutionary theory to arm-waving based on imperfect *Treatise* data.... Any subject that arouses such interest must be doing something right, even if later judgment rejects its conclusions" (1980b, 112). Almost a quarter of a century later the testing of observed versus sample-standardized data on diversity is still an active field (Alroy et al. 2001; Bush et al. 2004; Powell and Kowalewski 2002; Bush and Bambach 2004).

The degree to which Steve built his entire career around his interest in evolutionary theory was revealed to me during Steve's survey. We had a friendly debate about my responses one afternoon when I was visiting him in Cambridge. Functional morphology (one of the top twelve topics, appearing on nine ballots, six positive, three negative) was a topic I favored and Steve felt was unimportant. I persisted in my advocacy and Steve eventually erupted with an impassioned, "But Dick, it doesn't tell us anything about evolution!" Evolution was always and always his focus. This sidelight, coupled to his discussion of community reconstruction and mechanistic analyses of functional morphology as examples of areas that "will furnish no new or expansive generalities" (at least as he felt they were studied at the time), also demonstrates the shift in Steve's scope in the decade since he had said, "I see in these methods the promise that we can found a science of paleo-ecology." His interest in paleoecology had evaporated (see also Allmon et al., this volume), but the goal, first advocated in his 1970 paper, of using rigorous analytical methods to develop paleo-biology as a discipline that would "explain according to general principles," was now the main theme of his 1980 vision of paleobi-ology as a "nomothetic, evolutionary discipline."

In concluding the paper, Steve drew on his experience with stochastic modeling of diversity and on several examples of diver-sity studies by others to supply background for the discussion of the potential of paleontology as a nomothetic discipline. The examples are of the time, but Steve's statements about the goal of achieving a nomothetic discipline still form a credo for our field:

> Science is nomothetic insofar as its descriptions include particulars of given times and individual objects only as boundary conditions, not as intrinsic referents in the laws themselves.... Since our tradi-tional focus has been so idiographic, the nomothetic aspects of pale-ontology are now in greater need of attention—not because they are more important, but because they have been neglected.... The traditional view admits intrinsic historicity of events and applies nomothetic principles to abstracted aspects. The radical view asks if there might not be a sense—at an appropriate scale of analysis—in which the events themselves are essentially nomothetic in char-acter. That is, might a biological object be treated without reference either to its taxon or to the time in which it lived. Might species be like the molecules of classic gas laws.... Random models should be viewed as an appropriate tool for paleontologists of all persua-sions. First of all, they provide the first explicit set of null hypoth-eses for assessment of legitimate uniqueness. Since random systems generate a large amount of apparent order, we need to define the

bounds of pattern that random systems produce. More ordered patterns or lower degrees of order occurring too often must, with respect to the model that generated a null hypothesis, be regarded as nonrandom. (1980b, 113–4)

Steve concludes that, "paleontology is not a pure historical science; it resides in the middle of a continuum stretching from idiographic to nomothetic disciplines. It possesses a body of idiographic data virtually unparalleled in interest and importance among the sciences—for it is, after all, the history of life....The foundations for a nomothetic paleontology have been set—and there is so much more to do" (1980b, 116).

While at the crest of his enthusiasm for seeing his profession, paleontology, become a nomothetic science, Steve also wrote *The Mismeasure of Man* (1981l) his most important book on a nonevolutionary subject. Steve Gould, despite the wide range of his talents and interests, was a consistent, focused person. He regularly used the intellectual advances he found in his professional work to illuminate his views on nonpaleontological topics. Just as Steve argued that scientists are always theory-laden, something one must be aware of to conduct proper analysis within paleontology, he argued that it was the same for socially related topics. Dealing correctly with quantitative analysis and using appropriate models to do proper scientific investigation are central to Steve's view of nomothetic paleontology and of evolutionary theory. He felt the same should hold for evaluating people. Steve laid out his purpose for *The Mismeasure of Man* clearly: "this book discusses, in historical perspective, a principal theme within biological determinism: the claim that worth can be assigned to individuals and groups by *measuring intelligence as a single quantity*....This book seeks to demonstrate both the scientific weaknesses and political contexts of determinist arguments" (1981l, 20–21). He saw the value of proper scientific consideration of social issues in the following light: "Science cannot escape its curious dialectic. Embedded in surrounding culture, it can, nonetheless, be a powerful agent for questioning and even overturning the assumptions that nurture it. Science can provide information to reduce the ratio of data to social importance. Scientists can struggle to identify the cultural assumptions of their trade and to ask how answers might be formulated under different assertions. Scientists can propose creative theories that force startled colleagues to confront unquestioned procedures" (1981l, 23). While I do not claim that Steve's developing image of the nomothetic possibilities for paleontology was his motivation for looking at a difficult social issue with analytical rigor, his approach clearly followed from the idea that correct understanding, a prerequisite for successful science, was a valid model for looking at any basic question.

IV. 1980–1986: Hierarchical structure of evolutionary phenomena

The phases of Steve Gould's evolving intellectual focus do not have sharp "punctuated" boundaries. The seeds of each phase are generally detectable in earlier work. The interval from 1980 to 1986, a time of nearly continuous transition in Steve's thinking, starts with the nearly seamless shift of back-to-back papers in *Paleobiology*: the "nomothetic" paper (1980b) summarizes Steve's decade of attention to analytic methodology and "Is a new and general theory of evolution emerging?" (1980c) serves as the "bell-ringer" for his developing emphasis on the hierarchical nature of evolutionary phenomena that characterizes his work in the early 1980s. Steve's growing interest in the importance of mass extinctions, starting with his own look at one aspect of the Permian extinction (1980j), and stimulated greatly over the next few years by new discoveries across the profession, led to his formulation of a three-tiered hierarchy of evolutionary phenomena. The outcome of this transitional phase was his commitment to write *The Structure of Evolutionary Theory* (2002c), a project that lasted for twenty years.

Is a new and general theory of evolution emerging? (1980c)
Steve chose the issue of excessive extrapolation as a springboard for reconsidering evolutionary theory: "I think I can see what is breaking down in evolutionary theory—the strict construction of the modern synthesis with its belief in pervasive adaptation, gradualism, and extrapolation by smooth continuity from causes of change in local populations to major trends and transitions in the history of life" (1980c, 128). In several places, such as the concluding section of his "eternal metaphors" paper (1977b),

Steve had written about the hierarchical aspect that microevolution, including the direct action of natural selection, and macroevolution, embodied in speciation and large-scale patterns such as trends, bring to evolution. Hierarchy, however, now became central, as he took on the idea of a new view of evolutionary theory. Although Steve felt that in moving beyond simple microevolution as exemplified by Darwinian natural selection he did "not know what will take its place as a unified theory," he was willing to opine that, "the new theory will be rooted in a hierarchical view of nature" (1980c, 128–29). Over the next five years Steve's sense of what should be incorporated in evolutionary theory clarified as he began work on *The Structure of Evolutionary Theory* (2002c).

Ships passing in the night (1980j)

Diversity patterns figure prominently in Gould and Calloway's comparison of evolutionary dynamics in bivalves and brachiopods. The authors examine the question of competitive replacement and conclude that the rise to dominance of bivalves did not cause the decline of brachiopods: "The supposed replacement of brachiopods by clams is not gradual and sequential. It is a product of one event: the Permian extinction (which affected brachiopods profoundly and clams relatively little). When Paleozoic and post-Paleozoic times are plotted separately, numbers of clam and brachiopod genera are positively correlated in each phase. Each group pursues its characteristic and different history in each phase—clams increasing, brachiopods holding their own. The Permian extinction simply reset the initial diversities" (1980j, 383). They also emphasized the value of the nomothetic approach: "We are, nonetheless, intrigued that this pattern emerges in a simple analysis of residuals from an idealized curve based on data of the broadest kind. The corroboration of traditional knowledge by such nomothetic techniques should not be viewed with scorn as a sign that conventional procedures are all we need in paleontology but as an indication that the basic data of numerical diversity, for all their admitted inadequacy and inconsistency, probably do capture major patterns in the history of life" (1980j, 389).

Mass extinction became a centerpiece in Steve's evolving hierarchical view of evolutionary theory, and the nature of hierarchy

led to a warning against the simplicity of extrapolation in hierarchical systems: "The evolutionary ideas that most paleontologists employ are extensions by simple extrapolation, to vast times
and large groups, of processes that operate in local populations
over a few generations. But our tendency for uncritical extrapolation is bound often to lead us astray, unless reductionism is a
valid approach and evolutionary processes display no hierarchical
structure (with different styles of explanation emerging at higher
levels)" (1980j, 393–94). The authors also emphasized the importance of hierarchy for the role of paleontology in developing
evolutionary theory: "In framing their thoughts about macroevolution, paleontologists should consciously explore the ways
in which uncritical extrapolationism limit and channel thought.
Evolution works on a hierarchy of levels, and some causes at
higher levels are 'emergent.' These causes must be sought in
phenomena—like speciation—that cannot be rendered as an
extrapolation of sequential changes in gene frequencies within
local populations. The claim that paleontology can have an independent theory (within a unified system of evolutionary thought)
is not mere sectarian politics on our part, but a reflection of a
world arranged as a hierarchy of levels, not entirely as a smooth
continuum" (1980j, 395).

Validating a hierarchical approach to macroevolution (1982f)

Punctuated equilibrium was a touchstone for much of Steve's
thinking about the difference between microevolution and macroevolution. In a paper targeted for the neontological community
he used it to illustrate his developing ideas about trends: "Under
the alternate geometry of punctuated equilibrium...there is no
phyletic component to direct a trend; species arise (in geological time) with their differences established at the start, and
do not change substantially thereafter. Trends must therefore
be the product of a higher-order sorting that operates via the
differential birth and death of species considered as entities (the
same role that individual organisms, which do not change evolutionarily during their life, play in microevolution). This higher
order sorting of species, produced by differential origin and
extinction...must direct evolutionary trends within clades just as

natural selection, acting by differential birth and death of bodies, directs evolutionary change within populations (microevolution)" (1982f, 92). From this conclusion he advocates the independence of macro- and microevolution: "The key issue for the independence of macroevolution is not whether species selection operates in all trends (it does not), but whether the necessity, under punctuated equilibrium, of regarding trends as a higher-level sorting of species implies a new level in a hierarchy of evolutionary explanation.... It is in this sense that punctuated equilibrium is crucial to the independence of macroevolution—for it embodies the claim that species are legitimate individuals, and therefore capable of displaying irreducible properties" (1982f, 94). The importance of hierarchy in his emerging view of evolutionary theory follows: "The need for hierarchy in evolutionary theory is a contingent fact of the empirical world, not a mere issue of semantics or methodological styles.... The issue is larger than the independence of macroevolution. It is not just macroevolution vs. microevolution, but the question of whether evolutionary theory itself must be reformulated as a hierarchical structure with several levels—of which macroevolution is but one—bound together by extensive feedback to be sure, but each with a legitimate independence" (1982f, 97).

Diversity issues do not play a major role in this early statement of the central form of *The Structure of Evolutionary Theory* (2002c), except that in Steve's view trends are produced by differential diversification. For the first time Steve emphasizes the possibility of generating trends by higher speciation rates, rather than just by species selection through differential extinction: "I now believe that... evolutionary trends powered by differential birth are both common and more interesting in their unconventional implications. A trend can occur via differential speciation in two ways, only one of which may represent species selection. In "birth bias"... increased representation of one kind... arises only from its higher speciation rate (probability of extinction across all kinds may be constant). Trends are a product of differential origin" (1982f, 101–2). With evolutionary processes capable of generating trends in several different ways, a detailed knowledge of diversity dynamics would be important for understanding the nature of any trend.

The paper concludes with a hint that Steve was formalizing the views to be fleshed out later in *Structure*: "When a proper hierarchical theory is fully elaborated, it will not be entirely Darwinian in the strict sense of reduction to natural selection acting upon organisms. Yet I suspect that it will embody the essence of Darwinian argument in a more abstract and general form. We will have a series of levels with a source for the generation of variation and a mode (or set of modes) for selection among individuals at each level (Arnold and Fristrup 1982). The superseding of strict Darwinism may establish the Darwinian style of argument in its most general form as the foundation for a truly synthetic theory of evolution" (1982f, 104). Note here that Steve was not advocating abandoning Darwin, but of using "the Darwinian style of argument" separately tailored for different levels in the hierarchy, rather than extrapolating a continuum from microevolution to all levels.

Darwin centennials and catastrophes on earth (Gould 1983j, 1984h)

Stimulated by the then new emphasis on catastrophic mass extinction, Steve expounded on the importance of differentiating between micro- and macro-evolution in contributions to two symposium volumes, one commemorating the centennial of Darwin's death and the other on catastrophes in earth history. In each paper he used diversity pattern as a major example of information available about macroevolution in contrast to microevolution. In the Darwin centennial paper (1983j), Steve uses "the terms micro and macroevolution in the purely descriptive sense to designate the phenomenology, whatever its cause, of evolutionary change within versus among species," and he comments that "paleontology is a principal source of independent macroevolutionary theory that cannot be simply extrapolated from the evolutionary processes operating among organisms within populations" (1983j, 353). Diversity issues supply justification for his position: "Palaeontological phenomena described by differential rates of origin and persistence of species, rather than by differential rates of change within anagenetic lineages, provide a pool of events for an independent macroevolution," and he asserts that "patterns well described by stochastic models provide the best cases that palaeontology has offered during the last decade" [of patterns not directly predictable from microevolutionary events], concluding that "nothing in microevolutionary theory predicts whether or not stochastic models will apply" (1983j, 359).

Steve's paper in the catastrophes symposium is another argument for the "vindication of punctuational change" (1984h). Steve describes the trend to replace older gradualistic arguments related to a variety of aspects of the fossil record with new ideas supporting more abrupt, in some cases even catastrophically sudden, change. Again, diversity change is used for illustration: "Likewise, the most popular theories of my student days (the mid '60s, not so long ago) tried to deny—or, at least, to dilute substantially—the Cambrian explosion and the Permian extinction by labeling them 'preservational artifacts.' But the more these events are studied, the more they resist attempts to spread them out into gradualistic oblivion.... The history of marine invertebrate diversity is set primarily by its rare punctuations. The geological time scale is a mirror of these punctuations. No unit is without its 'boundary problems,' but the units are reasonably objective packages, not the arbitrary divisions of a gradual and uniform flow of life. Strength of punctuation established the hierarchy: the largest mark eras (Cambrian explosion, as well as Permian and Cretaceous extinctions); the smaller eras, periods" (1984h, 26).

The study of mass extinction has become a growth industry. The rate of publication on mass extinction held steady at a low rate from the nineteenth century until the mid-1950s, but has increased at an exponential rate for the last half century (Bambach 2006). In his paper in the catastrophes volume Steve made an interesting observation about the growth of interest in "punctuational" change that may be one reason the topic of mass extinction has become so popular: "I do not know how much of this new fascination for punctuational change resides in the stresses of our general culture. Information theory and general systems theory—with their concepts of equilibrium, steady state, homeostasis, feedback loops, and positive feedback leading to rapid autocatalytic change—certainly evoke a set of punctuational metaphors. Our uncertain world of nuclear armaments and deteriorating environments must also encourage a departure from gradualism" (1984h, 31).

The cosmic dance of Siva (1984x)

Steve Gould became a world-famous essayist through his column "This View of Life" which appeared in 300 consecutive issues of *Natural History* between January 1974 and January 2001. These essays generally were original scholarship, not rehashed material simplified for the general public. In a footnote in his 1989 discussion of "punctuated equilibrium in fact and theory" Steve stated his motive for writing the *Natural History* essays: "I use this series as a device to enhance personal learning, and therefore write about subjects that require probing and research. If I write about my own technical work, I learn nothing new" (1989e, 119; see also Allmon, this volume). In the preface to the last volume of his collected essays he was explicit about the scholarly content of his essays:

> Moreover, I refuse to treat these essays as lesser, derivative, or dumbed-down versions of technical or scholarly writing for professional audiences, but insist upon viewing them as no different in conceptual depth (however distinct in language) from other genres of original research. I have not hesitated to present, in this format, genuine discoveries, or at least distinctive interpretations, that would conventionally make their first appearance in a technical journal for professionals. I confess that I have often been frustrated by the disinclinations, and sometimes the downright refusals, of some (in my judgment) overly parochial scholars who will not cite my essays (while they happily quote my technical articles) because the content did not see its first published light of day in a traditional, peer-reviewed publication for credentialed scholars. Yet I have frequently placed into these essays original findings that I regard as more important, or even more complex, than several items that I initially published in conventional scholarly journals. (2001m, 6–7)

"The Cosmic Dance of Siva" is a perfect example of Steve's essays as original scholarly contributions. In this essay he first presented his idea that mass extinctions represent events at a level separate from regular micro- and macroevolutionary phenomena because they not only disrupt established evolutionary patterns, but unexpectedly and unpredictably alter the history of life. He later brought this idea into his professional work (see "The Paradox of the First Tier" below); it was a major source for his later advocacy of the importance of contingency in the history of life, and it remained at the heart of his thinking, being the focus of much of the last chapter of *The Structure of Evolutionary Theory* (2002c, especially 1320–40).

The suggestion by Luis Alvarez and colleagues that a massive bolide impact could have triggered the end-Cretaceous extinction dramatized the possibility of catastrophic extinction by providing possible evidence of colossal impact, documentation of a global iridium anomaly at the Cretaceous/Tertiary boundary (Alvarez et al. 1980). Attention was focused on extinction as never before. Dave Raup and Jack Sepkoski presented another intriguing idea on extinction at several professional meetings in 1983: the possibility that there had been a twenty-six million year periodicity to peaks of extinction intensity between the later Permian and now (Raup and Sepkoski 1984). Stimulated by this idea, several people who had heard Raup and Sepkoski's presentations wrote speculative papers that were published as articles in *Nature* in April, 1984, suggesting that the Sun might have a dark, distant companion that could trigger periodic meteor or comet showers from the outer reaches of the solar system and generate the proposed periodicity of extinction. In his August, 1984 *Natural History* column Steve discussed these developments and several others, but noted that he "would be performing no service in presenting a straight exposition" of the idea of periodic extinction or of the influence of the hypothesized dark companion star because so much had already been reported in the news media. Rather, "I wish, instead, to explain why this new theory of mass extinction might be so vitally important in altering our basic conception of the causes of pattern in life's history" (1984x; 1985y, 440).

Two important statements summarize Steve's views of the large-scale implications of the effects of unpredictable mass extinctions: (1) "In this view, external triggers of changing environment must drive the history of life on. But they drive it in unconventional directions: Where can we find the upward advance we seek so assiduously (to put ourselves on top of a struggling mass) if life only tracks a capriciously changing environment? Where can we locate predictable order at all if the primary environmental triggers are periodic cometary showers?" and (2) "In short, if mass extinctions are so frequent, so profound in their effects, and caused fundamentally by an extraterrestrial agency so catastrophic in impact and so utterly beyond the power of organisms to anticipate, then life's history either has an irreducible randomness or operates by new and undiscovered rules for perturbations, not (as we have always thought) by laws that regulate predictable competition during normal times" (1984x; 1985y, 446). Steve didn't ignore the fact that extinction also opened opportunity for the survivors. "Mass extinctions are not unswervingly destructive in the history of life. They are a source of creation as well, especially if the second view of external triggering is correct and the Red Queen of internal connection does not drive life inexorably forward. Mass extinction may be the primary and indispensable seed of major changes and shifts in life's history. Destruction and creation are locked in a dialectic of interaction" (1984x; 1985y, 448).

Steve had already written much in the professional literature on microevolutionary processes, such as natural selection in ecological time, and on macroevolutionary processes, such as speciation in geological time. But in this *Natural History* essay Steve pointed out that infrequent and unpredictable large-scale (and catastrophic) events could be seen as independent in time and effect from micro- and macroevolutionary processes. He would formalize these ideas as three tiers in the evolutionary hierarchy and emphasize their independence by pointing out the potential of events at a higher tier for frustrating progress made at a lower tier in a paper in *Paleobiology* in 1985. But the role of mass extinctions (later the "third tier") and their effects on evolutionary pattern was first discussed in this *Natural History* essay, as was the sense that the magnitude of extinction events may make nomothetic study of the history of life difficult or impossible.

The paradox of the first tier (1985f)

The professional paper that caps Steve's phase of emphasis on hierarchy of evolutionary phenomena, "The paradox of the first tier: an agenda for paleobiology" (1985f), clearly grew out of the *Natural History* essay, "The cosmic dance of Siva." Early in the paper Steve lays out his concepts of tiers of time: "The first tier includes evolutionary events of the ecological moment [natural selection and microevolution]. The second encompasses the evolutionary trends within lineages and clades that occur during millions of years in "normal" geological time between events of mass extinction.... The most exciting subject in paleobiology today, and the source (I suspect) of its principal agenda for the 1980s, lies in our recent recognition that one of our best-recognized and most puzzling phenomena, mass extinction, is not merely more and quicker of the same, but a third distinct tier with rules and principles of its own" (1985f, 2–3). Then he defines the paradox of the title: "I believe that our failure to find any clear vector of fitfully accumulating progress, despite expectations that processes regulating the first tier should yield such advance, represents our greatest dilemma for a study of pattern in life's history. I shall call it the *paradox of the first tier*" (1985f, 4). This is followed by a discussion of punctuated equilibrium and the second tier: "Punctuated equilibrium is a theory for the second tier—it studies the deployment of species and the origin of trends in normal geological time between episodes of mass extinction" (1985f, 4). Finally, Steve argues that the third tier is independent of the others: "As a summary statement, identifying the third tier as distinct, we may say that mass extinctions are more *frequent*, more *rapid*, more *extensive in impact*, and more *qualitatively different in effect* than our uniformitarian hopes had previously permitted most of us to contemplate" (1985f, 8). He also uses the third tier to resolve the paradox of the first: "It goes almost without saying that such a theory of mass extinction would largely resolve the paradox of the first tier. If anything like progress accumulates during normal times (and punctuated equilibrium casts doubt even upon this proposition), the vector of advance may be derailed often and profoundly enough to undo any long term directionality" (1985f, 10).

Although we can now see weaknesses in some of his views (periodicity of extinction did not become an important issue, the number of catastrophic mass extinction events may not be as large as Steve thought, and the effects of large-scale events are seldom as pervasive as Steve opined), study of mass extinctions and their effects dominated the 1980s and 1990s and still hold great interest (Bambach et al. 2004; McGhee et al. 2004; Bambach 2006).

Several items in the "paradox" paper foreshadow major themes of the final phase of Steve's career and make it an appropriate bridge to that later work. He denies directionality (and progress) in evolution. He also clearly looks ahead to the question of morphological disparity, as different from diversity, in the concluding section of the paper: "Questions are bursting out all over; I could fill the rest of this issue with a list, if the editor would only give me space. Just one, as an example of what we can now ask. Consider the history of mollusks and echinoderms and its general pattern of marked reduction in class-level diversity and great expansion at lower taxonomic levels within surviving groups. I believe that this reduction of diversity and expansion at stable points of design—and not progress—is the major pattern of life's history, at least for marine invertebrates" (1985f, 11). In these items we see the topics that would dominate *Wonderful Life* (1989d) and *Full House* (1996d), the two major books on evolutionary topics Steve would write prior to completing *The Structure of Evolutionary Theory* (2002c).

V. 1987–2002: Evolutionary theory expanded, progress out, contingency rules

The last phase of Steve Gould's evolving view of life's history contrasted with his earlier approach. He now frequently generalized about aspects of the entire history of life rather than focusing on specific topics such as speciation or differentiating levels of the evolutionary hierarchy. In one respect this was a natural culmination, a synthesis of the range of ideas he had been developing for the previous twenty years, and he capped it with his massive tome, *The Structure of Evolutionary Theory* (2002c). He did not abandon his earlier interests, however. In 1986 he published a monograph on *Cerion* (1986d) and he would write more papers on his favorite genus during the next decade. Heterochrony continued as a favorite topic, as well (1999d, 2000f). But rather than emphasizing a nomothetic approach to paleontology or explicating phenomena that might be used to identify or define independent levels in the evolutionary hierarchy, Steve emphasized holistic conclusions in many later works, even when the major portion of a publication focused on an example limited in space or time. Also, several of his most memorable forays were books written for the general public rather than research results presented in the technical professional literature. *Wonderful Life* (1989d) and *Full House* (1996d) are the documents from this last phase that most fully present his holistic overview of life's history. The first line of the "Preface and Acknowledgements" of *Wonderful Life* ("This book...attempts to tackle one of the broadest issues that science can address—the nature of history itself" [1989d, 13]) establishes Steve's ambition for that work (the book emphasizes contingency, not directionality, as the controlling feature underpinning history). Likewise, his comment, *"Full House* presents the general argument for denying that progress defines the history of life or even exists as a general trend at all" (1996d, 4) is unambiguous in declaring Steve's scope for that work.

These two books, both intended for a general audience, also reveal one reason why Steve's work was often considered controversial. Steve was a committed secular humanist, and both *Wonderful Life* and *Full House* argue that humans are an accident of evolution, not its purpose. This paper isn't the place to discuss this aspect of Steve's career (see Kelley, this volume), but it is important to understand that, reasoned as his statements are, they do generate a sense of aggressive challenge. It is clear from his rhetoric that Steve intended to create that sense of challenge. He wrote to get attention, and he hoped to stimulate debate.

The direction of evolutionary time (1987b)

Steve's last venture into the direct analysis of diversity patterns returned to the analysis of clade shapes first done ten years before (1973j, 1977a). Now, however, Steve (with his co-authors, former students Norman Gilinsky and Rebecca German) campaigns to drop the idea of progress from the history of life: "we wish to replace the grand but vague and noisome notion of progress with a question almost risibly limited by comparison—but imbued with the twin virtues of definition and testability: if you were handed a chart of clade diversity diagrams with unlabeled axes, would you know whether you were holding the chart upside down or right side up?" (1987b, 1437). Patterns of diversity are used to attack the question: "We pose an operational definition for arrows of time: does any asymmetry exist, statistically defined over large numbers of lineages, in the vertical dimension of clade diversity diagrams? If bottoms of lineages are definably different from tops, then evolutionary time has a direction, and the morphology of clade diversity diagrams can specify whether life's tape is running properly forward or illegitimately backward." (1987b, 1437). The authors determined the centers of gravity ("defined as the relative position in time of a clade's mean diversity, not as its time of maximum diversity" [1987b, 1438]) for a large number of extinct clades and looked for a pattern in their average position. Gould et al. found that early-arising clades had centers of gravity below 0.5 (earlier in time than the mid-point of the duration of the clade), whereas later-arising clades had centers of gravity generally close to 0.5 or, in the case of mammals, even higher, a result first noted a decade earlier (1977a). This was true for genera within families (1977a), families within orders (calculated three different ways to eliminate potential bias from variation in length of intervals), and for Cenozoic mammals (a new and more complete analysis than in 1977a). The authors argued for the reliability of their conclusions: "In practice, we attain greatest confidence when results are contrary to a primary bias. Since quality of preservation and quantity of available sediment increase as we approach the present, the known fossil record must impose a strong artifact favoring top-heavy clades. Since we have found an actual asymmetry of bottom-heaviness—and in older clades at that—we feel confident that we have detected a real pattern (weakened if anything by biases of the fossil record)" (1987b, 1440). The abstract of the paper summarizes its results: "Evolutionary time has a characteristic direction as demonstrated by the asymmetry of clade diversity diagrams in large statistical samples. Evolutionary groups generally concentrate diversity during their early histories, producing a preponderance of bottom-heavy clades among those that arise early in the history of a larger group. This pattern holds across taxonomic levels and across differences in anatomy and ecology (marine invertebrates, terrestrial mammals). The quantitative study of directionality in life's history (replacing vague, untestable, and culturally laden notions of 'progress') should receive more attention from paleontologists" (1987b, 1437).

Two aspects of this paper have always puzzled me. One is that Norm Gilinsky and I had done a bootstrapping study of families within orders (Gilinsky and Bambach 1986) and found that the centers of gravity of most real clades have a very strong tendency (fourteen of seventeen cases tested) to occur prior to that of the mean of multiple simulations for each clade when the same diversity dynamics are used in the simulation as occur in that real clade. In this sense, most clades, not just those originating early in time, have an asymmetry from "random" that marks a temporal direction. Yet that work was not referenced by Gould, Gilinsky, and German (1987b). Perhaps the authors only wanted a direct measure of asymmetry in the general form of clades to illustrate the general idea of a test independent of progress (the analysis Gilinsky and I had done was somewhat indirect, requiring comparison of the real clade with multiple simulations for each clade individually). The other puzzling aspect is why there was no discussion of the lack of bottom-heaviness of later-arising clades. Since Steve was against claiming progress as a characteristic of the history of life, it should have been appealing to note that, except for early-arising groups, no consistent arrow was reliably detectable in the general shape of later-arising clades (although the difference between simulated and real individual clades noted above is generally present).

Trends (1988c)

Steve was president of the Paleontological Society in 1987. His presidential address to the Paleontological Society summarized his ideas about trends. Along the way it allowed him to display his well-known interest in baseball as he reviewed his conclusion (first published in *Vanity Fair* in 1983 [1983a]) that the disappearance of 0.400 batting averages in the major leagues was a trend of decrease in variance as the skills of ball players for pitching and defense increased, not a directional trend toward lesser batting skills, and he used that example as one of the ways into a favorite topic of his later career, the lack of actual direction in many claimed trends. Steve's position was that, "Trends are the primary phenomenon of macroevolution, but they have often been misinterpreted because an old and deep conceptual error has induced us to misread, as anagenesis in abstracted entities, a pattern that actually records changes in variance by increase or decrease in diversity or disparity among species within clades" (1988c, 319). In his later career Steve emphasized change in variance in distributions—and the lack of selection or sorting—as producing apparent trends determined from "abstracted" entities (such as means). His new idea (inspired, as he acknowledged, by Steve Stanley's insights on Cope's Rule [Stanley 1973]), was that many claimed trends were not driven changes. Differential speciation, which would be important in changing variance even if it didn't result in a selection-driven trend, remained important as an attribute of macroevolution: "Macroevolution—including the central phenomenon of trends—must be conceptualized as the differential success of species. If microevolution results from the sorting of organisms within populations, then macroevolution occurs by sorting of species within clades (1986h). Nature operates an entity making and breaking machine in the processes of speciation and extinction. If differential birth and death produce anagensis in populations, then differential speciation and extinction forge trends within clades" (1988c, 320).

Gould's published presidential address also brought Bruce Runnegar's idea of disparity of form into his work for the first time: "Runnegar (1987) makes the valuable distinction between *diversity*, or number of species, and *disparity*, or average difference among species. If number remains constant and disparity increases or decreases (as new kinds of species replace old), the same effects may arise—for example, means and extremes may increase while modes remain constant if the mode lies near one end of a potential range and average disparity among a constant number of species grows" (1988c, 321). Steve discussed a variety of examples of trends as change of variance and even brought in the Burgess Shale late in the paper. The issue of trends was summarized as: "For trends, the key to understanding is often not the perceived excursion itself (for the 'entity' supposedly in motion may be a misleading abstraction, or a forced and secondary effect of starting points or limits of ranges), but an increase or decrease in number of disparity of species within clades" (1988c, 528–29).

These ideas would be the basis for much expanded discussion in *Wonderful Life* and *Full House*.

Fact and theory (1989e)

In a review, "Punctuated equilibrium in fact and theory" (1989e), Steve made some insightful remarks about his career interests. He reminisced a bit on his graduate school experience: "Norman

Newell was the most biologically oriented of 'old guard' palaeontologists and had personally fomented this restructuring [the shift
in paleontology from a dominant emphasis on biostratigraphy
to what is now called paleobiology] by urging students to tackle
evolutionary problems.... Niles [Eldredge] and I went to study with
Newell because we were primarily interested in evolution.... We
were both interested in small-scale, quantitative research on species
and lineages (a concern fostered by our other advisor, John Imbrie,
who taught us multivariate statistical analysis)" (1989e, 117–18).
He also revealed his commitment to hierarchy in evolutionary
theory: "I believe that hierarchy theory in causal perspective will
provide the most fundamental reconstruction of evolutionary
theory since the Modern Synthesis of the 1930s and 40s.... For
punctuated equilibrium, we may at least say that it directed the
attention of macroevolutionists to the causally distinct status of
species, and that this insight helped to lead the larger discipline of
evolutionary biology toward the expanded Darwinism of hierarchy
theory." (1989e, 126–27). He also elaborated at length on the
need to use the term species sorting, rather than species selection:
"our misnamed 'species selection' (of Stanley 1975 and Gould and
Eldredge 1977) was really a claim about species sorting" (1989e,
122), and he credited Elizabeth Vrba with bringing the distinction to the fore. Although he backed away from species selection
as a topic (both here and in the *Structure of Evolutionary Theory*)
he held on to species sorting as a macroevolutionary method of
making trends. Despite Steve's caution, Coyne and Orr, in a major
new review of speciation (Coyne and Orr 2004), state, "those
who continue to debate the possibility of species selection fail to
realize that comparative studies have already settled the issue.
What remains is to determine how often this type of selection has
shaped evolutionary trends" (Coyne and Orr 2004, 445). Steve had
claimed early on that species sorting and trends were associated.
Apparently there is still work to be done on Gouldian concepts.

Wonderful Life (1989d)

Wonderful Life won the Rhone-Poulenc Prize for science writing,
was a finalist for the Pulitzer Prize, and won the Forkosch Award.
Although intended for general readership, the book is the

central document about Steve's "post-nomothetic" emphasis on contingency in evolution. It is also where he launched his view contrasting early disparity with later diversity. Steve's thesis was that great disparity of form evolved early and the later increase of diversity that characterizes the overall history of life occurred within a much smaller number of anatomical body plans, groups that may have survived just by chance. This coupled Steve's long-held interest in the origins of morphology to his conviction that no progressive directionality had characterized the history of life. He explained the concept of disparity and the difference between it and diversity (also as used throughout this paper) thus: "Biologists use the vernacular term *diversity* in several different technical senses. They may talk about 'diversity' as number of distinct species in a group...But biologists also speak about 'diversity' as difference in body plans....Several of my colleagues (Jaanusson 1981; Runnegar 1987) have suggested that we eliminate the confusion about diversity by restricting this vernacular term to the first sense—number of species. The second sense—difference in body plans—should then be called *disparity*" (1989d, 49). The strange assortment of taxa found in the Burgess Shale of mid-Cambrian age and the contrast between the interpretation of their affinities by Charles Walcott, who originally described most of the fauna, and the reinterpretation of the fauna by Harry Whittington and his students, especially Simon Conway Morris and Derek Briggs, are the focus for the book, but generality about the nature of evolution is the real topic.

The contrast between diversification, which Steve acknowledged had increased greatly since the mid-Cambrian, and the variety of body plans of animals, which Steve argued was more varied in the mid-Cambrian than now, creates a dissonance between diversity and disparity: "The revision of the Burgess Shale rests upon its diversity in this second sense of *disparity* in anatomical plans. Measured as number of species, Burgess diversity is not high. This fact embodies a central paradox of early life: How could so much disparity in body plans evolve in the apparent absence of substantial diversity in number of species?" (1989d, 49).

Steve implies that the overall history of diversity distracts from understanding the nature of evolution: "Within these constraints of *monophyly* and *divergence*, the geometric possibilities for evolutionary trees are nearly endless. A bush may quickly expand to maximal width and then taper continuously, like a Christmas tree. Or it may diversify rapidly, but then maintain its full width by a continuing balance of innovation and death. Or it may, like a tumbleweed, branch helter-skelter in a confusing jumble of shapes and sizes. Ignoring these multifarious possibilities, conventional iconography has fastened upon a primary model, the 'cone of increasing diversity,' an upside-down Christmas tree. Life begins with the restricted and simple, and progresses ever upward to more and more and, by implication, better and better" (1989d, 38). He then charges:

> The iconography of the cone made Walcott's original interpretation of the Burgess fauna inevitable. Animals so close in time to the origin of multicellular life would have to lie in the narrow neck of the funnel. Burgess animals therefore could not stray beyond a strictly limited diversity and a basic anatomical simplicity.... I know of no greater challenge to the iconography of the cone—and hence no more important case for a fundamentally revised view of life—than the radical reconstructions of Burgess anatomy presented by Whittington and his colleagues... they have turned the traditional interpretation on its head. By recognizing so many unique anatomies in the Burgess, and by showing that familiar groups were then experimenting with designs so far beyond the modern range, they have inverted the cone. The sweep of anatomical variety reached a maximum right after the initial diversification of multicellular animals. The later history of life proceeded by elimination, not expansion. The current earth may hold more species than ever before, but most are iterations upon a few basic anatomical designs.... But the Burgess pattern of elimination also suggests a truly radical alternative, precluded by the iconography of the cone. Suppose that winners have not prevailed for cause in the usual sense. Perhaps the grim reaper of anatomical designs is only Lady Luck in disguise.... Perhaps the grim reaper works during brief episodes of mass extinction, provoked by unpredictable environmental catastrophes.... Groups may prevail or die for reasons that bear no relationship to the Darwinian basis of success in normal times. (1989d, 45–48)

Steve concentrates on the idea that much of the history of life had been determined simply by which of the original disparate body plans survived. He felt that most had gone extinct primarily by chance, not through failure in competition, and he called this loss decimation: "When I speak of decimation, I refer to reduction in the number of anatomical designs for life, not numbers of species" (1989d, 49). The chance of any particular body plan surviving and having the chance to diversify depended on many unpredictable events, making survival a contingent event. Hence the history of what did happen is just one of many possible histories that could have happened. Steve argues that if diversification is what impresses you, a different interpretive scheme will emerge than the one you construct if early disparity and later contingent survival is what determined which groups became diverse: "I believe that the reconstructed Burgess fauna, interpreted by the theme of replaying life's tape, offers powerful support for this different view of life: any replay of the tape would lead evolution down a pathway radically different from the road actually taken. But the consequent differences in outcome do not imply that evolution is senseless, and without meaningful pattern; the divergent route of the replay would be just as interpretable, just as explainable *after* the fact, as the actual road. But the diversity of possible itineraries does demonstrate that eventual results cannot be predicted at the outset. Each step proceeds for cause, but no finale can be specified at the start, and none would ever occur a second time in the same way, because any pathway proceeds through thousands of improbable stages. Alter any early event, ever so slightly and without apparent importance at the time, and evolution cascades into a radically different channel. This...represents no more nor less than the essence of history. Its name is contingency—and contingency is a thing unto itself, not the titration of determinism by randomness" (1989d, 51).

Steve described the fauna of the Burgess Shale at length and then stated his view of its importance: "The Burgess Shale includes a range of disparity in anatomical design never again equaled, and not matched today by all the creatures in all the world's oceans. The history of multicellular life has been dominated by decimation of a large initial stock, quickly generated in the Cambrian explosion. The story of the last 500 million years has featured restriction followed by proliferation within a few stereotyped designs, not general expansion of range and increase in complexity as our favored iconography, the cone of increasing diversity, implies" (1989d, 208).

He also summarized the larger meaning of his argument:

Contingency is both the watchword and lesson of the new interpretation of the Burgess Shale. The fascination and transforming power of the Burgess message—a fantastic explosion of early disparity followed by decimation, perhaps largely by lottery—lies in its affirmation of history as the chief determinant of life's directions.... The new view... is rooted in contingency. With so many Burgess possibilities of apparently equivalent anatomical promise—over twenty arthropod designs later decimated to four survivors, perhaps fifteen or more unique anatomies available for recruitment as major branches, or phyla, of life's tree—our modern pattern of anatomical disparity is thrown into the lap of contingency. The modern order was not guaranteed by basic laws (natural selection, mechanical superiority in anatomical design), or even by lower-level generalities of ecology or evolutionary theory. The modern order is largely a product of contingency. (1989d, 288)

Steve's focus had changed since he had written hopefully, in 1980, that "paleontology is not a pure historical science; it resides in the middle of a continuum stretching from idiographic to nomothetic disciplines."

Steve used the "bottom-heavy" clade analysis of Gould, Gilinsky, and German (1987b) as a starting point for a summarizing general statement. Because of early disparity, Steve felt that evolution had done most of its creative work at the beginning of diversification in "filling the ecological barrel," when clades were bottom-heavy. After that, life was a directionless series of interesting, but not fundamentally important, events:

"The early history of multicellular life is marked by a bottom-heavy signature for individual lineages, later times feature symmetrical lineages.... We may interpret this bottom-heavy pattern in several ways. I like to think of it as 'early experimentation and later standardization.' Major lineages seem able to generate remarkable disparity of anatomical design at the outset of their history—early experimentation. Few of these designs survive an initial decimation, and later diversification occurs only within the restricted anatomical boundaries of these survivors—later standardization. The number of species may continue to increase, and may reach maximal values late in the history of lineages, but these profound diversifications occur within restricted anatomies" (1989d, 304).

Destruction by the "third tier: of mass extinction and species sorting mechanisms, not natural selection on the genomes of individuals, which produce radiations and trends at the second tier, intervene to make selective evolution within species almost a sideshow. For G. Evelyn Hutchinson's "ecological theater and evolutionary play" (Hutchinson 1965), if you restaged the play, you could not predict what even the basic plot of the show might be:

> "However we interpret this bottom-heavy pattern, it strongly reinforces the case for contingency, and validates the principal theme of this book. First, the basic pattern is a disproof of our standard and comfortable iconography—the cone of increasing diversity. The thrall of this inconography and its underlying conceptual base prevented Walcott from grasping the true extent of Burgess disparity, and has continued to portray the controlling pattern of evolution in a direction opposite to its actual form. Second, maximal initial disparity and later decimation give the broadest possible role to contingency, for if the current taxonomic structure of life records the few fortunate survivors in a lottery of decimation, rather than the end result of progressive diversification by adaptive improvement, then a replay of life's tape would yield a substantially different set of surviving anatomies and a later history making perfect sense in its own terms but markedly different from the one we know" (1989d, 304).

Why we must strive to quantify morphospace (1991i)

Steve's claim of a uniquely broad range of early disparity was, and remains, controversial. Although not as sustained or widespread as the attention lavished on punctuated equilibrium, discussion about disparity has been serious and is not yet finished. Steve responded to some of the early criticisms in typical rousing form: "Three major arguments have been raised against the crucial claim...that disparity of anatomical design reached an early maximum in the history of multicellular life...I show that all these arguments are either false or illogical." (1991i, 411). However, in shifting attention from diversity, for which a variety of analytic techniques were available, to disparity,

with its implications of different degrees of morphological difference, Steve recognized that a new set of techniques would be needed
to rigorously analyze the issue. "I have argued that the critics of
greater early disparity are wrong in their central claims about cladistics and retrospective fallacies. But these critics, while choosing false
targets, are motivated by a serious and entirely legitimate malaise.
The problem, however, is caused by absent technique, not incorrect
argument or even inherent ambiguity" (1991i, 419).

So Steve set us an agenda, once again. Paleontology has developed a lot in the last thirty years, but working out this agenda
will surely be the way to get to a new level of understanding. It is,
however, a tough task: "What, then, do we need? . . . . We need . . . to
define a full range of the abstract (and richly multivariate) space
into which all organisms may fit (the morphospace). We must then
be able to characterize individual organisms and plot them within
this encompassing space. Finally, we need to measure density,
range, clumping, and a host of other properties that determine
differential filling of this totality; and we must be able to assess
the variation in this differential filling through time. (The claim
for greater initial disparity is, effectively, a statement that the
Cambrian range exceeds modern occupation.)" (1991i, 420).
Asking for the complete quantification of morphospace and its
occupation is a tall order. Even Steve didn't know how to get it
done: "These questions are dauntingly difficult, and I do not
pretend to have a solution. . . . I do confess some fears that, *in toto*,
the question of morphospace may be logically intractable, not
merely difficult. . . . Such issues may make a general solution intractable" (1991i, 420–421). George McGhee has written an introductory review of theoretical morphology (McGhee 1999), and Karl
Niklas has discussed the various ways morphology in plants may
respond to a range of needs (Niklas 1997), but Gould's dream is
not likely to become reality any time soon.

A fundamental challenge in emphasizing disparity, rather
than diversity, is determining what we should be measuring. As
Runnegar defined disparity, it is a comparative concept: "There
are two main ways of measuring evolutionary distance in living
and fossil organisms, although the degree of complexity may also
provide a rough indication of the amount of evolution that has
occurred. The first is *taxonomic diversity* (the number of kinds . . .)

and the second is *morphologic disparity* (the amount of difference between related phyla, classes, species, individuals, proteins, genes etc.)" (Runnegar 1987, 41). Runnegar specifies that, "innovation—the crossing of a functional threshold—gives rise to morphological disparity" and notes that "the main source of new higher taxa (macroevolution) lies in the pathway that leads to morphological disparity. Such developments involve the crossing of one or more new functional thresholds and the organisms that approach these thresholds develop morphologies that make this possible" (Runnegar 1987, 41). If disparity is really a comparative concept ("the amount of difference")—and if disparity is not just the range of variation in morphological characters but something of larger scope (Runnegar says innovation, not variation "gives rise to morphological disparity" and Steve called disparity "difference in body plans"), then the problem Steve recognized starts with the question: what are we actually trying to specify? It isn't something we have ever dealt with in a quantitative manner. Runnegar even argued, "morphological disparity is often best described in qualitative terms" (Runnegar 1987, 41).

Full House (1996d)

Steve was skilled at making arguments that were logical and convincing, but, at the same time, raised hackles. Although he was careful to make his statements so he could defend them, many seem to be planned to generate reactions. *Full House* is an interesting, but pugnacious, recasting of the traditional view of the history of life and regularly invites argument. Steve emphasizes two main phenomena in *Full House*: (1) many perceived trends are not directional sequences but just changes in variance (an idea expanded from his presidential address to the Paleontological Society; 1988c), and (2) the fact that bacteria not only are among the earliest evolved forms of life, but are extraordinarily abundant and important for biogeochemical processes today. He uses these to build a case for denying progress in evolution.

Steve opens with a blast: *"Full House* presents the general argument for denying that progress defines the history of life or even exists as a general trend at all. Within such a view of life-as-a-whole, humans can occupy no preferred status as a pinnacle or culmination. Life has always been dominated by its bacterial mode" (1996d, 4). His rhetorical attack continues:

> *Homo sapiens* is not representative, or symbolic, of life as a whole...how can this invention [human consciousness and intelligence] be viewed as the distillation of life's primary thrust or direction when 80 percent of multicellularity (the phylum Arthropoda) enjoys such evolutionary success....Why, then, do we continually portray this pitifully limited picture of one little stream in vertebrate life as a model for the whole multicellular pageant?....I shall argue in this book that our unquestioning approbation of such a scheme provides our culture's most prominent example of a more extensive fallacy in reasoning about trends—a focus on particulars or abstractions...egregiously selected from a totality because we perceive these limited and uncharacteristic examples as moving somewhere—when we should be studying variation *in the entire system* (the 'full house' of my title) and its changing pattern of spread through time. I will emphasize the set of trends that inspires our greatest interest—supposed improvements through time. And I shall illustrate an unconventional mode of interpretation that seems obvious once stated, but rarely enters our mental framework— trends properly viewed as results of expanding or contracting variation, rather than concrete entities moving in a definite direction. This book, in other words, treats the "*spread* of *excellence*," or trends to improvement best interpreted as expanding or contracting variation. (1996d, 15–16).

Steve presents a primer on skewness in distributions, noting that skewness usually develops when one side of a distribution is limited by minimal ("left wall") or maximal ("right wall") conditions. He also investigates the problem of extracting single examples out of a range of cases or of treating the overall average of a distribution (the mean) as an entity and concludes, "I therefore submit that the history of any entity (a group, an institution, an evolutionary lineage) must be tracked by changes in the variation of all components—the full house of their entirety—and not falsely epitomized as a single item (either an abstraction like a mean value, or a supposedly typical example) moving in a linear direction" (1996d, 72–73). Steve discusses many aspects of changing variance and its affect on distributions, and especially how change of variance influences the value of extremes at different times. Almost sixty pages are devoted to his favorite example, the disappearance of the 0.400 hitter in baseball. He demonstrates that the hitting skills of the best batters has not decreased; instead, overall pitching and fielding skills have increased in quality so that batters with skills equal to the 0.400 hitters of the past simply cannot achieve such high absolute levels of success. Curiously, in a book dedicated to "denying that progress defines the history of life or even exists as a general trend at all," Steve uses an example in which there is an actual trend of overall progress (improvement in several aspects of play in baseball) as the main case to illustrate the methodology of his argument!

Steve then brings up bacteria and their importance in the history of life, pointing out that they are very diverse and make up possibly a majority of living biomass in the world today. The distributional argument, as related to life's history, is stated as:

> "life had to begin next to the left wall of minimal complexity....As life diversified, only one direction stood open for expansion. Nothing much could move left and fit between the initial bacterial mode and the left wall. The bacterial mode itself has maintained its initial position and grown continually in height....Since space remains available away from the left wall and toward the direction of greater complexity, new species occasionally wander into this previously unoccupied domain, giving the bell curve of complexity for all species a right skew, with capacity for increased skewing through time" (1996d, 171).

> As the main claim of this book, I do not deny the phenomenon of increased complexity in life's history—but I subject this conclusion to two restrictions that undermine its traditional hegemony as evolution's defining feature. First, the phenomenon exists only in the pitifully limited and restricted sense of a few species extending the small right tail of a bell curve with an ever-constant mode at bacterial complexity—and not as a pervasive feature in the history of most lineages. Second, this restricted phenomenon arises as an incidental consequence—an "effect," in the terminology of Williams (1966) and Vrba (1980), rather than an intended result—of causes that include no mechanism for progress or increasing complexity in their main actions. (1996d, 197)

Note how the whole diversification of eukaryotes, both protests and multicellular animals and plants, becomes "pitifully limited" to "a few species" in "the small right tail" of a skewed bell curve with a bacterial mode. Steve uses carefully chosen rhetoric to downplay the topics that not only have been a primary focus of most paleontologists, but that interested him, too, through much of his career.

The point of view about the history of life expressed in *Full House* is that evolutionary branching may have produced the wide range of organisms in the modern world, including complex mammals as well as simple amoebas among the eukaryotes and even less morphologically complex prokaryotes, but that this is simply a product of the expansion of life from a "left wall" of minimal complexity, not a driven trend to increased complexity. Left walls in this model are irreducible minimalist boundaries, permitting expansion in only one direction. However, a right wall, by the same logic, should be an impenetrable barrier, against which a distribution would "pile up" if it were to expand to meet it. In that case the only way to expand further would be to climb the right wall, to evolve some new ability that makes the ascent possible. And here is the place where Steve's approach glosses over some really important steps. Bacteria, as prokaryotes, simply cannot attain the complexity of eukaryotes. If eukaryotes and multicellularity had not evolved, bacteria could not and would not have become coral reefs, giant sequoias, whales, or people—they were not produced by increase in bacterial variance. Although Steve's argument about passive expansion from left walls is fundamentally sound, it is not the whole story.

The biosphere is an entity only at the grossest scale. Evolution has formed many entities (evolutionary lineages—clades) in the hierarchically structured evolutionary tree of life. Each branch of the tree had a history that, while influenced by the other branches, was also independent of them. Each clade has a history, and those may be quite different from the overall pattern of life-as-a-whole. Achieving the range of life in the biosphere has not been a simple singular expansion from minimal complexity. Many different expansions are combined, some necessarily in specific sequence, in the overall biosphere, and each has its own separate story. For example, Maynard-Smith and Szathmáry (1995) and Knoll and Bambach (2000) touch on these issues and the question of directionality in the history of life. Early in the book, however, Steve is careful to note he was adopting the perspective of viewing "life-as-a-whole" and considering "variation *in the entire system* (the "full house" of my title) and its changing pattern of spread through time." He was not looking at the internal organization of that whole or what might be necessary to permit the whole story to actually unfold. In the limited sense Steve took care to define, the pattern considered only at the holistic level, his story is correct.

We need to remember that many of Steve's works were almost political in nature. They were intended to define a philosophical position or point the way to future study. *Full House* does not look at the trees, just the forest. Steve was interested in the conceptual viewpoint, not the detail of description.

The Structure of Evolutionary Theory (2002c)

Steve Gould's last major professional contribution was his massive *The Structure of Evolutionary Theory* (2002c). I will not parse this large book—its subject is not diversity. The only items under the term "diversity" in the index are two references to Darwin's admitted failure in solving "the problem of diversity" (2002c, 47, 50). But all the features of the hierarchy of evolutionary phenomena that Steve had discussed over the years using diversity as an exemplar are incorporated in the text. For instance, as noted above, much of the last chapter, "Tiers of time and trials of extrapolationism, with an epilog on the interaction of general theory and contingent history," is devoted to concepts initially developed earlier (1984x, 1985f).

The Structure had a long gestation. Steve said he worked on the project for twenty years (2002c, 1339), an interval starting in the early 1980s when he first expressed his commitment to a hierarchical view of evolution. Yet, in the preface of his first professional book, *Ontogeny and Phylogeny* (1977e), Steve noted he had started that project in 1970 as a prelude to his planned book on macroevolution. Steve had it in mind to write a major tome on evolution for his entire career.

VI. The ideas mattered; it was philosophy, not description

I conclude this lengthy review with three different observations: a note on the unity of Steve Gould's commitment in all that he did to evolutionary thinking; a brief general review of his reputation related to the shape of his evolving career; and an observation on the breadth of his influence.

The Good People of Halifax (2001f)

Steve Gould used evolutionary concepts in all his writings, general as well as technical. One of his finest public statements was written after the tragedy of September 11, 2001. When the attacks occurred in New York and Washington, Steve was flying to New York from Milan. His plane was grounded in Halifax, Nova Scotia, and he was there for a week before he could leave. His first writing after the event, published in Canada's national newspaper, *The Globe and Mail*, was a thank-you for the kindness the citizens of Halifax had shown all 9,000 stranded air travelers who had been headed for New York around that fateful time. Part of the message was derived from his thinking about the tiers of evolution (1985f), but with a twist. While acknowledging the disruptive effect of catastrophes at a level analogous to the third tier, he turned the argument so that features analogous to selectional events at the first tier actually win out. I don't think it was just for the essay; I think it is a different, but legitimate, view of the structure of evolution as he had formulated it.

> In an important, little appreciated and utterly tragic principle regulating the structure of nearly all complex systems, building up must be accomplished step by tiny step, whereas destruction need occupy but an instant. In previous essays on the nature of change, I have called this phenomenon the Great Asymmetry (with uppercase letters to emphasize the sad generality). Ten thousand acts of kindness done by thousands of people, and slowly building trust and harmony over many years, can be undone by one destructive act of a skilled and committed psychopath. Thus, even if the *effects* of kindness and evil balance out in the course of history, the Great Asymmetry guarantees that the *numbers* of kind and evil people could hardly differ more, for thousands of good souls overwhelm each perpetrator of darkness.
>
> I stress this greatly underappreciated point because our error in equating a *balance of effects* with *equality of numbers* could lead us to despair about human possibilities, especially at this moment of mourning and questioning, whereas, in reality, the decent

multitudes, performing their ten thousand acts of kindness, vastly outnumber the very few depraved people in our midst. Thus, we have every reason to maintain our faith in human kindness, and our hopes for the triumph of the human potential, if only we can learn to harness this wellspring of unstinting goodness in nearly all of us, (2001f; 2001m, 390–91).

Conclusions on career shape and reputation

Steve Gould's reputation and the regard with which his work was held changed as his career progressed. From the late 1960s to the early 1980s Steve occupied a central position in paleontology, formulating ideas and concepts to which the whole profession directly responded. Then, as he began work on *The Structure of Evolutionary Theory* in the early 1980s and he changed from concentrating on traditional paleontology to thinking about evolutionary biology as a whole, fewer practicing paleontologists found as much direct relevance in his work for their daily tasks. Also, his style in presenting his overarching generalizations seemed to oversimplify or to gloss over situations that were of interest to workers in specialized areas. This is always one of the dangers of theorizing, and the impressions made by his style in works intended for a general audience made misunderstanding easier, especially if the language wasn't read very carefully.

But his ideas of hierarchy and interactions between levels are real issues. There is a difference between gene sequencing and era-bounding extinction events. Steve never denied that evolutionary change in lineages was anything but change in gene frequencies. But he did, correctly, insist that there are patterns of evolutionary change to be seen at other scales of time and taxonomy, which are not predictable from gene sequencing. On that, I believe, Steve was right. His interest, however, was focused on the general concepts; he left us to work out the details, they simply weren't his concern.

In his later career he emphasized contingency and used the "third tier" of mass extinctions to deny progress. He also changed his treatment of diversity. Where he had once used it as his exemplar for macroevolution being independent from microevolution and to argue for high status for paleontology as a discipline, his emphasis changed; in his holistic presentations, diversity was characterized almost as a barrier to understanding what was important. This was mostly a rhetorical device to focus attention on the concepts of contingency and disparity that he felt were important new viewpoints needing emphasis, but it also de-emphasized the importance of even his own earlier work. To be sure, details of diversity history are contingent (most obviously in the effect of mass extinctions); but recent work on alpha diversity (Powell and Kowalewski 2002; Bush and Bambach 2004), on changes in the balance of major ecotypes through time (Bambach et al. 2002; Bush et al. 2007), and on the nature of the sequence of major evolutionary breakthroughs in the evolution of the biosphere (Knoll and Bambach 2000) suggest that there may be more structure to the history of life than the impression Steve created of an unpredictable, directionless evolutionary sequence. The full story is likely to be a path between Steve's *Wonderful Life/Full House* and Simon Conway Morris' *The Crucible of Creation/Life's Solution* (Conway Morris 1998, 2003). Real life may be more a mix of contingency and determinism than Steve intimated.

Stephen Jay Gould's pervasive influence in modern evolutionary biology and paleontology (*Nature*, October 13, 2003)

In the week before the Geological Society of America meeting at which this paper was originally presented in a symposium on Steve Gould's career, *Nature* for October 13, 2003 contained two items that would have interested Steve greatly. They both demonstrate the depth of his influence on evolutionary studies. One was a news report by Paul Harvey and Andy Purvis (Harvey and Purvis 2003) commenting on Robert Ricklefs's study of diversification in clades of passerine birds (Ricklefs 2003). They reported that Ricklefs found no phylogenetic association between morphological or behavioral innovation and species-rich clades: "Ricklefs proposes that unusual expansion of geographical ranges might explain the unusually species-rich clades. . . . But what about species-poor taxa? Here the explanation is that tribes with few species simply never had a chance to radiate because, for instance, they are restricted to remote locations away from the continental landmasses or are dietary specialists" (Harvey and Purvis 2003, 676). The other report is a brief note documenting that a particular single gene mutation in the genus *Euhadra*, a Japanese land snail, reverses its coiling direction and forces the bearers to only breed among themselves (Ueshima and Asami 2003). Ueshima and Asami conclude, "Our results indicate that single-gene speciation is possible, at least in hermaphroditic snails, contrary to the traditionally held view" (Ueshima and Asami 2003, 679). Here we have contingency, not key innovations, controlling evolutionary success and instant speciation, punctuation at its best.

Steve is not cited in either report (or in the full report on passerine birds by Ricklefs). But he doesn't need to be. Punctuated equilibrium and contingency are well-established concepts now; as ideas they are no longer news. But Steve made both types of study central to modern biology and paleontology. Now we all work in Gould's world.

References

Alroy, J., C. R. Marshall, R. K. Bambach et al. 2001. Effects of sampling standardization on estimates of Phanerozoic marine diversification. *Proceedings of the National Academy of Sciences USA* 98: 6261–66.

Alvarez, L. W., W. Alvarez, F. Asaro, and H. V. Michel. 1980. Extraterrestrial cause for the Cretaceus-Tertiary extinction. *Science* 208: 1095–1108.

Arnold, A. J., and K. Fristrup. 1982. The theory of evolution by natural selection: a hierarchical expansion. *Paleobiology* 8: 113–29.

Bambach, R. K. 2006. Phanerozoic biodiversity mass extinctions. *Annual Review of Earth and Planetary Science* 34: 127–55.

Bambach, R. K., A. H. Knoll, and J. J. Sepkoski, Jr. 2002. Anatomical and ecological constraints on Phanerozoic animal diversity in the marine realm. *Proceedings of the National Academy of Sciences USA* 99: 6854–59.

Bambach, R. K., A. H. Knoll, and S. C. Wang. 2004. Origination, extinction, and mass depletions of marine diversity. *Paleobiology* 30: 522–42.

Bush, A. M., and R. K. Bambach. 2004. Did alpha diversity increase during the Phanerozoic? Lifting the veils of taphonomic, latitudinal, and environmental biases. *Journal of Geology* 112: 625–42.

Bush, A. M., R. K. Bambach, and Gwen M. Daley. 2007. Changes in theoretical ecospace utilization in marine fossil assemblages between the mid-Paleozoic and late Cenozoic. *Paleobiology* 33(1): 76–97.

Bush, A. M., M. J. Markey, and C. R. Marshall. 2004. Alpha, beta, gamma: the effects of spatially organized biodiversity on sampling-standardized diversity analysis. *Paleobiology* 30: 666–86.

Conway Morris, S. 1998. *The crucible of creation: The burgess shale fauna and the rise of animals.* New York: Oxford University Press.

————. 2003. *Life's solution: Inevitable humans in a lonely universe.* Cambridge, UK: Cambridge University Press.

Coyne, J. A., and H. A. Orr. 2004. *Speciation.* Sunderland, MA: Sinaure Associates, Inc.

Eldredge, N. 1971. The allopatric model and phylogeny in Paleozoic invertebrates. *Evolution* 25: 156–67.

————. 1972. Systematics and evolution of *Phacops rana* (Green 1832) and *Phacops iowensis* (Delo 1935) (Trilobita) from the Middle Devonian of North America. *Bulletin of the American Museum of Natural History* 147(2): 45–113.

————. 1995. *Reinventing Darwin: The great debate at the high table of evolutionary theory.* New York: John Wiley and Sons.

Eldredge, N., and S. J. Gould. 1972. Punctuated equilibrium: an alternative to phyletic gradualism. In T. J. M. Schopf, ed., *Models in paleobiology,* 82–115. San Francisco: Freeman, Cooper and Company.

Gilinsky, N. L. 1994. Volatility and the Phanerozoic decline of background extinction intensity. *Paleobiology* 20: 445–58.

Gilinsky, N. L., and R. K. Bambach. 1986. The evolutionary bootstrap: a new approach to the study of taxonomic diversity. *Paleobiology* 12: 251–68.

————. 1987. Asymmetrical patterns of origination and extinction in higher taxa. *Paleobiology* 13: 427–45.

Gilinsky, N. L., and I. J. Good. 1989. Analysis of clade shape using queing theory and the fast Fourier transform. *Paleobiology* 15: 321–33.

Harvey, P. H., and A. Purvis. 2003. Opportunity versus innovation. *Nature* 425: 676–77.

Hutchinson, G. E. 1965. *The ecological theater and the evolutionary play.* New Haven: Yale University Press.

Jaanusson, V. 1981. Functional thresholds in evolutionary progress. *Lethaia* 14: 251–60.

Jones, D. S., and S. J. Gould. 1999. Direct measurement of age in fossil *Gryphaea*: the solution to a classic problem in heterochrony. *Paleobiology* 25: 158–87.

Knoll, A. H., and R. K. Bambach. 2000. Directionality in the history of life: diffusion from the left wall or repeated scaling of the right? In D. H. Erwin and S. L. Wing, eds., *Deep time: Paleobiology's perspective,* 1–14. *Paleobiology,* suppl. to Vol. 26, No. 4.

Maynard Smith, J., and E. Szathmáry. 1995. *The major transitions in evolution.* Oxford: W. H. Freeman Spektrum.

Mayr, E. 1963. *Animal species and evolution.* Cambridge, MA: Harvard University Press.

McGhee, G. R., Jr. 1999. *Theoretical morphology: The concept and its applications.* New York: Columbia University Press.

McGhee, G. R., Jr., P. M. Sheehan, D. J. Bottjer, and M. L. Droser. 2004. Ecological ranking of Phanerozoic biodiversity crises: ecological and taxonomic severities are decoupled. *Palaeogeography, Palaeoclimatology, Palaeoecology* 211: 289–97.

Niklas, K. J. 1997. *The evolutionary biology of plants*. Chicago: University of Chicago Press.

Oppel, A. 1856–1858. Die Juraformation Englands, Frankreichs und des Südwestichen Deutschlands (1856: 1–438; 1857: 439–694; 1858: 695–857). Stuttgart: *Württembergische Naturwissenschaft Verein Jahress-chrift* vols. 11–14.

Orbigny, A. D. d.' 1842. *Paléontologie francaise, terranes jurassiques, pt 1. Cephalopodes*. Paris: Masson.

Powell, M. G., and M. Kowalewski. 2002. Increase in evenness and sampled alpha diversity through the Phanerozoic: comparison of early Paleozoic and Cenozoic marine fossil assemblages. *Geology* 30: 331–34.

Raup, D. M., S. J. Gould, T. J. M. Schopf, and D. S. Simberloff. 1973. Stochastic models of phylogeny and the evolution of diversity. *Journal of Geology* 81: 525–42.

Raup, D. M., and S. J. Gould. 1974. Stochastic simulation and the evolution of morphology—towards a nomothetic paleontology. *Systematic Zoology* 23: 305–22.

Raup, D. M., and J. J. Sepkoski Jr. 1984. Periodicity of extinctions in the geologic past. *Proceedings of the National Academy of Sciences USA* 81:801–5.

Ricklefs, R. E. 2003. Global diversification rates of passerine birds. *Proceedings of the Royal Society of London* B 270:2285–91.

Runnegar, B. 1987. Rates and modes of evolution in the Mollusca. In K. S. W. Campbell and M. F. Day, eds., *Rates of evolution*, 39–60. London: Allen and Unwin.

Schopf, T. J. M., D. M. Raup, S. J. Gould, and D. S. Simberloff. 1975. Genomic versus morphologic rates of evolution: influence of morphologic complexity. *Paleobiology* 1: 63–70.

Sepkoski, J. J., Jr. 1981. A factor analytic description of the Phanerozoic marine fossil record. *Paleobiology* 7: 36–53.

Stanley, S. M. 1975. A theory of evolution above the species level. *Proceedings of the National Academy of Sciences USA* 72: 646–50.

Ueshima, R., and T. Asami. 2003. Single-gene speciation by left-right reversal. *Nature* 425: 679.

Valentine, J. W. 1973. *Evolutionary paleoecology of the marine biosphere*. Englewood Cliffs, NJ: Prentice-Hall.

Vrba, E. S., and S. J. Gould. 1986. The hierarchical expansion of sorting and selection: sorting and selection cannot be equated. *Paleobiology* 12: 217–28.

Wright, Sewall. 1967. Comments on the preliminary working papers of Eden and Waddington. In P. S. Moorehead and M. M. Kaplan, eds., *Mathematical challenges to the neo-Darwinian interpretation of evolution*, 117–20. Monograph No. 5. Philadelphia: Wistar Institute Press.

3

The Legacy of Punctuated Equilibrium

Dana H. Geary

Introduction

The idea for which Steve Gould is best remembered might well be punctuated equilibrium, which he coauthored with Niles Eldredge. A great many thousands of pages have been written about punctuated equilibrium since 1972; it is clearly impossible to characterize all of the discussion in a short essay. My intent here is to summarize my view of what about punctuated equilibrium has proven correct or successful and what has changed or remains uncertain. I will also discuss various aspects of the legacy of punctuated equilibrium.

What punctuated equilibrium is and is not

Regarding George Washington's 1796 Farewell Address, Joseph Ellis observes that its "main themes...are just as easy to state succinctly as they are difficult to appreciate fully" (Ellis 2000, 128). One might well say the same about punctuated equilibrium. Punctuated equilibrium (PE) describes a particular pattern of morphological change in the fossil record; it is about the "origin and deployment of species in geological time" (Gould 2002c, 765). The basic predictions PE are that most species, when observed on geologic time scales, exhibit insignificant change during their lifetimes. When a daughter species arises, it does so abruptly (again, on geological time scales) and is typically accompanied in the record by its ancestral species. Thus, evolutionary change is predominantly associated with branching events rather than accumulating in an anagenetic mode.

Though seemingly simple enough, the literature harbors a diversity of opinion on what exactly PE is (e.g., Somit and Peterson 1992). In addition, creationists have misrepresented PE in a host of self-serving ways (see Kelley, this volume). I will restrict myself here to the views of the scientific community.

I will not take up the question of species definitions in this essay. The fossil record is our only source of information on long-term patterns of evolutionary tempo and mode, and the fossil record gives us "morphospecies." Fossil morphospecies are not necessarily defined by morphology only; morphologic data are commonly supplemented with data on temporal and geographic distribution and variation, and paleoecological and paleoenvironmental context. When fossil morphospecies have been checked against genetic and experimental data, their validity has been confirmed (e.g., Jackson and Cheetham 1990).

The Empirical Record

Many paleontologists have conducted studies intended to gauge the fit of fossil data to the predictions of PE. Despite this large body of empirical work (or perhaps because of its volume and complexity), no one has yet carried out a comprehensive critical review. Such an assessment would be both interesting and challenging.

The published record is highly variable with respect to several key parameters. Available stratigraphic completeness, microstratigraphic acuity, and temporal scope vary significantly among studies (Schindel 1980, 1982; Sadler 1981), as do the attention paid to quantifying them. The geographic coverage offered by a particular study ranges from a single locality to extensive sampling across a species' range. The attention given to establishing phylogenetic relationships varies widely. Morphometric coverage of specimens ranges from a few simple characters to multivariate or other more holistic approaches to morphology (see Erwin and Anstey

[1995] for discussion of potential biases involved with continuously varying vs. discrete character states, multivariate techniques, and/or data transformations). Finally, published studies are highly variable with respect to sample sizes, as well as the degree of statistical rigor applied to the data. Thus, as argued by Jackson and Cheetham (1999), "reports of gradual or punctuated speciation have not been subjected to consistently rigorous critical evaluation despite great differences in the quality of evidence available."

An additional concern is that the fossil record is very strongly biased against species that are rare or temporally or geographically restricted (Koch 1978; Schopf 1982). Convincing demonstration of change or stasis, however, requires abundant and widely distributed species. Thus, in choosing taxa that meet these criteria, we are limiting our overall sample to those relatively unusual taxa that happen to be abundant. The degree to which this biases the empirical record, or even the direction of bias, is unclear. One might argue that this bias works against PE, in that the host of short-lived, geographically restricted taxa that are so underrepresented in the fossil record both come and go in less than a geologic instant. On the other hand, the same problem can be seen as a bias toward stasis (Erwin and Anstey 1995). Citing quantitative genetic models that predict an increased likelihood of stasis with increasing population size, Erwin and Anstey point out that when we can track many individuals over time, they must have been so numerous that stasis is likely to have prevailed. It is obvious that a paleontologist cannot study what does not exist, and so we will make our generalizations about those fossil lineages whose change (or lack thereof) can be measured, but the issue warrants acknowledgement.

What can be controlled is the selection of taxa to document. Ideal is consideration of all preserved taxa within a clade, or perhaps an entire biota, rather than selection of particular "interesting" lineages. As Steve emphasized (2002c), a considerable portion of the existing literature is probably focused on these interesting cases, chosen because something (i.e., change) rather than "nothing" (i.e., stasis) could be documented (referred to as "publication bias" or "Cordelia's dilemma" [2002c, 763–65]).

Erwin and Anstey (1995) tallied the results of fifty-eight published studies (many of them likely subject to this publication bias), taking each author's conclusions at face value. They conclude that the collective results of these studies differ from the initial predictions of Eldredge and Gould (1972e). A wide variety of patterns have been documented, with no single pattern predominate among them (see also Barnosky 1987). Erwin and Anstey suggest that the greatest departure from expectation is in the preponderance of studies exhibiting both stasis and gradualism in the history of a single lineage. These authors argue that just as neobiologists recognize a pluralism of species definitions and speciation mechanisms, paleobiologists should recognize a pluralism of viewpoints on species-level change. In particular, the fossil record appears to provide evidence that speciation can be both a long-term continuous process as well as an episodic one.

Many other commentators characterize the overall pattern differently, arguing that a pattern of stasis and punctuations prevails among fossil metazoans (Gould 1992i, 2002c; Prothero 1992; Stanley 1992; Jackson and Cheetham 1999; Benton and Pearson 2001). Most of these authors also conclude that planktonic foraminifera exhibit gradualism more frequently than do metazoans (or are at least more variable with respect to tempo). Jackson and Cheetham (1999) argue that a relatively small number of studies are in fact sufficiently detailed to assess the frequency of punctuated equilibrium, but those that are sufficient support a punctuated pattern. Hunt (2007) describes a novel quantitative approach that finds support for directional change in fewer than 5 percent of assessed sequences, with the remaining 95 percent nearly evenly divided between stasis and a random walk model.

Steve always emphasized that the only way to assess relative frequency is among entire clades or biotas (e.g., 1977c). The classic studies of large numbers of species include Hallam's bivalves (1978); Kelley's bivalves (1983, 1984); Cheetham and Jackson's extensively documented bryozoans (Cheetham 1986, 1987; Cheetham and Hayek 1988; Cheetham and Jackson 1995; Cheetham et al. 1983, 1984; Jackson and Cheetham 1990, 1994); Stanley and Yang's bivalves (1987); Prothero and Heaton's mammals (1996); and the trilobites, corals, and brachiopods of the Hamilton group (Eldredge 1971; Pandolfi and Burke 1989; Lieberman et al. 1995; Brett and Baird 1995); all of which indicate an overwhelming predominance of stasis and rapid change. Even in Lake Pannon, where we have found many cases of gradualism among gastropods and bivalves, most mollusc species arise abruptly (Geary 1990, 1992, 1995; Geary et al. 2002).

Thus, while an interesting range of tempos and modes has been described—and I agree with Erwin and Anstey that we would do well to consider a variety of species-level processes—it does appear that when large, relatively unbiased samples are compiled, the pattern of PE is dominant. Although I think Steve was a bit soft on some case studies, I will give him the last word here. Writing about the predominance of the PE pattern, he expresses surprise that this "clear signal has not been more widely appreciated as the most decisive result in a quarter century of research and debate about punctuated equilibrium" (2002c, 854).

Lessons

Anyone who ever read "This View of Life" in *Natural History* magazine knows that for Steve, every creature (from siphonophore to seahorse, hyena to hen) carried with it lessons on many levels, often a lesson on how science works as well as a lesson on how evolution works. A lesson on *how science works* was a featured component of Eldredge and Gould's 1972 debut paper, which begins: "In this paper we shall argue: 1) The expectations of theory color perception to such a degree that new notions seldom arise from facts collected under the influence of old pictures of the world. New pictures must cast their influence before facts can be seen in a different perspective" (1972e, 83). In that paper, Niles and Steve set out to give us a new picture (gaps may be real, stasis is data) and to remind us to consider our own biases. Tom Schopf, the editor of *Models in Paleobiology* (the 1972 book in which punctuated equilibrium first appeared), called the idea that theory dictates what one sees the "larger and more important lesson" of Eldredge and Gould (1972e; for further discussion of perceptual bias, see Fortey 1985, 1988; Sheldon 1993; Erwin and Anstey 1995). One of the things that distinguished Steve was his emphasis on science as a human endeavor and his consistent self-placement within, rather than apart from, the fray: "We readily admit our bias toward it [PE] and urge readers…to remember that our interpretations are as colored by our preconceptions as are the claims of the champions of phyletic gradualism by theirs" (1972e, 98).

The larger lesson of PE for *how evolution works* has always centered on the implications of the PE pattern for macroevolutionary theory; "if species originate in geological instants and then do not alter in major ways, then evolutionary trends cannot represent a simple extrapolation of allelic substitution within populations. Trends must be the product of differential success among species" (Gould 1980c, 125). More than twenty years later, but in very much the same vein, Steve writes: "Punctuated equilibrium validates the hierarchical theory of selection. This hierarchical theory establishes the independence of macroevolution as a theoretical subject…thereby precluding the full explanation of evolution by extrapolation of microevolutionary processes to all scales and times" (2002c; also 1982f, 1995d; see Stanley 1975, 1979 for important early discussion of this topic; also Lieberman, this volume).

It is possible, of course, to believe in the predominance of PE as a pattern, without necessarily accepting the entire set of implications that Steve so forcefully laid out (Futuyma 1987, 1989; Kitcher 2004). Conversely, it is possible to reject the pattern of PE, but still find meaning in a hierarchical approach (e.g., Schopf 1982; Maynard Smith 1983a, b, 1984; Hoffman 1989, 1992).

The last thirty-plus years of discussion give PE itself a history, which some authors view as transpiring in three stages (Gould and Eldredge 1986i; Gould 2002c; Dawkins 1986; Hoffman 1989, 1992; Ruse 1992; Erwin and Anstey, 1995). In this view, the history of PE parallels von Baer's (1866) characterization of the reception of unconventional ideas in science (Gould and Eldredge 1986i): first, the idea is dismissed as false; second, the idea is rejected as against religion (in this case, Darwinism); and third, the idea is seen as the logical outcome of what was, in fact, known all along.

During Stage One, the basic pattern of rapid speciation and long-term stasis was described and debated (e.g., Gingerich 1974, 1976). Not everyone was convinced of the ubiquity or importance of the PE pattern (e.g., Levinton 1988; Hoffman 1989, 1992), but other issues moved to the fore after the 1970s.

By the 1980s, much of the focus had shifted to the implications of PE for evolutionary biology (the lesson for macroevolution). For detractors, a focal point for Stage Two was Steve's (1980c) paper titled "Is a new and general theory of evolution emerging?" Herein, Steve emphasized a three-tiered hierarchy marked by discontinuities between gradual, adaptive allelic substitution within populations and change at speciation (the "Goldschmidt break") and between speciation and macroevolutionary trends (the "Wright break"). Steve characterized traditional views of the synthesis with the following statement from Ernst Mayr: "The proponents of the synthetic theory maintain that all evolution is due to the accumulation of small genetic changes, guided by natural selection, and that trans-specific evolution is nothing but an extrapolation and magnification of the events that take place within populations and species" (Mayr 1963). Steve went on to argue that "if Mayr's characterization of the synthetic theory is accurate, then that theory, as a general proposition, is effectively dead, despite its persistence as textbook orthodoxy." (1980c, 120). Mentioning death and the synthesis in the same sentence raised some hackles, to say the least, although critics took no notice of the ways in which Steve had qualified his remarks (see further analysis of this statement in Allmon, this volume). Offput by this and other claims, Dawkins (1986) refers to the early 1980s as PE's "grandiloquent era," Hoffman (1992) characterizes Steve's views as "indefensible," and many others consider them at the very least overblown or extreme (e.g., Ruse 1992; Erwin and Anstey 1995).

During the late 1980s and into the 1990s, Steve softened his rhetoric and spent considerable effort trying to explain what he had and had not meant. "Punctuated equilibrium is now quite acceptable, rather smaller in scope than once suspected, and a good, comfortable part (or perhaps, at best, a mild extension) of neo-Darwinism" (1986i, 145). The notion that PE now implied a modest expansion of neodarwinian theory rather than its burial, led his critics to characterize this final stage as a retreat (Ruse 1992; Erwin and Anstey 1995). The following extract from Halstead (1985) typifies the rancor with which his work was treated by some critics:

> He seems to be setting up a face-saving formula to enable him to retreat from his earlier aggressive saltationism, having had a bit of a thrashing, his current tack is to suggest that perhaps we should keep the door open in case he can find some evidence to support his pet theories so let us be "pluralist." (Halstead 1985, 318)

Steve refers to this three-fold history as an "urban legend" that he regards as "about as close to pure fiction as any recent commentary by scientists has ever generated" (2002c, 1006). Never at a loss for a pithy phrase, even when characterizing the rap against him, Steve suggests that various accounts of his own writings are in line with a classic theme of Western sagas: the "growth, exposure, and mortification of hubris." Or again according to his critics, the history of PE has proceeded as "modest origin, bombastic rise, and spectacular fall" (2002c, 1007). Steve's deep frustration with this literature stems from what he sees as a misrepresentation of his views, particularly the 1980 paper. The misrepresentations, when repeated often enough, made his later denials sound like a retreat. For example, based on the quotations provided above (from Mayr 1963, 1980c), it is often claimed that Steve declared Darwinism dead (e.g., Dawkins 1986). Steve's subsequent efforts at clarification (e.g., he intended only that "the synthesis can no longer assert full sufficiency to explain evolution at all scales" (2002c, 1003) probably still fall on some unsympathetic ears. Similarly, critics pounced upon Steve's supposed endorsement of Goldschmidt saltationism (his original discussion was certainly more nuanced), claiming that he was also attempting to revise our notions of microevolutionary mechanics.

In my view, Steve did not substantially change his message on the implications of PE. Much of his rhetoric may have been intended as a force for "disruptive selection," if you will, to move a more critical mass of practitioners off of the conventional mode, in order to expand our collective view of the ways in which evolution works. Steve provoked his critics to be more explicit about the variety of processes acceptable under a neo-Darwinian umbrella and what one could and could not extrapolate from microevolutionary processes. So did Steve go too far and then retreat? Or did the field itself move and then claim it had been there all along? Did Steve set up a neo-Darwinian straw man, or were his own arguments so shallowly miscast as to constitute a ridiculous caricature? This is not an issue one can decide by reading the derivative literature. Wherever one falls along this spectrum, however, we can acknowledge Steve's contribution to putting these issues on center stage and forcing all sides to articulate their positions more clearly.

Causes

It seems clear that the patterns predicted by PE have not changed, and most would agree that they have largely proven correct. Furthermore, I believe that Gould and Eldredge were consistent in their view of the lessons for how both science and evolution work. In important ways, however, there are basic elements of PE that have changed, or at the least, components of the theory about which we are less certain now than Gould and Eldredge were thirty years ago. In *Structure* (2002c), Steve concedes that the original formulations of PE were wrong in several ways. I focus here on two issues: whether or why significant change happens exclusively at speciation, and the mechanisms behind stasis (in other words, the *causes of punctuations* and the *causes of stasis*).

Eldredge and Gould (1972e) viewed the cause of punctuations as "a simple consequence of the allopatric theory." Mayr's allopatric model was the most widely accepted of the day, and it followed logically that evolutionary change in small, peripheral populations would transpire in a time frame that appeared instantaneous on geologic scales.

Furthermore, "the importance of peripheral isolates lies in their small size and the alien environment beyond the species border that they inhabit—for only here are selective pressures strong enough and the inertia of large numbers sufficiently reduced to produce the "genetic revolution" [Mayr 1963, 533] that overcomes homeostasis" (1972e, 114). Mayr's "genetic revolutions" were thus a critical component of early versions of PE. Mayr's argument was that the increased homozygosity typical of a founder population meant that "every gene may or will have a new selective value in the drastically altered genetic environment of the founder population" (Mayr 1963, 534). During the resulting "genetic revolution, the population will pass from one well-integrated and stable condition through a highly unstable period to another period of balanced integration" (Mayr 1963, 538). Thus, Eldredge and Gould (1972e; also 1977c) were explicit about the fact that change, both morphologic and genetic, occurs predominantly at speciation.

By 2002, however, Steve frankly admitted that "advocating for a direct acceleration of evolutionary rate by the processes of speciation" was a major error (2002c, 798–99). He sees the problem resolved, however, in a 1987 paper by Doug Futuyma, in which Futuyma makes the case that *if* there is a correlation between punctuations and speciation it is *not* due to an acceleration of rate at speciation. Instead, Futuyma argues that change can occur at any time during the life of a species, as evidenced by extensive observations of rapid adaptive change in local populations. Unless this change is locked in by the establishment of reproductive isolation, however, it will be ephemeral. Change is only retained, or stabilized, when speciation via reproductive isolation occurs. In Futuyma's words: "Reproductive isolation confers sufficient permanence on morphological changes for them to be discerned in the fossil record" (1987, 465).

Whether or not Futuyma's explanation ultimately proves correct (or the frequency with which it does), the description of a mechanism that incorporates both the punctuated pattern that geologists see and the prodigious variability and change that biologists see can be viewed as a positive outcome of PE. Writing in 2002, Steve celebrates this as progress in his characteristically upbeat fashion: "Futuyma's incisive macroevolutionary argument...offers a far richer, far more interesting, and theoretically justified rationale for correlating episodes of evolutionary change with speciation" (2002c, 77).

Stasis, on the other hand, has clearly been easier to document than to explain. Steve's views on the cause(s) of stasis changed in important ways. The early focus was squarely on developmental constraint and internal homeostatic mechanisms: "The coherence of a species, therefore, is not maintained by interaction among its members (gene flow). It emerges, rather, as an historical consequence of the species' origin as a peripherally isolated population that acquired its own powerful homeostatic system" (1972e, 114). Thus, early views on the causes of stasis were linked to the causes of punctuations, and both had their roots in Mayr: "A genuine genetic revolution is characterized by a breakdown of genetic homeostasis through a loss or a reconstitution of previously existing balancing systems" (Mayr 1963, 539). (Mayr, however, has acknowledged that he never put such emphasis on long-term stasis [Mayr 1992].)

By 2002, however, Steve admits that criticism of these ideas was justified and that he no longer believes that constraint plays a critical causal role in stasis. He enumerates several possible causes of stasis, including "conventional" stabilizing selection (e.g., Charlesworth et al. 1982), habitat tracking (i.e., Eldredge 1995, 1999), and the restrictions imposed by subdivided populations (Lieberman and Dudgeon 1996) (2002c). Lieberman and Dudgeon's proposal is that species are commonly subdivided into temporary, semi-autonomous populations, each of which may adapt to its local surroundings or be subject to drift. Over the entire lifetime of the species, however, local changes will be buffered by input from elsewhere in the species range and stasis will prevail (see Eldredge et al. 2005; Lieberman, this volume, for further discussion). Steve was favorably inclined toward this latter type of explanation for stasis because he considers it more explicitly and irreducibly macroevolutionary (2002c). He admitted, however, that for the time being, the relative importance of various proposed causes remains unknown.

The movement away from constraint and homeostasis shifted our suite of explanations in an important way; we have gone from a view that "species cannot change" (except under unusual circumstances) to a view that "species can change (at least locally), but generally do not." A number of authors (e.g., Williams 1992; McKinney and Allmon 1995; Lieberman and Dudgeon 1996; Sheldon 1996) view a lack of overall change (stability) as a "response" to a fluctuating environment (or, as Steve put it, a "form of compromise" [2002c]).

Overall then, while the reality of the pattern is acknowledged, the causes of punctuations and stasis are not well understood. Yet this is not a failure of PE. On the contrary, one could argue that these uncertainties in our understanding are now more apparent and the questions better focused because of the debate over PE. Yet the situation forces us to take a careful look at the implications of PE. If the PE pattern prevails, but its original explanations were wrong or at best the causes are now uncertain, where does that leave us? Were Eldredge and Gould right for the wrong reasons? Or were they right (about the pattern) and the reasons don't matter? In particular, do the reasons matter to a macroevolutionary theory?

I would argue that the reasons do matter, that understanding the causes of stasis and punctuations is critical to the development of a causal macroevolutionary theory. It is undeniable that many generations of naturalists could study living organisms with no indication of the reality of the PE geometry, and this in itself defines a higher-level pattern. But the nature of the distinction between micro- and macrolevel processes must stem from causes. And while the PE pattern in and of itself serves to individuate species in geologic time, I think that our changing view of the causes of punctuations and stasis renders this individuation more descriptive and less causal than it was thirty years ago. The underlying biological nature of the micro–macro discontinuity was clearly stronger in earlier versions of PE, and has yet to be replaced.

PE does indeed validate a distinct macroevolutionary theory because it was not, nor could it have been, predicted from studies of microevolution. As argued by Francisco Ayala (a population geneticist):

> the theory of population genetics is compatible with both punctualism and gradualism; and, hence, logically it entails neither. Whether the tempo and mode of evolution occur predominantly according to the model of punctuated equilibria or according to the model of phyletic gradualism is an issue to be decided by studying macroevolutionary patterns, not by inference from microevolutionary processes. (Ayala 1982)

In my view, however, our understanding of the distinction between micro- and macroevolutionary processes lags behind our rich, yet largely descriptive articulation of macroevolutionary pattern (e.g., Erwin 2000).

Most would agree, however, that Steve and Niles were successful in bringing a hierarchical approach to the table, into the language and thinking of evolutionary biologists and paleontologists, and that there is no question that this is an important legacy of punctuated equilibrium. Overall, the degree to which irreducible higher-level processes have been important in the history of life remains to be seen.

The legacy of punctuated equilibrium

Although Steve argued repeatedly that a hierarchical view of evolution was PE's most important implication, there are many other important components to the legacy of PE. One is the body of empirical evidence collected with the intent of proving or disproving PE. In their papers, Steve and Niles were always direct and honest about the limitations of the fossil record, but also clear and forceful in arguing for a role for paleontology in formulating our thoughts about speciation. Many interesting questions raised by PE and subsequent empirical work on the topic of species-level change have yet to be understood; important parts of this empirical legacy are yet to come. For example, anagenetic gradualism, even if relatively rare, is common enough to warrant better explanation (Geary 1995). Steve appreciates this in *Structure* (2002c): "I do not think we have even begun to explore the range of potential explanations for the puzzling phenomenon of anagenetic gradualism. I, at least, find the subject very confusing and challenging."

There are many other unresolved issues regarding species-level change for which paleontological data are both well suited and necessary (e.g., McKinney and Allmon 1995).

Another important part of PE's legacy is the dialogue that it fostered between evolutionary biologists and paleontologists. Of course, much of this dialogue could be characterized as contentious. I would say, however, that ground has been given on all sides, and that this must represent progress. To put it more positively, the engagement itself (between biologists and paleontologists) is healthy, and as reasonable people have argued for many decades, is critical to the development of evolutionary thinking.

Although PE is specifically about change at the species level, a general acceptance of nonuniform rates of evolution across all levels is part of PE's legacy. In *Structure* (2002c), Steve describes a broad range of studies, extending across evolutionary biology and on to many other disciplines, in which periods of stasis and relatively rapid episodes of transition have been described and their recognition at least in part attributed to PE. Steve could not and did not claim to have invented the idea of rapid transformations, but certainly the extensive discussion and debate about PE helped put this notion into the popular lexicon.

Finally, in assessing the overall impact of his ideas, it has long seemed to me that much of Steve's enthusiasm came from a deep satisfaction in simply stirring things up. From behind his "modest" proposal based on "conventional" ideas, came major challenges to evolutionary thinking. Steve clearly valued the unconventional and the heterodox, and in the end seemed quite willing to do whatever was necessary to shake things up. As so nicely put by Allen Orr (2002): "Gould might well then represent something new in the historical strata of science: the first self-consciously revolutionary scientist—the first scientist who set out to create a revolution at least in part because he felt that the field just needed one." We can all read history and enjoy the perspective of hindsight, but it takes a special person to see one's own moment clearly enough to be a real force for change.

A final note

People like Steve don't come along very often. I'm quite sure that I've never known anyone else even in his ballpark. A historical figure comes to mind as an apt comparison, who can, at the least, facilitate a few final remarks about Steve. Although it seems unlikely that Steve will ever have an aircraft carrier named after him, there are nonetheless a number of similarities between Teddy Roosevelt and Steve Gould. An obvious one is that each had a long-term association with New York City, with important forays to Harvard University. Both men burst upon their respective political or academic scenes at very young ages. Both were simultaneously well situated within the elite of their political or academic circles, but carried unwavering sympathy for those less advantaged than themselves. Each man was a prolific writer; not everyone liked their writing styles. Their urban backgrounds were apparent in their morphologies, rendering them unlikely candidates for fieldwork, but they persevered. Most importantly, each man was a force: a force of personality backed up by a real force of intellect. Each possessed abundant energy, an energy with deep roots tapped into a love of life, a love of the natural world, of history and biology, and a love of engagement (they weren't afraid to mix it up a bit). Both men had exceptional personal courage, extending to a near disregard for their own physical suffering or danger. In the early 1980s, Steve was diagnosed with a highly virulent form of cancer, believed at the time to be invariably fatal. Steve approached his disease with characteristic zeal; he might have been the type specimen for a "positive attitude." In the midst of his brutal treatment, when his body was uncharacteristically thin and immunologically compromised, he traveled to Arkansas to testify against creationism. Not surprisingly, the infection he contracted there nearly killed him. As we all know, however, he fought both the infection and the mesothelioma and survived another twenty years.

One hundred years ago, Teddy Roosevelt laid the groundwork for our system of national parks, codifying a view of humankind's relationship to the wilderness that was novel, indeed antithetical to the views of most of his contemporaries. Today's conservation movement is an unavoidable part of the political landscape, and owes many of its roots, acknowledged or not, to Teddy Roosevelt. I am confident that Steve's contributions, some acknowledged and some perhaps with their roots forgotten, will still be a fundamental part of the landscape of evolutionary biology 100 years from now.

Acknowledgments

Thanks to Rob Bleiweiss, Gene Hunt, and Jeremy Jackson for comments.

References

Ayala, F. 1982. Microevolution and macroevolution. In D. S. Bendall, ed., *Evolution from molecules to man*, 387–402. Cambridge: Cambridge University Press.

Barnosky, A. D. 1987. Punctuated equilibrium and phyletic gradualism: Some facts from the Quaternary mammalian record. *Current Mammalogy* 11: 109–47.

Benton, M. J., and P. N. Pearson. 2001. Speciation in the fossil record. *Trends in Ecology and Evolution* 16: 405–11.

Brett, C. D., and G. C. Baird. 1995. Coordinated stasis and evolutionary ecology of Silurian to Middle Devonian faunas in the Appalachian Basin. In D. H. Erwin and R. L. Anstey, eds., *New approaches to speciation in the fossil record*, 285–317. New York: Columbia University Press.

Charlesworth, B., R. Lande, and M. Slatkin. 1982. A neo-Darwinian commentary on macroevolution. *Evolution* 36: 474–98.

Cheetham, A. H. 1986. Tempo of evolution in a Neogene bryozoan: rates of morphologic change within and across species boundaries. *Paleobiology* 12: 190–202.

———. 1987. Tempo of evolution in a Neogene bryozoan: are trends in single morphologic characters misleading? *Paleobiology* 13(3): 286–97.

Cheetham, A. H., and L. C. Hayek. 1988. Phylogeny reconstruction in the Neogene bryozoan Metrarabdotos: a paleontologic evaluation of methodology. *Historical Biology* 1: 65–84.

Cheetham, A. H., J. B. C. Jackson, and L. C. Hayek. 1993. Quantitative genetics of bryozoan phenotypic evolution. I. Rate tests for random change versus selection in differentiation of living species. *Evolution* 47: 1526–38.

———. 1994. Quantitative genetics of bryozoan phenotypic evolution. II. Analysis of selection and random change in fossil species using reconstructed genetic parameters. *Evolution* 48: 360–75.

Dawkins, R. 1986. *The blind watchmaker.* New York: W. W. Norton.

Eldredge, N. 1971. The allopatric model and phylogeny in Paleozoic invertebrates. *Evolution* 25: 156–67.

———. 1995. *Reinventing Darwin: The great debate at the high table of evolutionary theory.* New York: John Wiley.

———. 1999. *The pattern of evolution.* New York: W. H. Freeman.

Eldredge, N., J. N. Thompson, P. M. Brakefield, S. Gavrilets, D. Jablonski, J. B. C. Jackson, R. E. Lenski, B. S. Lieberman, M. A. McPeek, and W. Miller III. 2005. The dynamics of evolutionary stasis. *Paleobiology* 31: 133–45.

Ellis, J. J. 2000. *Founding brothers: The revolutionary generation.* New York: Vintage Books.

Erwin, D. H. 2000. Macroevolution is more than repeated rounds of microevolution. *Evolution and Development* 2: 78–84.

Erwin, D. H., and R. L. Anstey. 1995. Speciation in the fossil record. In D. H. Erwin and R. L. Anstey, eds., *New approaches to speciation in the fossil record,* 12–38. New York: Columbia University Press.

Fortey, R. A. 1985. Gradualism and punctuated equilibria as competing and complementary theories. Special Papers in Palaeontology, 33: 17–28.

———. 1988. Seeing is believing: gradualism and punctuated equilibria in the fossil record. *Science Progress* 72: 1–19.

Futuyma, D. J. 1987. On the role of species in anagenesis. *American Naturalist* 130: 465–73.

———. 1989. Macroevolutionary consequences of speciation: Inferences from phytophagous insects. In D. Otte and J. A. Endler, eds., *Speciation and its consequences,* 557–78. Sunderland, MA: Sinauer Associates, Inc.

Geary, D. H. 1990. Patterns of evolutionary tempo and mode in the radiation of *Melanopsis* (Gastropoda; Melanopsidae). *Paleobiology* 16: 492–511.

———. 1992. An unusual pattern of divergence between two fossil gastropods: ecophenotypy, dimorphism, or hybridization? *Paleobiology* 18: 93–109.

———. 1995. The importance of gradual change in species-level transitions. In D. H. Erwin and R. L. Anstey, eds., *New approaches to speciation in the fossil record.* New York: Columbia University Press.

Geary, D. H., A. W. Staley, P. Müller, and I. Magyar. 2002. Iterative changes in Lake Pannon *Melanopsis* reflect a recurrent theme in gastropod morphological evolution. *Paleobiology* 28: 208–21.

Gingerich, P. D. 1974. Stratigraphic record of early Eocene *Hyopsodus* and the geometry of mammalian phylogeny. *Nature* 248: 107–9.

————. 1976. Paleontology and phylogeny: patterns of evolution at the species level in early Tertiary mammals. *American Journal of Science* 276: 1–28.

Hallam, A. 1978. How rare is phyletic gradualism and what is its evolutionary significance? Evidence from Jurassic bivalves. *Paleobiology* 4: 16–25.

Halstead, B. 1985. The evolution debate continues. *Modern Geology* 9: 317–26.

Hoffman, A. 1989. *Arguments on evolution.* Oxford: Oxford University Press.

————. 1992. Twenty years later: Punctuated equilibrium in retrospect. In A. Somit and S. A. Peterson, eds., *The dynamics of evolution: The punctuated equilibrium debate in the natural and social sciences,* 121–38. Ithaca: Cornell University Press.

Hunt, G. The relative importance of directional change, random walks, and stasis in the evolution of fossil lineages. *Proceedings of the National Academy of Sciences, U.S.A.* 104 (47): 18404–8.

Jackson, J. B. C., and A. H. Cheetham. 1990. Evolutionary significance of morphospecies: A test with cheilostome bryozoa. *Science* 248: 579–83.

————. 1994. Phylogeny reconstruction and the tempo of speciation in cheilostome Bryozoa. *Paleobiology* 20: 407–23.

————. 1999. Tempo and mode of speciation in the sea. *Trends Ecology Evolution* 14: 72–77.

Kelley, P. H. 1983. The role of within-species differentiation in macroevolution of Chesapeake Group bivalves. *Paleobiology* 9: 261–68.

————. 1984. Multivariate analysis of evolutionary patterns of seven Miocene Chesapeake Group mollusks. *Journal of Paleontology* 58: 1235–50.

Kitcher, P. 2004. Evolutionary theory and the social uses of biology. *Biology and Philosophy* 19: 1–15.

Koch, C. F. 1978. Bias in the published fossil record. *Paleobiology* 4: 367–72.

Levinton, J. S. 1988. *Genetics, paleontology, and macroevolution.* Cambridge: Cambridge University Press.

Lieberman, B. S., C. E. Brett, and N. Eldredge. 1995. A study of stasis and change in two species lineages from the Middle Devonian of New York State. *Paleobiology* 21: 15–27.

Lieberman, B. and S. Dudgeon. 1996. An evaluation of stabilizing selection as a mechanism for stasis. *Palaeogeography, Palaeoclimatology, Palaeoecology* 127: 229–38.

Maynard Smith, J. 1983a. The genetics of stasis and punctuation. *Annual Review of Genetics* 17: 11–25.

————. 1983b. Current controversies in evolutionary biology. In M. Grene, ed., *Dimensions of Darwinism,* 273–86. Cambridge: Cambridge University Press.

————. 1984. Palaeontology at the high table. *Nature* 309: 401–2.

Mayr, E. 1963. *Animal species and evolution.* Cambridge, MA: Harvard University Press.

————. 1992. Speciational evolution or punctuated equilibria. In A. Somit and S. A. Peterson, eds., *The dynamics of evolution: The punctuated equilibrium debate in the natural and social sciences,* 21–53. Ithaca: Cornell University Press.

McKinney, M., and W. D. Allmon. 1995. Metapopulations and disturbance: from patch dynamics to biodiversity dynamics. In D. H. Erwin and R. L. Anstey, eds., *New approaches to speciation in the fossil record,* 123–83. New York: Columbia University Press.

Orr, H. A. 2002. The descent of Gould. *The New Yorker,* Sept. 20, 132–38.

Pandolfi, J. M., and C. D. Burke. 1989. Shape analysis of two sympatric coral species: Implications for taxonomy and evolution. *Lethaia* 22 183–93.

Prothero, D. R. 1992. Punctuated equilibrium at twenty: A paleontological perspective. *Skeptic* 1: 38–47.

Prothero, D. R., and T. H. Heaton. 1996. Faunal stability during the Early Oligocene climatic crash. *Palaeogeography, Palaeoclimatology, Palaeoecology* 127: 257–83.

Ruse, M. 1992. Is the theory of punctuated equilibria a new paradigm? In A. Somit and S. A. Peterson, eds., *The dynamics of evolution: The punctuated equilibrium debate in the natural and social sciences,* 139–67. Ithaca: Cornell University Press.

Sadler, P. M. 1981. Sediment accumulation rates and the completeness of stratigraphic sections. *Journal of Geology* 89: 569–84.

Schindel, D. E. 1980. Microstratigraphic sampling and the limits of paleontologic resolution. *Paleobiology* 6: 408–26.

————. 1982. Resolution analysis: a new approach to the gaps in the fossil record. *Paleobiology* 8: 340–53.

Schopf, T. J. M. 1982. A critical assessment of punctuated equilibria I. Duration of taxa. *Evolution* 36: 1144–57.

Sheldon, P. R. 1993. Making sense of microevolutionary patterns. In D. R. Lees and D. Edwards, eds., *Evolutionary patterns and processes,* 19–32. Linnean Society Symposium Series No. 14. New York: Academic Press.

————. 1996. Plus ça change—a model for stasis and evolution in different environments. *Palaeogeography. Palaeoclimatology, Palaeoecology* 127: 209–27.

Stanley, S. M. 1975. A theory of evolution above the species level. *Proceedings of the National Academy of Sciences, U.S.A.* 72: 646–50.

————. 1979. *Macroevolution.* San Francisco: W. H. Freeman and Company.

————. 1992. The empirical case for the punctuational model of evolution. In A. Somit and S. A. Peterson, eds., *The dynamics of evolution: The punctuated equilibrium debate in the natural and social sciences,* 85–102. Ithaca: Cornell University Press.

Stanley, S. M., and X. Yang. 1987. Approximate evolutionary stasis for bivalve morphology over millions of years: a multivariate, multilineage study. *Paleobiology* 13: 113–39.

Von Baer, K. E. 1866. Über Prof. Nic. Wagner's Entdeckung von Larven die sich fortpflanzen, und über die Padogenesis überhaupt. *Bulletin Acad. Imp. Sciences St. Petersburg* 9:63–137.

Williams, G. D. 1992. *Natural selection: Domains, levels and challenges.* New York: Oxford University Press.

A Tree Grows in Queens

Stephen Jay Gould and Ecology

*Warren D. Allmon, Paul J. Morris,
and Linda C. Ivany*

Introduction

Our interests and fascinations can be much more than just our idle diversions; they can affect what we do with our lives. If you like the color red, or playing the oboe, or driving a car, for example, these passions might lead your career in those directions. If your personal interest is popular or economically valuable, you may succeed in applying it vocationally. If it isn't, but you're a virtuoso, you may still succeed. But what if you're a scientist and your interests must be put against an independent separate reality? The answer is that it's not much different. Scientists largely pursue what they are interested in. Yet what if you set out to answer a scientific question and you're just not interested in and/or not technically able to do what turns out to be what is necessary to find the solution? It doesn't matter how good you are if you're wrong.

Steve Gould's scientific and literary handling of ecology can be examined in this context. He just wasn't personally interested in it. Given that he developed macroevolutionary theories that conspicuously excluded a major role for ecology, it is thus worth considering to what degree this personal inclination may have colored his theorizing in this direction, and to what degree it affected his reactions to data and theories which seemed to suggest ecology did matter in macroevolution. Secondarily, Steve's role as a popular natural history essayist eventually compelled him to address environmental themes. Yet his lack of personal interest in the science of ecology, combined with his overarching humanistic passions, led him to create a "humanistic ecology" in his writing that lacked urgency in its calls for environmental conservation.

Steve's technical macroevolutionary thinking did not leave much room for ecology: his emphasis on morphological stasis, internal constraint, and nonadaptationism de-emphasized the long-term effects of natural selection and biotic interactions in causing long-term evolutionary trends, which he attributed largely to the sorting of more or less static entities driven by processes only distantly related to local ecological factors; any trends that might become established, were likely to be derailed by the caprice of mass extinction. Here, however, we would like to explore the reverse argument, that Steve's thoughts on ecology predated and strongly affected—or at least set the stage for—his later evolutionary theorizing. It is possible, in other words, that one reason that Steve came to some of his conclusions about whether ecology "matters" in macroevolution was because he didn't really give it much thought at all, and when he did think about it, it did not impress him. It would be ironic if this turned out to be the case, for Steve was conspicuously aware of the effects of personal bias in science (see discussion in Allmon, this volume), and even specifically criticized colleagues for committing "a classic psychological fallacy" in denying significance to a phenomenon (in this case, species selection) "by confusing personal interest with general importance" (2002c, 711–12).

We, however, lack the hubris to see Steve as simply having fallen into the fallacy of confusing personal interest with general importance. The explanation is more complex. All three of us have struggled for many years, during and after our time as his students, with why he was not impressed with ecological interactions, which to us seemed to be of evident importance for evolutionary theory. Although trying to understand the psychological underpinnings of someone's scientific view point is always a dangerous endeavor, we believe that there is a reasonable explanation for Steve's views that is grounded in both his personal interests and in the fundamental underpinnings of his approach to scientific problems. We believe that the near absence of ecology from Steve's evolutionary theories was due to a combination of three factors: his personal disinterest in ecology, his view of ecology as limited to interactions on geologically short temporal scales, and to his perception of congruence between environmentalism and the biases that he sought to expunge from science.

I. A tree grows in Queens

As anyone who has read Steve's popular essays knows, he grew up as a "street kid" in "the middle of New York City" (e.g., 1984e), Queens, more precisely, and he remained "a city boy at heart"(2000k, 152). Growing up in the city, he had little direct access to natural environments, and once described "nature to New York kids" as Central Park and Jones Beach (1995m; 1995k, 109). "My youthful 'splendor in the grass' was the bustle of buildings of New York. My adult joys have been walks in cities, amidst stunning human diversity of behavior and architecture...more than excursions in the woods....I am not insensible to natural beauty, but my emotional joys center on the improbable yet sometimes wondrous works of that tiny and accidental evolutionary twig called *Homo sapiens*" (1991a, 13). In this context, the following story strongly suggests that Steve's New York City childhood was a major influence on his later views of ecology and the environment.

Kathy Hoy Burgess was a PhD student at Harvard studying insect ecology and also serving as a teaching assistant in Steve's large lecture course, "The History of the Earth and of Life" (see Ross, this volume). In 1987 or 1988 (she cannot recall which), she met with Steve to discuss her thesis. Here is Kathy's recollection of that meeting (we are very grateful to Kathy for sharing this memory with us):

> Steve said, "Let's get a cup of tea." We were passing in the yellowed light and creaking wood of the hallway of the MCZ, and I must have looked surprised. He went on that he liked to go out with each of his students to hear about their work, and why didn't we walk together down to [Harvard] Square? I don't know whether I was more

shocked that Steve had time, interest, or [that he] considered me one of his students. I had taught in Steve's class...a couple of years earlier, and had thoroughly enjoyed making use of my experience in geology and paleontology. As an undergraduate at Dartmouth I had been a biology major, and had narrowly missed the chance for a double major with geology. Being part of the epic class was all I could ever have dreamed. Each graduate assistant could offer a study section of his or her own design, and Steve supported my "History of Plant and Animal Interactions"...Steve met with his [teaching assistant] team regularly to discuss teaching strategies and problem students, and I really couldn't have had more fun. But off for tea, that was a treat.

The walk was long and so was the tea. I was researching the ecology of flower feeding, and focusing on insects that ate flowers rather than leaves. I was now well into my dissertation work and had theory and fieldwork to present. What is important to know is how active a listener Steve was, even on a topic that I was aware was not of keen interest to him. Without the shorthand of my field, I had the rare challenge of proving the importance of my work to Stephen Jay.

The details of the day are dear to me.... I rested the monologue and asked something that I had been wondering: "From your work and your writing, it seems you are not very interested in ecology. Why is that?" I had a list of possible reasons. Unexpectedly, Steve told me a story, one he seemed to be just remembering as he told it. When he was a boy, perhaps eight or ten, he had found a tree seedling growing in a crack in the sidewalk in Queens. He knew that was a bad place for it so he carefully dug it out and got dirty in the process. He found a better place in a wide swath next to the sidewalk, pushed in the roots, and ran to get water for it. He cared for it for several days, bringing it cups of water after school. Then one afternoon he came to see the seedling crushed to the ground and broken. "That was it. It was too much effort." After that, [he said,] he was not very interested in observing nature. The impressionistic vignette lingered, then I asked, if ecology—the interactions of living things and their environment—is not very interesting, where does that leave paleoecology? Steve considered a few moments, and said he would have to think about that.... The connection of ecology and evolution was an important issue for me as I moved between labs and faculty at Harvard. Even if Steve had no enduring interest in caterpillars, it was good to know that day that I made him think.

We find this to be an immensely interesting story; for Steve to give a very personal answer to a question in the midst of an intellectual discussion was not, at least in our experience, typical. Much like his essays, stories on a tangent about intellectually interesting things he had encountered were usual starting points for an answer, but this sort of intense personal history, in our experience, was not. Each of the three of us frequently heard from Steve variations on the theme of "ecology isn't interesting because it does little to inform evolutionary theory." Kathy Hoy's story, however, suggests an additional level to this argument: there was something about ecology in relation to individual organisms that simply didn't interest Steve, that there was indeed a personal element of disinterest influencing his view of ecology.

II. Paleontology plus ecology as paleobiology

Early in his career Steve accepted that natural selection in local environments was important in guiding morphological form and trends in evolution (1970c) and was actively interested in how ecological communities were organized (1969h; see Bambach, this volume). With the publication of the theory of punctuated equilibrium in 1972, however, Steve began to doubt the hegemony of ecology in macroevolution. Still, in a paper published in 1976, and revised in 1981 (1976c; 1981z), he did explore how ecology could matter to paleontology, if not necessarily to all long-term evolutionary patterns. Comparison of the two versions of this paper is instructive, for they span the period in which some of his most radical thinking and writing on macroevolution was taking place.

In the original paper, he credited much of the "transformation" of paleontology that took place in the 1970s to the influence of theoretical ecology, including population dynamics, island biogeography, life history studies, and species interactions. Ecology, he said, can "encourage palaeontologists to study adaptation for its immediate significance. Traditional palaeontology rarely worked at this level; it focused instead on the meaning of adaptation as a contribution to long-term evolutionary trends. . . . If we are now treating immediate significance seriously with these new themes, then almost every empirical study in palaeontology will be conducted differently in the future" (1976c, 233). He even pointed to some of his own work (1977g) as "personal testimony" of the utility of this approach, and suggested that "adaptive morphology must be consulted for the solution of many problems in diversity" such as key innovations.

Yet even as these themes were now open to paleontologists, their long term significance to macroevolution was less clear. Evolutionary trends, Steve ventured, are a "phenomenon that probably requires explanation at its own level." This explanation was likely to be rooted in the implications of punctuated equilibrium (which, Steve frequently noted, he and Niles Eldredge had not completely grasped in 1972): "Events of speciation provide an essentially stochastic input to evolutionary trends. The trends themselves represent differential preservation and success of a subset of speciations. Trends cannot be explained in the ecological time of speciation itself, but only in the evolutionary time of a higher-order 'selection' of speciation events. . . . The laws of ecology will not encompass trends, but they do set the speciations that serve as their building blocks" (1976c, 235–36). Thus, while Steve perceived important contributions from ecology to paleobiology, he was clearly limiting them to the first tier, to the immediate adaptive context of individual organisms. A metaphor in *The Origin of Species* that Steve frequently highlighted with his students is relevant here. On the penultimate page, Darwin writes: "It is interesting to contemplate an entangled bank. . . and to reflect that these elaborately constructed forms. . . have all been produced by laws acting around us" (1859, 489). We think that the metaphor of "the entangled bank" was very reflective of Steve's views of the role of ecology in nature—the complex web of interactions between individuals of species with their own independent contingent histories, histories that led to those particular individuals with their particular character suites happening to live together in that place and time, a result of their evolutionary histories.

Steve was particularly struck by the inappropriateness of the application of the ecological idea of succession to fossil sequences, both in discussions with his students (e.g., LCI and PJM both clearly recall Steve explicitly pointing out the inappropriate application of ecological succession to a stratigraphic sequence of corals) and in print: "Extrapolation from ecological to evolutionary time remains a dangerous game," he said, and the "concept of succession does not include replacement by evolution. The laws of scaling require different explanations at contrasting levels" (1976c, 236). These cautions were strengthened in the 1981 version, and he cited the work of Hansen (1978), Vrba (1980), and himself (1980c) in emphasizing "the inadvisability of simple extrapolation across levels of analysis" (1981z, 299). The use of succession as a "model for microtemporal change" in the fossil record would not work: "palaeobiologists must not uncritically extrapolate to their domain the models that operate so well in ecological time. Several postulated 'successions' in the recent palaeontological literature probably span too long a time for proper application. . . . May we speak of true succession if most species disappear by total extinction rather than by local replacement?" (1981z, 303).

The 1981 version also included more on results of his involvement in the so-called MBL studies (1973j, 1974i, 1977a; named for some of the work having been carried out at the Marine Biological Laboratory at Woods Hole, Massachusetts; see also Bambach, this volume) on stochastic factors in evolution. The "predominant impression" from these studies, he said "is one of conformity between patterns in real and random worlds" (1981z, 312–13). In other words, deterministic forces such as natural selection might— at least occasionally—be less important in shaping long-term evolutionary patterns than had been conventionally believed.

Particularly notable in the 1981 version is the emphasis on the end-Cretaceous extinction. He added a long paragraph that included the beginnings of what would eventually be a major theme: Explanations such as the Alvarez impact hypothesis (Alvarez et al. 1980), he said, "may point to a fundamental difference between reasons for local extinction in ecological time (primarily density-dependent) and mass extinction in geological time." Steve implied, however, that there did need to be an ecological component to such analysis (countering, at least at this stage, later criticism (e.g., Erwin, 2004, 25) that he was completely uninterested in the ecology of mass extinction): "these catastrophic theories will not compel ascent until they can provide an ecologically reasonable scenario for the differential patterns that characterize mass extinctions—a consideration notably lacking in papers by non-biologists published so far" (1981z, 309).

III. The hardening of his synthesis

The 1981 revision of "Ecology plus paleontology" marked the last time that Steve explicitly discussed in print any significant evolutionary role for ecological interactions. Throughout the 1980s his views on the irrelevance of ecology for long-term evolutionary trends appeared to solidify and became more clearly integrated into his evolutionary theorizing. This seems to have been the result of at least two scientific (as opposed or in addition to personal interest) factors. First, he was clearly very impressed by the results of the MBL studies, which appeared to show that a large amount of pattern could be produced by non-deterministic processes. The inappropriateness of the application of succession to the fossil record also continued to impress him and to suggest that all application of ecology to long-term evolutionary patterns was suspect.[1] Second, his increasing articulation of a hierarchical view of evolution logically excluded ecology as a major causal factor of macroevolution.

Many of his technical papers in the early 1980s express this nonecological theory of macroevolution. In a landmark paper with Brad Calloway, for example, Steve argued that the long-term consequences of competition for major evolutionary patterns, such as the switch in dominance between brachiopods and bivalves following the Paleozoic, were minimal (1980j). "I have long felt," he later wrote in one of his popular essays, "that images of balance and optimizing competition have been greatly oversold, that important and effectively random forces buffet the history of life, that most groups of organisms make their own way according to their own attributes, and that interactions among most groups are, on the broad scale of time in millions of years, more like Longfellow's 'ships that pass in the night' than the Book of Ruth's 'whither thou goest, I will go'" (1994n; 1995k, 100–101).

As Flannery (2002, 54) noted, Steve didn't like Dobzhansky's (1937) notion of adaptive peaks "because it cedes too much ground to natural selection and ecology." Adaptive peaks, according to Wright (1931), Dobzhansky, and many others, explain the existence of discrete morphologies, such as dogs and cats, "because ecological niches exist for dog-like and cat-like creatures, but not for in-between kinds. Steve, however, argued that the two discrete types owe their existence to historical constraint in the form of their own, separate inheritances from an ancestral dog-like and cat-like creature" (Flannery, 2002, 54). Indeed, Steve argued for the dominance of history over ecology as a controlling force because convergent evolution due to selection was not, in his view, very common. (Steve didn't deny convergence happened, but he argued that when it did, it was also possibly explicable by structural and fortuitous factors [e.g., 1971i].) Steve eventually came to view the adaptive landscape, if he thought about it at all, as the consequence of the contingent histories of species, and thus a product of their evolution rather than a cause. He argued this with particular enthusiasm of course in the case of the Burgess Shale (1989d). His views were countered by, among others, Conway Morris (2003), who argued explicitly that convergence due to natural selection (i.e., ecology) was a major theme in the history of life.

Steve maintained that it was unlikely that ecology could affect long-term evolutionary patterns in part because he found it unlikely that environments and their selective regimes would remain constant over evolutionary time. This was one reason he thought that stabilizing selection was unlikely to be the main cause of stasis. An "evolutionary trajectory through a temporal series of environments encounters so many random effects of great magnitude," he wrote, "that I expect historical individuality to overwhelm coordination...I regard each species as a contingent item of history with an unpredictable future" (1994n; 1995k, 103). Steve acknowledged the ecological patterns of biotic interaction documented by Vermeij (e.g., 1977; 1987), but he did not see how they could be translated into evolutionary/geological time (1990e); the trends were both too ponderously slow and too tied to multiple contingent histories to be explicable in this way. Steve also questioned whether we really know how ecological communities are structured (1995u; 1998x, 404). If we don't understand ecology, he reasoned, it is unlikely that we can apply it successfully to evolution.

This view reached its logical extreme conclusion with Steve's three-tiered hierarchical theory of macroevolution (1985f): ecological interactions (which he termed "Darwin's wedge," in reference to Darwin's Malthusian metaphor of the wedge [Darwin 1859, 67]) clearly mattered in ecological time, but probably had little effect on speciation (which he believed to be largely caused by stochastic accidents). Speciation, via punctuated equilibrium and sorting of lineages, was the most important source of long-term evolutionary trends, and even these trends were upset by mass extinctions, which might at least occasionally (and in the mid-1980s it was thought that it might be much more frequently) be caused by extraterrestrial events clearly independent of the relative merits of ecological interactions or adaptations that characterized "normal times" during intervals of "background extinction." His final exposition of a hierarchical view of evolution (2002c) had toned down the effects of mass extinction, but still argued that trends due to sorting of species limited the long-term influence of species interactions with each other and the environment occurring on ecological time scales.

Steve's invocation of the metaphor of the wedge is particularly revealing. The metaphor invokes an image of many independent entities, each competing for limited resources, each with its position determined by a contingent history of hammer blows. Interactions between wedges are fluid and fleeting, with only the long term fact of limited resources remaining unchanged. This is a metaphor of a brutal Malthusian ecology, fit for the fate of the seedling on the sidewalk, a metaphor where the only ecological constant is limited resources, and where ecology can be relegated to a result rather than a cause of evolution. This seemed to be Steve's view of ecology: the immediate biological context for the survival of individual organisms on short time scales.

Each of us recalls instances from our time as graduate students when Steve mentioned his lack of personal interest in ecology, and his pervasive doubts about the macroevolutionary significance of ecology. In 1987, for example, Steve made his first trip to Panama to visit Barro Colorado Island (as part of his duties as a member of the Smithsonian Council, on which he served from 1976 to 1988). Upon hearing about the upcoming trip, one of us (WDA[2]) excitedly encouraged him to see as much of the rainforest as possible while he was there. Steve replied to the effect that yes of course he would see the forest, but he was really going because Panama offered a better view of Halley's comet (see 1993l, 179–80). (He admittedly did have a keen interest in astronomy from childhood and several times went to great lengths to observe eclipses [see

1994t] and the collision of Shoemaker-Levy with Jupiter [1994v].)
In seminars and reading courses, Steve frequently commented
that he thought paleoecology was little more than the descriptive
application of modern concepts at the wrong temporal scales to
the fossil record. We still wonder how much of these comments
about "paleoecology as not important" was a reflection of what
he really thought and how much it was a pedagogical position to
provoke a reaction from us (which it did; see below), but our gut
feeling is that it was an expression of what he really thought.

We thus see Steve's view of ecology as being a mixture of both
personal disinterest and a relegation of ecology to operation on
geologically short time scales, confined to first tier interactions
among organisms in the entangled bank, its organisms placed
together in one place by their own historically contingent evolutionary and biogeographic histories.

IV. A wolf at the door: Steve and environmentalism

In his technical writing, Steve was thus a professional doubter
of the importance of ecology in the history of life. Yet his other
professional life as popular natural history essayist made him a
de facto spokesperson for not just evolutionary biology, but also
biology and even science in general. In this second role, he was
less than enthusiastic about using his bully pulpit for preaching
about environmental issues, yet his position in the literary and
popular science constellation eventually compelled him to address
environmental concerns.

The first obstacle to his embracing environmentalism in his
popular writing was that he did not like, or even know how, to
write conventional essays about the beauty and wonder of nature
(what he called the "strange ways of the beaver tradition" of essay
writing [1995u; 1998x, 394]). He was, in contrast, more interested in a "humanistic natural history" (see further discussion
in Allmon, this volume). Steve's "personal theory about popular
writing in science" divided the genre into two modes, which he
called "Galilean, for intellectual essays about nature's puzzles,
and Franciscan, for lyrical pieces about nature's beauty." He
was, he said, "an unrepentant Galilean" working "in a tradition
extending from the master himself [Galileo], to Thomas Henry

Huxley in the last century, down to J. B. S. Haldane and Peter Medawar in our own. I greatly admire Franciscan lyricism, but I don't know how to write in that mode" (1994t; 1995k, 10). "I do love nature," he wrote, perhaps somewhat defensively, "as fiercely as anyone who has ever taken up a pen in her service. But I am even more fascinated by the complex level of analysis just above and beyond...that is, the history of how humans have learned to study and understand nature. I am primarily a 'humanistic naturalist' in this crucial sense....That is, I am enthused by nature's constitution, but even more fascinated by trying to grasp how an odd and excessively fragile instrument—the human mind—comes to know this world outside, and how the contingent history of the human body, personality, and society impacts the pathways to this knowledge" (1998x, 5).

Aesthetics also played an important role in this lack of interest in ecology. Although he said that "nothing matched the insight and satisfaction gained from seeing the whole earth as a resplendent sphere in space" (1995f, 512), his tastes were clearly more attuned to humanism and less to nature: "I even believe—though I would not push the point, for the concept can too easily cede to human arrogance and a discounting of natural forms—that intelligent reconstruction can 'improve' upon natural design (though only by the criterion of human aesthetic preference, the most parochial of all possible judgements)"[3] (1998x, 3).

Second, ecology perhaps reflected too many elements of uniformity (of state) and determinism for Steve to be comfortable with the discipline. Near the beginning of each term in his "History of Earth and Life" course, Steve would write four words on the blackboard: Progress, Determinism, Gradualism, and Adaptationism. He described these four words as pervasive biases in science, and much of the course involved examination of episodes in the history of the earth and the history of science that illustrated these biases and the errors that arose from them. Indeed much of Steve's writing also focused on refuting these four biases. Some elements of ecology, in particular ecology as viewed by environmentalists, embody all of these biases. For example, a view of environmental protection that holds any change to be a bad thing (e.g., any introduced species is an inherently a bad thing, any environmental degradation by human constructions is inherently a bad thing) implies that the present is an optimal end point of progressive evolution. Exalting species as the epitome of specialized design produced over millions of years of evolution embodies adaptationism and progress. Projecting ecology into geological time scales can readily involve elements of adaptationism, gradualism and progress, and attempting to forecast the future state of environmental systems is fundamentally a deterministic exercise. We are not claiming that ecology as a discipline is, or was, dominated by these biases (some parts of the discipline, such as non-equilibrium dynamics or the application of cascades from complex systems theory, are explicitly not gradual, progressive, or deterministic). We do, however, suggest that Steve could easily have seen too many elements of ecology as reflecting these biases, and thus they were anathema to him.

Third, and perhaps more important, Steve had genuine doubts about some of the substantive claims of modern environmentalism, and therefore seemingly could not convince himself that the environmental crisis was real, or at least that it was/is as big a crisis as some environmentalists suggested. For example, he did acknowledge that rainforests were "the appropriate focus of the environmental movement" because of their high species diversity[4] (1993m; 1995k, 383), and stated that with the extinction of "too many" species, "the gloriously arborescent tree of life becomes a tawdry set of branches, far too sparse and all awry, with a layer of rot and decay at the base" (1995f, 514). Yet he also said he had trouble getting worked up about current biodiversity loss because we didn't know how many species there are (e.g., 1993m). He devoted an entire essay (1997g) to the issue of invasive plants, remarkably concluding that they were not entirely a bad thing. That essay was criticized by Flannery (2002, 54) who noted that "what is astonishing about this argument is that it takes no account of ecology."

Steve finally admitted that environmental themes had been "curiously underplayed, I must say in self-criticism" in his popular essays (1995k, x), and in the early 1990s he emphasized environmental themes in several pieces. For example, he said that if we could recognize that we shared a very close genealogical connection to other species, especially apes, then "We will be a bit freer, a bit more enlightened, a bit readier to work for planetary preservation with the rest of kindred life" (1992r; 1995k, 400). He noted with sadness the loss of the blaauwbock, or blue antelope, which became extinct in 1799, just thirty-three years after its formal description—the first known extinction of a large-bodied terrestrial mammal species in historic times (1993q; 1995k, 272). And wrote with some measure of disgust about those who had hunted animals like the dodo, the passenger pigeon, and the Galapagos tortoises to extinction or near extinction simply because they were easy prey and no thought was given to their precipitous decline (1991, reprinted in 1993). Yet, in describing the first documented extinction of a marine invertebrate animal, a group thought to be relatively immune to extinction, he seems to take satisfaction in pointing out that here we were not to blame. It was simply a case of extinction of a specialized animal, the limpet *Lottia alveus*, whose habitat was radically altered by a natural contingent turn of events (infestation and die-back of their seagrass host). The limpet was simply unlucky, and went the way of many before it. Admittedly, though, at the close of the essay he suggests that the extinction of this limpet may be a harbinger for things to come, a "warning against [our] complacency" because it lived "in a realm of supposed invulnerability" and yet met with extinction even without our help (1991o).

His major statement on environmentalism, however, came in a 1990 *Natural History* essay (1990s; restated in part in a later short book chapter, 1995f) in which he made a twofold argument. First, he said that, despite the claims of the environmental movement, humans will and can have no *long-term* effect on the Earth, because planetary and human time scales are so different. The notion that "[h]umans must learn to act as stewards for this threatened world" is, he said, "rooted in the old sin of pride and exaggerated self-importance. We are one among millions of species, stewards of nothing. By what argument could we, arising just a geological microsecond ago, become responsible for the affairs of a world 4.5 billion years old, teeming with life that has been evolving and diversifying for at least three-quarters of this immense span?" Such an idea makes sense, Steve maintained,

> if we, despite our late arrival, now held power over the planet's future. But we don't, despite popular misconceptions that we might. We are virtually powerless over the earth at our planet's

own geological timescale.... [after the K-T extinction] the earth survived...and, in wiping out the dinosaurs paved the road for the evolution of large mammals, including humans. We fear global warming, yet even the most radical model yields an earth far cooler than many happy and prosperous times of a prehuman past. We can surely destroy ourselves, and take many other species with us, but we can barely dent bacterial diversity and will surely not remove many species of insects and mites. On geological scales, our planet will take good care of itself and let time clear the impact of human malfeasance. (1990s; 1993l:47–48)

Perhaps for Steve, as for many other Earth scientists (including ourselves), this fatalistic but long-ranging view actually can offer a bit of solace in what otherwise could be seen as an ultimately hopeless situation for global biodiversity. While it may make him sound insensitive to the ongoing human crisis, the statement is most certainly correct on long time scales.

Second, he said that environmental degradation mattered precisely because it is occurring at *our* time scale: "We cannot threaten at geological scales, but such vastness has no impact upon us. We have a legitimately parochial interest in our own lives, the happiness and prosperity of our children, the suffering of our fellows" (1990s; 1993l, 47–48). We should not protect species and their environments, Steve said, "because we fear for global stability in a distant future not likely to include us. We are trying to preserve populations and environments because the comfort and decency of our present lives, and those of fellow species that share our planet, depend upon such stability" (1990s; 1993l, 47–48).

From these two arguments, Steve reached a conclusion and recommendation for the future that revealed much about his views of environmentalism, and perhaps also about his reasons for not embracing it more enthusiastically. Steve, the "street kid" and grandson of poor immigrants clearly found the socioeconomic background of many modern environmentalists deeply distasteful:

We must squarely face an unpleasant historical fact. The conservation movement was born, in large part, as an elitist attempt by wealthy social leaders to preserve wilderness as a domain for patrician leisure and contemplation.... We have never entirely shaken the legacy of environmentalism as something opposed to immediate human needs, particularly of the impoverished and unfortunate....Environmental movements cannot prevail until they convince people that clean air and water, solar power, recycling, and reforestation are best solutions (as they are) for human needs at human scales—and not for impossibly distant planetary futures. (1990s; 1993l, 47–48)

Steve's phrase "we fear for global stability in a distant future" embodies a perspective of ecology as deterministic (with the ability to predict states in the distant future) and static (with change from a perceived current stable state as being worrisome), a perspective clearly at odds with the fundamental underpinnings of his approach to science. If ecology is simply the entangled bank of interactions that emerges from the converging history of species living in one place, then it must be so complexly dependent on contingent histories as to be unpredictable when scaled into geologic time. From the human centered bias of the present as the peak of natural progress, any change from stable ecological states must be a bad thing, but from the perspective of geologic time, ecological change and extinction are the normal and expected processes. Environmentalism, from Steve's perspective, was in many ways fundamentally at odds with the biases he spent his career trying to expunge from science.

V. Coordinated stasis

In the spring of 1992, two of us (PJM and LCI) participated in a reading course with Steve, together with fellow graduate students David Kendrick, Ken Schopf, Loren Smith, and Peg Yacobucci. The readings focused on important papers in paleobiology published between 1900 and 1980, and included, concomitant with our interests, several examples from the early work in paleoecology. At this juncture of the course, it became clear that Steve's view of paleoecology was that it was little more than the descriptive application of concepts drawn from modern ecology theory, usually at the wrong temporal scales, to the fossil record—a sort of "we can do it too" mentality that contributed no new insights into evolutionary process or cause.

Coincidentally, over this same interval of time, Ivany, Morris, and Schopf had been joining Carlton Brett, then at the University of Rochester, and his students on a number of field trips to the Paleozoic rocks of New York State. On these occasions as well as during several formal invited lectures at Harvard, we were exposed to Carl's thoughts about the nature of faunal assemblages in these rocks and how they change over time. Carl was convinced that ecological assemblages and the taxa that comprise them exhibited what he called coordinated stasis (see, e.g., Brett and Baird 1995), wherein change seemed to be intermittent, abrupt, and separated by long intervals of stasis across the vast majority of taxa. The juxtaposition of Steve saying that paleoecology had nothing to offer evolutionary theory and Carl's observations about coincident punctuated equilibrium among many coexisting taxa in ecological assemblages fomented the development of what we called "ecological locking" (thanks to Loren Smith for the term!) (Morris et al. 1995). We argued that ecological interactions within well-established assemblages were capable of limiting evolutionary change in the component taxa. In developing ecological locking, we were trying in particular to address the need (spurred by Steve's comments) for mechanistic links between macroevolution and selection, and this was one of the main drivers behind the "spaghetti diagram" in the Morris et al. paper.

Over the course of the spring of 1992, Ivany, Morris, and Schopf spent lots of time whispering in corners (at least that is how it felt to us; Steve's students were almost always very open about what they were working on), sketching out ideas about how ecological systems might scale into geological time to explain the apparent patterns of coordinated stasis. Late in the spring we brought ecological locking as a relatively complete concept to Steve (just the four of us in the seminar room with a blackboard) by means of an extended version of the "spaghetti diagram." A key point we made in that meeting (again driven by his earlier comments about ecological time scales and geological time scales being different) was that there could be and probably were ecological dynamics working on population-level time scales that were capable of producing long term patterns that would be visible on the time scales preserved in the fossil record. A very strong undercurrent in that meeting was the three of us telling Steve that he was wrong— that paleoecology *was* important and that it had important things to say about evolutionary processes.

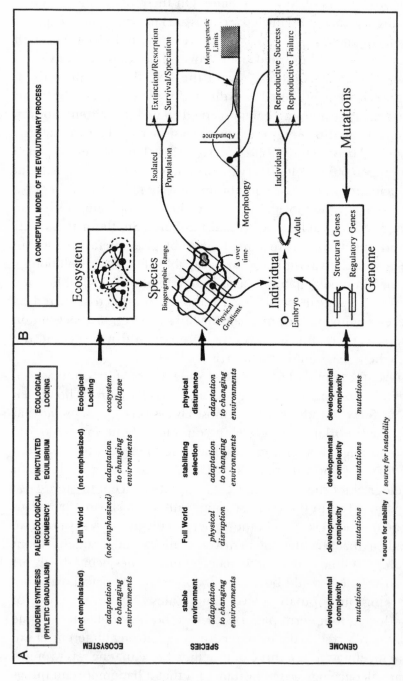

Figure 4.1. Morris et al. The Challenge of Paleoecological Stasis: Reassessing Sources of Evolutionary Stability. *Proceedings of the National Academy of Sciences USA*, Vol. 92, 11269–273, Copyright 1995 National Academy of Sciences, U.S.A.

After that meeting, the concepts of coordinated stasis and ecological locking were more generally discussed around the lab. Steve was very encouraging about our working out how to publish our ideas. He suggested *Proceedings of the National Academy of Sciences* as a venue, and offered to put forward the paper as a member. He was also very encouraging to us in thinking about how ecological locking might help to explain other macroevolutionary patterns (such as onshore/offshore patterns; e.g., Jablonski and Bottjer 1990). This encouragement contributed to the ultimate organization and publication of a symposium on coordinated stasis and related ideas at the 1994 national meeting of the Geological Society of America (Ivany and Schopf 1996). During all of this, we had the sense that he didn't think that we were right in the details, but that there were ecological sorts of things he had to go off and think about some more (and work into his tiered view of the world). Judging from his later discussions of ecology, it seems that he did indeed think about it, and decided that ecology was still something to relegate to the first tier. Nonetheless, he devoted some space to it in *Structure* (2002c, 920).

VI. Does ecology matter?

Steve thus had both personal and technical reasons for de-emphasizing the importance of ecology in his world view and evolutionary theorizing. Whatever the origins of these views, however, they do not currently appear to be very popular, even among colleagues generally friendly to Steve's ideas. Steve's co-advocate for the importance of stasis, for example, has long argued for a major role for stabilizing selection in maintaining that apparently pervasive pattern (e.g., Eldredge 1985, 1989; Eldredge et al. 2005). Other authors who are generally amenable to the punctuated view of evolution are also comfortable granting to ecology a larger role in macroevolution (e.g., Stanley 1979; Vrba 1980; Jackson 1988; Bambach 1993, 1999; Allmon, 1994; Kelley and Hansen 2001; Jablonski 2005). Although mass extinction is now widely recognized as a major factor in the history of life, Steve's extreme view of mass extinction as an independent tier of macroevolutionary process largely detached from events and processes in ecological time has generally not fared well, and the study of recovery from mass extinction is almost completely an ecological subject (e.g., Hart 1996; Lockwood 2004, 2005; Payne 2005). As Erwin puts it:

> I have little qualm with Steve's analysis of the evolutionary significance of mass extinctions, but the ecological poverty of his analysis is particularly evident in this discussion [in *Structure*]. Steve has long had a particular aversion to the inclusion of G. Evelyn Hutchinson's ecological play in his evolutionary theater. Yet all macroevolutionary events, mass extinctions chief among them, play themselves out through macroecological processes. Changing biogeographic patterns, competition between species or clades, and the acquisition of resources and space matter for macroevolution as much as microevolution, albeit at a different spatial and temporal scale (hence macroecology). Without inclusion of these processes we have little likelihood of understanding either mass extinctions or the biotic rebounds that follow them, and their exclusion ultimately renders *Structure* fascinating but incomplete as a new theory of macroevolution. (2004, 25)

Although, as Steve long maintained, "evolutionary paleoecology" may not yet have a fully coherent body of its own theory, research results continue to appear regularly in the recent paleobiological literature on how ecology may affect macroevolutionary change (e.g., Dietl et al. 2000; Sepkoski et al. 2000; Connolly and Miller 2001, 2002; Allmon and Bottjer 2001; Kowalewski and Kelley 2002; Todd et al. 2002; Kelley et al. 2003; Peters 2004; Finnegan and Droser 2005; Jablonski 2005; Aberhan et al. 2006). Even within his own sphere of hierarchical macroevolution, Steve may have missed "a good bet by overlooking so much of ecology," says Jablonski (2002, 370): if, for example, "geographic range or rarity are species-level traits or are at least likely to confer emergent fitness, then the burgeoning field of macroecology is rich in potential cases of species selection."

Steve had a very clear understanding of the scope of geologic time, and for him, ecological processes occurred on time scales far too short to affect long term macroevolutionary patterns. Ecological time, while long on human terms, was fleeting in a geological context, and became simply part of the fluid selective context of individual species. For Steve, Darwin's metaphors of the entangled bank and the wedge suggested that although ecology was the immediate environmental context for individual organisms, it was the largely the result of their independent contingent evolutionary histories, rather than a cause of evolutionary change. The extension of ecology beyond such historically contingent interactions seemed to Steve to fall afoul of the biases of science that he spent his career fighting, such as determinism and uniformity. Furthermore, views held by some in the environmental movement of the ecological consequences of human environmental degradation in particular seemed to Steve to be futile deterministic attempts to predict the future state of an unpredictably complex system. That little tree that lived, and died, in Queens may well have inspired some of Steve's attempts to transform evolutionary theory. So far, however, that transformation has not included the relegation of ecology to the ash heap of macroevolutionary causes.

Acknowledgments

We are especially grateful to Kathy Hoy Burgess for sharing with us her memory of a long-ago conversation with Steve Gould, and to Norman Johnson and two anonymous reviewers for comments on previous drafts of the manuscript.

References

Aberhan, M., W. Kiessling, and F. T. Fürsich. 2006. Testing the role of biological interactions in the evolution of mid-Mesozoic marine benthic ecosystems. *Paleobiology* 32(2): 259–77.

Allmon, W. D. 1991, A plot study of forest floor litter frogs, Central Amazon, Brazil. *Journal of Tropical Ecology* 7: 503–22.

———. 1994. Taxic evolutionary paleoecology and the ecological context of macroevolutionary change. *Evolutionary Ecology* 8: 95–112.

Allmon, W. D., and D. J. Bottjer, eds. 2001. *Evolutionary paleoecology. The ecological context of macroevolutionary change.* New York: Columbia University Press.

Bambach, R. K. 1993. Seafood through time: changes in biomass, energetics, and productivity in the marine ecosystem. *Paleobiology* 19(3): 372–97.

———. 1999. Energetics in the global marine fauna: a connection between terrestrial diversification and change in the marine biosphere. *Geobios* 32: 131–44.

Brett, C. E., and G. C. Baird. 1995. Coordinated stasis and evolutionary ecology of Silurian to Middle Devonian faunas in the Appalachian Basin. In D. H. Erwin and R. L. Anstey, eds., *New approaches to speciation in the fossil record*, 285–315. New York: Columbia University Press.

Conway Morris, S. 2005. *Life's solution.* New York: Cambridge University Press.

Connolly, S. R., and A. I. Miller. 2001. Global Ordovician faunal transitions in the marine benthos: proximate causes. *Paleobiology* 27(4): 779–95.

———. 2002. Global Ordovician faunal transitions in the marine benthos: ultimate causes. *Paleobiology* 28(1): 26–40.

Darwin, C. 1859. *On the origin of species.* London: John Murray.

———. 1860. *The voyage of the Beagle.* 1962 printing, Natural History Library edition. New York: Doubleday.

Dietl, G. P., R. R. Alexander, and W. F. Bien. 2000. Escalation in Late Cretaceous-early Paleocene oysters (Gryphaeidae) from the Atlantic Coastal Plain. *Paleobiology* 26(2): 215–37.

Dobzhansky, T. 1937. Genetics and the origin of species. 2nd ed., 1941; 3rd ed., 1951. New York: Columbia University Press.

Eldredge, N. 1985. *Time frames. The rethinking of Darwinian evolution and the theory of punctuated equilibria.* New York: Simon and Schuster.

———. 1989. *Macroevolutionary dynamics. Species, niches, and adaptive peaks.* New York: McGraw-Hill.

Eldredge, N., J. N. Thompson, P. M. Brakefield, S. Gavrilets, D. Jablonski, J. B. C. Jackson, R. E. Lenski, B. S. Lieberman, M. A. McPeek, and W. Miller III. 2005. The dynamics of evolutionary stasis. In E. S. Vrba and N. Eldredge, eds., *Macroevolution. Diversity, disparity, contingency. Essays in honor of Stephen Jay Gould,* 133–45. Supplement to *Paleobiology* 31(2).

Erwin, D. H., 2004, One very long argument. Review of S. J. Gould, *The Structure of evolutionary theory. Biology and Philosophy* 19:17–28.

Finnegan, S., and M. L. Droser. 2005. Relative and absolute abundance of trilobites and rhynchonelliform brachiopods across the Lower/ Middle Ordovician boundary, eastern Basin and Range. *Paleobiology* 31(3): 480–502.

Flannery, T., 2002, A new Darwinism? Review of S. J. Gould, *The structure of evolutionary theory,* and *I have landed: The end of a beginning in natural history. New York Review of Books,* May 23, 52–54.

Hart, M. B., ed. 1996. Biotic recovery from mass extinction events. Geological Society Special Publication no. 102. London: The Geological Society.

Ivany, L. C., and K. M. Schopf, eds., 1996. New perspectives on faunal stability in the fossil record. *Palaeogeography, Palaeoclimatology, Palaeoecology,* Special issue, 127(1–4).

Jablonski, D. 2002. A more modern synthesis. Stephen Jay Gould 1941–2002. *American Scientist,* July-August, 368–72.

Jablonski, D. 2005. Evolutionary innovations in the fossil record: the intersection of ecology, development, and macroevolution. *Journal of Experimental Zoology* (Mol. Dev. Evol.) 304B: 504–19.

Jablonski, D., and D. J. Bottjer. 1990. Onshore-offshore trends in marine invertebrate evolution. In R. M. Ross and W. D. Allmon, eds., *Causes of evolution. A paleontological perspective*, 21–75. Chicago: University of Chicago Press.

Jackson, J. B. C. 1988. Does ecology matter? Review of G. J. Vermeij, *Evolution and escalation: an ecological history of life. Paleobiology* 14(3): 307–12.

Kelley, P. H., and T. A. Hansen. 2001. The role of ecological interactions in the evolution of naticid gastropods and their molluscan prey. In W. D. Allmon and D. J. Bottjer, eds., *Evolutionary paleoecology. The ecological context of macroevolutionary change*, 149–70. New York: Columbia University Press.

Kelley, P. H., M. Kowalewski, and T. A. Hansen, eds. 2003. *Predator-prey interactions in the fossil record.* New York: Kluwer Academic/Plenum Publishers.

Kowalewski, M., and P. H. Kelley, eds. 2002. The fossil record of predation. Paleontological Society Papers No. 8.

Lewin, R. 1983. No dinosaurs this time. *Science* 221: 1168–69.

Lockwood, R. 2004.The K/T event and infaunality: morphological and ecological patterns of extinction and recovery in veneroid bivalves. *Paleobiology* 30(4): 507–21.

———. 2005. Body size, extinction events, and the early Cenozoic record of veneroid bivalves: a new role for recoveries? *Paleobiology* 31(4): 578–90.

Morris, P. J., L. C. Ivany, K. M. Schopf, and C. E. Brett. 1995. The challenge of paleoecological stasis: Assessing sources of evolutionary stability in the fossil record. *Proceedings of the National Academy of Sciences, U.S.A.* 92: 11269–73.

Payne, J. L. 2005. Evolutionary dynamics of gastropod size across the end-Permian extinction and through the Triassic recovery interval. *Paleobiology* 31(2): 269–90.

Peters, S. E. 2004. Evenness of Cambrian-Ordovician benthic marine communities in North America. *Paleobiology* 30(3): 325–46.

Sepkoski, J. J., Jr., F. K. McKinney, and S. Lidgard. 2000. Competitive displacement among post-Paleozoic cyclostome and cheilostome bryozoans. *Paleobiology* 26(1): 7–18.

Stanley, S. M.. 1979. *Macroevolution. Pattern and process.* San Francisco: W. H. Freeman.

Todd, J. A., J. B. C. Jackson, K. G. Johnson et al. 2002. The ecology of extinction: molluscan feeding and faunal turnover in the Caribbean Neogene. *Proceedings of the Royal Society B: Biological Sciences* 269: 571–77.

Vermeij, G. J. 1977. The Mesozoic marine revolution: evidence from snails, predators and grazers. *Paleobiology* 3: 245–58.

———. 1987. *Evolution and escalation. An ecological history of life.* Princeton, NJ: Princeton University Press.

Vrba, E. 1980. Evolution, species, and fossils: how does life evolve? *South African Journal of Science* 76: 61–84.

Wright, S. 1931. Evolution in Mendelian populations. *Genetics* 16: 97–159.

Stephen Jay Gould's Winnowing Fork

Science, Religion, and Creationism

Patricia H. Kelley

Gould and the Creationists

One of the first things I remember seeing, when I moved into my graduate student office a few doors down the hall from Steve Gould's, was a clipping posted in the hallway with an alarmist title proclaiming that Gould had disproved evolution. I was a bit taken aback; I had come to Harvard to study paleontology with Stephen Jay Gould, and I began to wonder what I had gotten myself into!

I didn't waste too much time before asking Steve about the article. He greeted my query with a chuckle; I don't remember his exact reply, but it was something like, "Those silly creationists, thinking punctuated equilibrium supports their views. Of course it does nothing of the sort."

It was 1975, a few years after the publication of the seminal paper introducing the concept of punctuated equilibrium (1972e), and at the time Steve's reaction struck me as one of mild amusement and perhaps a little pride that his work had become well known enough to attract creationists' attention. And yet over the next few years Gould's attitude toward creationism seemed to go through its own evolution from this initial stage of amusement to one of irritation to outright animosity towards creationism. Indeed, Gould became one of creationism's harshest critics, testifying in the *McLean v. Arkansas* evolution-creationism balanced treatment case and lashing out against creationism in numerous popular essays, particularly through his monthly venue in *Natural History* magazine. Indeed, this acrimony is still evident in *The Structure of Evolutionary Theory* (2002c). Why such personal animosity?

Gould's rancor toward creationism was spurred by the misuse of punctuated equilibrium to support creationist tenets, either by conflating punctuated equilibrium with Goldschmidt's saltationism or by using it to claim that the fossil record contains no intermediates between higher taxa. Decades later, Gould (2002c) reflected on the treatment of his ideas by creationists, citing half a dozen creationist references that distorted or misquoted his work. For instance, Gould (2002c, 987) cited a pamphlet titled "Harvard scientists agree: Evolution is a hoax!!!" (perhaps the very item I saw posted outside Steve's office door?) that claimed "the facts of 'punctuated equilibrium'...fit the picture that Bryan insisted on, and which God has revealed to us in the Bible. Every species of organism was separately created during the six 'days' of creation." Another creationist work quoted by Gould (2002c, 987) claimed that the repudiation of gradualism by "the new theory of punctuated equilibrium brings the thinking of science remarkably closer to the biblical view. It is notable that the more evidence scientists discover (or fail to discover), the closer scientific theory moves toward the unchanging biblical pattern."

In a telling statement in *Discover Magazine*, Gould (1981f, 36) commented, "creationists rely upon distortion and innuendo to buttress their rhetorical claim. If I sound sharp or bitter, indeed I am—for I have become a major target of these practices." Gould responded to such attacks by taking every opportunity to expose creationist approaches as "modern pseudoscience" and "intellectual dishonesty." For instance, in a *Natural History* essay on the refutation of flood geology in the early 1800s (reprinted in *The Flamingo's Smile*), Gould (1985o, 124) stated, "Modern creationists...do no field work to test their claims (arguing instead by distorting the work of true geologists for rhetorical effect) and will not change one jot or tittle of their preposterous theory." Gould used an essay on Nebraska Man (reprinted in *Bully for Brontosaurus*) to contrast the self-correcting process of science with the approach of creationists, concluding that "The world of creationism is too imbued with irrefutable dogma, and they don't seem able even to grasp enough about science to put up a good show in imitation" (1991n, 447). Gould repeatedly drew a distinction between true science, which is based on hypothesis testing, and creation "science," the tenets of which are untestable and based on biblical literalism.

Gould's Winnowing Fork

His winnowing fork is in his hand, and he will clear his threshing floor and gather his wheat into the granary, but the chaff he will burn with unquenchable fire.

(Matt. 3:12)

John the Baptist's metaphor of impending judgment in the third chapter of Matthew's gospel strikes me as an apt image to describe Gould's differentiation of creationism and religion. Although his bitterness toward creationism might have led to an overall animosity toward religion, the unquenchable fire of Gould's vitriolic chastisement of creationism did not extend to religion in general. Perhaps remarkably for a nonbeliever, Gould was able to winnow the chaff of creationism from the wheat of religion. Gould realized from the start that creationism was not synonymous with religion, writing that creationists have been "disowned by leading churchmen of all persuasions, for they debase religion even more than they misconstrue science" (essay on "A Visit to Dayton," reprinted in 1983d, 275). Gould's essays repeatedly assert that the enemy of science is not religion, but irrationalism, dogma, and intolerance (see, for example, 1985o, 125).

Gould's attitude toward religion is best expressed in his 1999 book *Rocks of Ages: Science and Religion in the Fullness of Life.* In this short volume, Gould expanded on a 1997 *Natural History* article in which he defined the principle of "nonoverlapping magisteria" or NOMA (magisterium signifying a "domain of authority in teaching"; i.e., a "site for dialog and debate"). Gould (1999n) argued that science and religion represent separate and nonconflicting realms; science deals with the empirical realm, developing theories to explain the observed facts of nature. Religion, as described by Gould, represents the moral realm, addressing questions of

ultimate meaning and value. According to him, both magisteria
are necessary for human fulfillment.

Under the NOMA principle, the relationship between science
and religion should be characterized by "respectful noninterfer-
ence" as well as "intense dialogue" between the two magisteria.
According to this view, there should be no inherent conflict
between the two magisteria; instead, apparent conflict occurs
when the principle of NOMA is violated by either side, for instance
when scientists draw moral conclusions from their findings or
when creationists impose their dogma on the magisterium of
science. Gould (1999n, 209–10) castigated as violators of NOMA
both scientist colleagues "who tout their private atheism...as a
panacea for human progress against an absurd caricature of 'religion'" and creationists who "try to suppress the uncomfortable
truths of science."

Gould's *Rocks of Ages* might be viewed as a manifestation of a trend
of expanding interest in the relationship of science and religion. In
recent years, scientific societies, such as the American Association for
the Advancement of Science and the National Academy of Sciences,
opened science-religion dialogues; publications on the subject have
burgeoned; institutions have been established to address the issue
(e.g., the Chicago Center for Religion and Science and the Center
for Theology and Natural Sciences in Berkeley); and universities
such as Cambridge, Oxford, and Princeton have made joint faculty
appointments in science and theology (see Easterbrook 1997, and
Bryan 1999, for summaries). *Newsweek* magazine even went so far
as to publish an article titled "Science Finds God" (Begley 1998).
Widely differing views of the relationship have been propounded,
ranging from the notion that science and religion are inherently
conflicting to the idea that the two should be fully integrated.
Ian Barbour (1990; 1997) has provided a convenient, if perhaps
simplistic (Bryan 1999), taxonomy for existing views concerning
the relationship of science and religion. Barbour classified views of
science and religion into the four categories of conflict, independence, dialogue, and integration.

The view of science and religion as irreconcilably conflicting
is prevalent in American society today (Kelley et al. 1999). In my
own teaching experience, which has been mostly in the southern
United States, students entering my class frequently have had the

preconceived notion that science and religion are incompatible and that they must either make a conscious decision to reject science or to reject religion (Kelley 1999; see also Gibson 1999). As described by Gould (1999n), this notion of warfare between science and religion was popularized in the late nineteenth century by books written by Andrew Dickson White and John William Draper. Interestingly, this view has been perpetuated by both scientists (e.g., the scientific materialism of zoologist Richard Dawkins) and by young-earth creationists who require a supposedly literal reading of the Bible (Barbour 1990). Gould repeatedly argued against the view that science and religion are inherently conflicting, emphatically stating, "No battle exists between science and religion" (1991a, 400). Much of *Rocks of Ages* is devoted to reinforcing this point.

Gould's NOMA principle occupies a middle ground on the continuum of views concerning the relationship of science and religion. Using Barbour's (1990; 1997) taxonomy, the NOMA principle represents the "independence" category (Bryan 1999), in which the two fields are considered distinct, with science confined to the empirical realm and religion to the moral realm. Obviously this view is not unique to Gould; for instance, a resolution by the National Academy of Sciences in 1981 (see National Academy of Sciences 1984, 6) stated, "Religion and science are separate and mutually exclusive realms of human thought whose presentation in the same context leads to misunderstanding of both scientific theory and religious belief." This view of science and religion as independent has been criticized as a sort of "coward's way out," a means of avoiding controversy. Gould (who was never one to avoid controversy) anticipated this criticism, writing (1999n, 9–10), "NOMA represents a principled position on moral and intellectual grounds, not a merely diplomatic solution." Indeed, he chastised those who adopt an avoidance strategy in which neither side engages the other in debate (1999n, 220–21).

Under NOMA, Gould (1999n) advocated that the relationship between science and religion should be characterized by "intense dialogue." NOMA thus appears to also share some aspects of Barbour's third category, "dialogue." Barbour (1990, 17) suggested that "for many scientists, exposure to the order of the universe, as well as its beauty and complexity, is an occasion of wonder and reverence," which opens the scientist to religious answers. Barbour thus envisioned dialogue as taking place at the boundary of science and religion, a boundary that Gould (1999n) described as complexly interdigitating at all scales. For Gould, addressing of the "interesting questions" requires contributions from both magisteria, through the vehicle of "talking to each other in mutual respect" (1999n, 211). When this occurs, "science can then forge true partnerships with philosophy, religion, and the arts and humanities, for each must supply a patch in that ultimate coat of many colors, the garment called wisdom" (1999e, 2087).

Barbour (1990) suggested that the dialogue of science and religion can be fostered by consideration of their respective methodologies, and he reviewed claims of methodological parallels between science and religion. Despite arguments that science is not as objective as once thought and that it is "theory-laden" and dependent on prevailing paradigms (views with which Gould would have agreed wholeheartedly), Barbour (1990, 23) concluded that the methodological distinctions between science and religion are still valid: "The kinds of data from which religion draws are radically different from those in science, and the possibility of testing religious beliefs is more limited.... Religious belief must always be seen in the context of the life of the religious community and in relation to the goal of personal transformation." Gould (1999n, 210) supported the methodological distinctions between science and religion: "NOMA honors the sharp differences in logic between scientific and religious arguments." Nevertheless, dialogue must still occur.

Barbour's fourth category is integration, a view that advocates direct interaction, even fusion, between the realms of science and religion (an approach Gould, 1999n, termed "syncretism"). Syncretism often relies on arguments based on apparent design in nature (which in a previous century would have been referred to as "natural theology"). Modern incarnations of design arguments include "intelligent design," which argues that the "irreducible complexity" of nature requires the existence of a "blueprint" and the actions of an "intelligent agent" (see Scott, 2004b, for a concise summary of the main ID arguments). Despite its avoidance of overtly religious terminology, ID appears to many as simply another vehicle to slip creationist views into the public school curriculum (Numbers 1998; Bryan 1999; Forrest and Gross 2005; see also the decision in the *Kitzmiller et al. v. Dover Area School District* by Jones 2005).

A related concept is the anthropic principle, which exists in a bewildering array of forms (e.g, Weak Anthropic Principle, Strong Anthropic Principle, Participatory Anthropic Principle, Final Anthropic Principle) discussed at length by Barrow and Tipler (1986). The Weak Anthropic Principle (WAP) was defined by Barrow and Tipler (1986, 16) as "The observed values of all physical and cosmological quantities are not equally probable but they take on values restricted by the requirement that there exist sites where carbon-based life can evolve and by the requirement that the Universe be old enough for it to have already done so." Gould (1999n, 218–19) dismissed the WAP as "utterly trivial," stating, "The weak version only tells us that life fits well with nature's laws, and couldn't exist if the laws were even the tiniest bit different. Interesting, but I see no religious implications." The Strong Anthropic Principle (SAP) states "The Universe must have those properties which allow life to develop within it at some stage in its history" (Barrow and Tipler 1986, 21); it has been used to argue that the unique set of conditions and circumstances that makes our universe habitable requires the existence of an intelligent creator. Gould (1999n, 219) argued against the illogic of this conclusion drawn by SAP.

What Did Gould Get Right?

Although the NOMA principle has been criticized for presenting a simplistic view of the relationship between science and religion, I believe Gould to be fully correct in insisting that science and religion are not inherently incompatible. In an essay on the Huxley-Wilberforce debate reprinted in *Bully for Brontosaurus,* Gould (1991n, 400) observed that "many prominent evolutionists have been devout, and many churchmen have placed evolution at the center of their personal theologies." Indeed, a recent Paleontological Society Short Course (Kelley et al. 1999) on the evolution/creationism controversy included presentations by several geologists who are persons of faith (Bryan, Bambach, Young, Dodson, and Kelley are those known to me to be practicing

Christians). Several books have been published in recent years by other paleontologists and evolutionary biologists who find no conflict between their faith and science (e.g., K. R. Miller 1999; K. B. Miller 2003; Godfrey and Smith 2005). According to a recent poll, approximately 40 percent of biological and physical scientists (including mathematicians) consider themselves to be believers, a result remarkably consistent with that of a similar poll conducted in 1914 (Larson and Witham 1997, 1999). Public avowals of faith by prominent scientists are becoming commonplace (see summary in Easterbrook 1997). Thus religious belief is not inimical to the practice of science.

Gould correctly stated that most mainline religions support evolution, including the Roman Catholic Church, the American Jewish Congress, and various Protestant denominations. A number of religious organizations have released official statements either supporting evolution or opposing the teaching of creationism in public school science classes. A sampling of such statements is included in table 5.1; see Matsumura (1995) for additional comments.

In the same vein, Gould (1999n) recognized that clergy and religious scholars are typically supporters of NOMA and the First Amendment. As he pointed out (and as always shocks my students, even after I browbeat them that science and religion are not incompatible), a large number of the plaintiffs in the Arkansas "balanced treatment" court case were clergy or religious scholars. Included were the Arkansas leadership of the United Methodist, Episcopal, Roman Catholic, African Methodist Episcopal, and Presbyterian Churches; additional United Methodist, Presbyterian, and Southern Baptist clergy; and the American Jewish Congress, Union of American Hebrew Congregations, and the American Jewish Committee as organizational plaintiffs.

Gould would not be surprised by the success of The Clergy Letter Project, a grassroots movement spearheaded by biologist Michael Zimmerman of Butler University (www.butler.edu/clergy project/rel_evol_sun.htm). The project was initiated in response to antievolution policies passed by the Grantsburg, Wisconsin, school board in 2004 and has since become a nationwide project. As of November 2007, more than 11,000 clergy had signed the following statement:

Table 5.1. Official statements of religious organizations regarding evolution and/or creationism.

Organization	Date	Statement
Roman Catholic Church (Pope John Paul II)	1981	"The Bible . . . does not wish to teach how heaven was made but how one goes to heaven"
General Convention of the Episcopal Church	1982	"Be it Resolved, that the 67th General Convention affirm the glorious ability of God to create in any manner, whether men understand it or not, and in this affirmation reject the limited insight and rigid dogmatism of the "Creationist" movement"
Central Conference of American Rabbis	1984	"The principles and concepts of biological evolution are basic to understanding science."
United Church of Christ	1983	"We acknowledge modern evolutionary theory as the best present-day scientific explanation of the existence of life on earth; such a conviction is in no way at odds with our belief in a Creator God."
American Jewish Congress	1984	"It is our position that scientific creationism is a religious theory and that, therefore, the First Amendment's establishment clause prohibits its being taught as science in public school classes."
United Presbyterian Church USA	1983	"Calls upon Presbyterians . . . to resist all efforts to establish any requirements to teach. . . 'creation science'"
United Methodist Church	1984	"The Iowa Annual Conference opposes efforts to introduce "Scientific" creationism into the science curriculum of the public schools"

> We the undersigned, Christian clergy from many different tradi-
> tions, believe that the timeless truths of the Bible and the discov-
> eries of modern science may comfortably coexist. We believe that
> the theory of evolution is a foundational scientific truth, one that
> has stood up to rigorous scrutiny and upon which much of human
> knowledge and achievement rests....We believe that among God's
> good gifts are human minds capable of critical thought and that
> the failure to fully employ this gift is a rejection of the will of our
> Creator. To argue that God's loving plan of salvation for humanity
> precludes the full employment of the God-given faculty of reason
> is to attempt to limit God, an act of hubris. We urge school board
> members to preserve the integrity of the science curriculum by
> affirming the teaching of the theory of evolution as a core compo-
> nent of human knowledge. We ask that science remain science and
> that religion remain religion, two very different, but complemen-
> tary, forms of truth.

Beginning on February 12, 2006, The Clergy Letter Project has
also sponsored an annual Evolution Sunday involving hundreds
of congregations from the United States, Canada, Australia,
the United Kingdom, and Nigeria, addressing in sermons,
classes, or other venues the compatibility of science and religion. Denominations and faiths represented include Jewish,
Roman Catholic, Baptist, Christian Church/Disciples of Christ,
Congregational, Episcopal, Lutheran, Mennonite, Metropolitan,
Presbyterian, Swedenborgian, United Church of Christ, United
Methodist, Unitarian Universalist, and Unity churches, and a
variety of nondenominational congregations.

What Did Gould Get Wrong, and Why?

Although Gould understood that creationism is not synonymous
with religion, he vastly underestimated the extent and power of
creationism. Throughout his writings, Gould portrayed creationism
as a uniquely American phenomenon that was characteristically
southern, rural, and poor. He also viewed the legislative challenge
posed by creationism as essentially over. Gould was incorrect on
all these counts.

In an article originally published in *Time* magazine and reprinted
in his last collection of essays (2003e, 214), Gould wrote: "No other

Western nation has endured any similar movement, with any political clout, against evolution—a subject taught as fundamental, and without dispute, in all other countries that share our major sociocultural traditions." Gould viewed creationism as uniquely American because of its dependence on a literalist interpretation of the Bible, "a doctrine only well developed within the distinctively American context of Protestant church pluralism" (1999n, 130). Nevertheless, a perusal of recent issues of the *Reports of the National Center for Science Education,* which monitors creationist activities worldwide, indicates that creationism has become an issue in such countries as Australia, the United Kingdom, Germany, Hungary, Norway, and Poland (which share our Western traditions) as well as in Japan and Turkey. The anti-evolution movement is particularly strong in Turkey; for example, Numbers (1998) reported on the distribution of creation-science books to teachers by the ministry of education in Turkey. Perhaps the most obvious vehicle for Turkish anti-evolutionism is the *Atlas of Creation* (Yahya 2006), an extravagantly illustrated diatribe written by Adnan Oktar under the pen name Harun Yahya and distributed gratis to schools and universities in several countries, including the United States. The goal of the *Atlas* is to "show everyone that this theory [evolution] is a deception," which the author considers "the basis for all anti-spiritual philosophies" (Yahya 2006, 6).

Numbers (1998) also discussed the rise of creationist organizations internationally. Perhaps the most successful of these organizations are the Creation Science Foundation, established in Queensland, and the Korea Association of Creation Research, both founded in 1980. Each of these organizations established outreach branches in the United States (Numbers 1998). As long ago as 1980, books by leading creationist Henry Morris had been translated into Chinese, Czech, Dutch, French, German, Japanese, Korean, Portuguese, Russian, and Spanish, making creationism "an international phenomenon" (Numbers 1982, 544).

The international spread of creationism is also demonstrated by the Council of Europe's Parliamentary Assembly approval of a resolution in October 2007 titled, "The dangers of creationism in education." The resolution (full text available at: http://assembly.coe.int/Main.asp?link=/Documents/AdoptedText/ta07/ERES1580.htm#P16_91) states that, "Creationism, born of the denial of the evolution of species through natural selection, was for a long time an almost exclusively American phenomenon. Today creationist ideas are tending to find their way into Europe and their spread is affecting quite a few Council of Europe member states." The resolution urges member states to strengthen the teaching of evolution and to "firmly oppose the teaching of creationism as a scientific discipline on an equal footing with the theory of evolution and in general resist presentation of creationist ideas in any discipline other than religion" (Resolution 1580, art. 19.4).

Thus, Gould misjudged creationism as a "uniquely American" phenomenon. Within the United States, Gould (1999n) also misjudged creationism to be a marginal, minority form of Christianity that was characteristically southern, rural, and poor. This conclusion is perhaps understandable based on the notoriety southern states have gained for their antievolution/pro-creationism activities (e.g., Scopes trial in Dayton, Tennessee; balanced treatment laws enacted in Arkansas and Louisiana). Ronald Numbers (1998) has convincingly argued, however, that the popular stereotype of the American South as uniquely and traditionally hostile to evolution is incorrect.

Numbers investigated the history of Darwinism from the 1860s to late 1920s and found the South was far from unified in attitudes toward evolution. During this period, evolution was taught at many southern colleges, including Tulane University and the state universities of Mississippi, Alabama, Georgia, Virginia, North Carolina, and South Carolina. Evolution was also taught at colleges with religious affiliations, including Presbyterian (e.g., Davidson College, North Carolina), Southern Baptist (Baylor University, Texas), Methodist (Wofford College, South Carolina) and Quaker (Guilford, North Carolina) schools. Although anti-evolution laws were indeed passed in Tennessee, Mississippi, and Arkansas in the fervor of fundamentalism in the 1920s, anti-evolution bills were defeated in Alabama, Georgia, North and South Carolina, Kentucky, Louisiana, Texas, Florida, and Oklahoma, and in Virginia the bill failed even to find a sponsor (Numbers 1998). The fact that the anti-evolution movement progressed as far as it did in these states is evidence of how well accepted evolution had become in the South by the 1920s (Numbers 1998).

Nor are modern creationist activities restricted to the South (Numbers 1998). True, Arkansas and Louisiana both passed laws in the late 1970s mandating balanced treatment of "evolution science" and "creation science" (both laws were declared unconstitutional), and in 1995 Alabama inserted an anti-evolution disclaimer in all biology textbooks used by the state. However, beginning in the mid-1990s creationist activities spread far beyond Gould's putative "Southern, rural, and poor" regions to such states as Pennsylvania, New Hampshire, Ohio, Indiana, Michigan, Wisconsin, New Mexico, California, and Washington (Numbers 1998). More recently, according to the *Reports of the National Center for Science Education*, creationist activities have surfaced in Massachusetts, Minnesota, Iowa, Wyoming, and Montana, among others.

In addition to underestimating the extent of creationism, Gould misjudged the political power of creationism. He prematurely celebrated victory in 1987 when the U.S. Supreme Court in *Edwards v. Aguillard* found the Louisiana balanced treatment law unconstitutional. In an essay titled "Genesis and Geology," reprinted in *Bully for Brontosaurus*, Gould (1991n, 403) declared, "Our legislative victory...ended an important chapter in American social history...the legislative strategy of passing off dogma as creation science and forcing its instruction in classrooms has been defeated." However, creationist legislation continues to be introduced, at the national, state, and local levels.

At the federal level, the Santorum Amendment, which would have singled out evolution as a controversial topic requiring special treatment, was introduced as an amendment to the No Child Left Behind Act in 2001 but was ultimately not included in the bill that was signed into law. Creationist legislation has also continued to be introduced in such states as South Carolina, Georgia, Alabama, Mississippi, Louisiana, Missouri, Oklahoma, Texas, Kansas, Ohio, Michigan, Minnesota, and New Mexico, as detailed in recent issues of the *Reports of the National Center for Science Education* (see Scott 2004b, for the text of several such bills). And local school boards prove fertile ground for the current iteration of creationism, intelligent design. Gould would have applauded the decision by Judge Jones in the *Kitzmiller et al. v. Dover Area School District* intelligent design court case that played out in Dover, Pennsylvania: "the

Board's ID Policy violates the Establishment Clause. In making this determination, we have addressed the seminal question of whether ID is science. We have concluded that it is not, and moreover that ID cannot uncouple itself from its creationist, and thus religious, antecedents" (Jones 2005, 136). However, the long-term influence of this decision is yet to be seen.

Gould remained optimistic that creationist victories would be but temporary. In 1999 he wrote a piece for *Time* magazine decrying the Kansas school board's decision to delete evolution from the curriculum, and predicting that this strategy would be only temporarily successful. The reprinted article, appearing in his collection *I Have Landed* (2003e), included a footnote rejoicing that the subsequent Kansas school board election in 2000 defeated creationist candidates and resulted in the reinstatement of evolution in the curriculum. However, elections in 2004 tipped the balance of school board membership in favor of creationism, resulting in the approval of anti-evolution standards in 2005. School board composition again changed in 2006 following widespread criticism of these standards, and in February 2007 evolution was returned to the curriculum in a scientifically appropriate way. In Kansas, as elsewhere, creationism has proven to be a tougher adversary than Gould expected.

Why was Gould so wrong about the persistence and power of creationism? Is NOMA too simplistic a concept, and was Gould naïve in proposing it? Perhaps the intermediate positions of independence and dialogue between science and religion are untenable, and resolution will only be one of conflict or integration? Must an extreme solution prevail, in which "either science and religion must battle to the death, with one victorious and the other defeated; or else they must...be fully and smoothly integrated into one grand synthesis" (1999n, 51)?

I don't think so. Clearly, science and religion need not conflict, as those of us who are both scientists and persons of faith will testify (e.g., Dodson 1999; Miller 1999; Kelley 2000, in press; Miller 2003). Nor can or should science and religion be integrated; science tests hypotheses about the natural world, and hypotheses about the supernatural (the realm of religion) are not open to scientific testing (Scott 2004a). As a person of faith, I bristle at the idea that my faith should depend upon validation by science, as it seems to me that "scientific creationists" and advocates of intelligent design wish to do. Admittedly, each individual must achieve a balance in his or her own life between the distinct realms (magisteria, in Gould's terminology) of science and religion. As Gould (1999n, 59) stated, "these two domains hold equal worth and necessary status for any complete human life;...they remain logically distinct and fully separate in styles of inquiry, however much and however tightly we must integrate the insights of both magisteria to build the rich and full view of life traditionally designated as wisdom."

If NOMA is not an overly simplistic, overly idealistic view, then why did Gould so severely underestimate the power and extent of creationism? Gould saw creationism as a political issue, declaring that "the rise of creationism is politics, pure and simple" (1981f, 34). Certainly creationism is part of a political agenda, as evidenced by the insertion of creationist planks in the platforms of various state Republican parties beginning in the 1990s (Numbers 1998). However, creationism (though not synonymous with religion) is a religious concept, the power of which Gould, as a nonbeliever, could not understand.

In *Rocks of Ages*, Gould (1999n) described his upbringing as lacking in formal religious training; his parents, although retaining pride in their Jewish heritage, rejected any form of religious belief. Though Gould expressed great respect for religion, he considered himself an agnostic. Nevertheless, his knowledge of religious history was broad, and he knew the Bible well, often sprinkling his essays with biblical allusions. Indeed, he presented his own exegesis of Genesis 1 in several of his essays: "Genesis and Geology," reprinted in *Bully for Brontosaurus*, and "The First Day of the Rest of Our Life" and "The Narthex of San Marco and the Pangenetic Paradigm," reprinted in *I Have Landed*. (Compare Gould's insights with those of Genesis scholars such as Conrad Hyers 1984, 1999.)

Although Gould possessed considerable knowledge of religious and church history, as well as familiarity with the Bible, he lacked personal knowledge of the role that faith plays in the life of the believer. Despite his professed respect for religion, he could be very dismissive of others' religious beliefs. One incident in particular stands out in my mind. As part of a macroevolutionary course in which I was enrolled, Steve and colleague Dick Lewontin had scheduled a day-long symposium for a Sunday. Fellow student Anne Raymond and I approached Steve to see if the schedule could be altered so that we could attend church. Steve did not understand why corporate worship was important to us, and responded to our request by snapping, "I don't know what you people do!" He wasn't going to change the symposium so that we could do whatever it was that we did when we went to church. (We ended up going to an early service before the symposium.) Though Steve mellowed with time (he even thanked me when I told him he was in my prayers, a few weeks before his untimely death), he never understood the importance of faith in the believer's life. At best, he viewed it as a psychological crutch in a world that can be overwhelming with hardship, tragedy, and confusion (1999n).

Scott (1999) reported the results of several Gallup polls as well as surveys of college students that suggest that 90 percent of Americans believe in God but that less than half accept some form of evolution. In part, I believe this disparity is due to the fact that the American public (including school teachers; Scott 1999) has very little understanding of what science is (let alone what evolution is) and how it differs from religion. Individuals often feel forced to choose (unnecessarily) between their faith and accepting evolution. For the believer, a personal relationship with the divine brings meaning to human existence, a meaning to which the agnostic Gould was blind. Nevertheless, as paleontologist and devout Roman Catholic Peter Dodson (1999, 183) stated, "When forced to choose between a religion that enriches human experience and an evolutionary science that ignores human experience and minimizes humans as a species, people will unhesitatingly choose the religion that gives meaning to their daily struggles." Our task, as scientists, is to make it clear, as Steve did, that such a choice is not necessary.

References

Barbour, I. G. 1990. *Religion in an age of science: The Gifford Lectures.* Vol. 1. San Francisco: HarperSanFrancisco.

———. 1997. *Religion and science: Historical and contemporary issues.* San Francisco: HarperSanFrancisco.

Begley, S. 1998. Science finds God. *Newsweek* 132(3): 46–52.

Barrow, J. D., and F. J. Tipler. 1986. *The anthropic cosmological principle.* New York: Oxford University Press.

Bryan, J. R. 1999. Science and religion at the turn of the millennium. In Kelley et al., *Evolution-creation controversy II*, 1–15. Paleontological Society Papers 5.

Dodson, P. 1999. Faith of a paleontologist. In Kelley et al., *Evolution-creation controversy II*, 183–95. Paleontological Society Papers 5.

Easterbrook, G. 1997. Science and God: A warming trend? *Science* 277:890–93.

Forrest, B., and P. R. Gross. 2005. *Creationism's trojan horse: The wedge of intelligent design.* New York: Oxford University Press.

Gibson, M. A. 1999. Teaching biopoesis and evolution: Taking the controversy out of controversial topics. In Kelley et al., *Evolution-creation controversy II*, 226–33. Paleontological Society Papers 5.

Godfrey, S. J., and C. R. Smith. *Paradigms on pilgrimage: Creationism, paleontology, and biblical interpretation.* Toronto: Clements Publishing.

Gould, S. J. 1981. Evolution as fact and theory. *Discover Magazine*, May 1981, 34–37.

———. 1983. *Hen's teeth and horse's toes.* New York: W. W. Norton & Company.

———. 1985. *The flamingo's smile.* New York: W. W. Norton & Company.

———. 1992. *Bully for brontosaurus.* New York: W. W. Norton & Company.

———. 1999a. *Rocks of ages: Science and religion in the fullness of life.* Library of Contemporary Thought. New York: The Ballantine Publishing Group.

———. 1999b. Darwin's more stately mansion. *Science* 284:2087.

———. 2003. *I have landed.* New York: Three Rivers Press.

Hyers, C. 1984. *The meaning of creation: Genesis and modern science.* Atlanta: John Knox Press.

———. 1999. Common mistakes in comparing biblical and scientific maps of origins. In Kelley et al., *Evolution-creation controversy II*, 197–206. Paleontological Society Papers 5.

Jones, J., III. 2005. Kitzmiller et al. v. Dover Area School District. Memorandum Opinion (Dec. 20, 2005). US District Court for the Middle District of Pennsylvania. Case 4:04-cv-02688-JEJ Document 342. www.pamd.uscourts.gov/kitzmiller/kitzmiller_342.pdf

Kelley, P. H. 1999. An honors course on evolution and creationism: teaching experiences in the deep south. In Kelley et al., *Evolution-creationism controversy II*, 217–25. The Paleontological Society Papers 5.

———. 2000. Studying evolution and keeping the faith. *Geotimes* 45(12): 22–23, 41.

———. In press. Teaching Evolution During the Week and Bible Study on Sunday: Perspectives on Science, Religion, and Intelligent Design. In J. S. Schneiderman and W. D. Allmon, eds., *For the rock record: Geologists on intelligent design creationism.* Berkeley: University of California Press.

Kelley, P. H., J. Bryan, and T. A. Hansen, eds. 1999. *The evolution-creationism controversy ii: Perspectives on science, religion, and geological education.* The Paleontological Society Papers 5 (series editor, W. Manger).

Larson, E. J., and L. Witham. 1997. Scientists are still keeping the faith. *Nature* 386:435–36.

———. 1999. Scientists and religion in America. *Scientific American* 281(3): 88–93.

Matsumura, M. 1995. *Voices for evolution.* Berkeley: The National Center for Science Education, Inc.

Miller, K. B., ed. 2003. *Perspectives on an evolving creation.* Grand Rapids, MI: Wm. B. Eerdmans.

Miller, K. R. 1999. *Finding Darwin's God: A scientist's search for common ground between God & evolution.* Cliff Street Books, New York: Harper Collins.

National Academy of Sciences. 1984. Science and creationism: A view from the National Academy of Sciences. Washington, DC: National Academy Press.

Numbers, R. L. 1982. Creationism in 20th century America. *Science* 218:538–44.

———. 1998. *Darwinism comes to America.* Cambridge, MA: Harvard University Press.

Scott, E. C. 1999. Problem concepts in evolution: cause, purpose, design, and chance. In Kelley et al., *Evolution-creation Controversy II,* 169–81. Paleontological Society Papers 5.

———. 2004a. Comments on Richter's "What science can and cannot say." *Reports of the National Center for Science Education* 22(1–2): 23.

———. 2004b. *Evolution vs. creationism: An introduction.* Westport, CT: Greenwood Press.

Yahya, H. 2006. *Atlas of creation.* Istanbul, Turkey: Global Publishing.

Top-Tier

Stephen Jay Gould and Mass Extinctions

David C. Kendrick

Only thirty-six pages out of 1,343, about 2.7 percent of the total, in *The Structure of Evolutionary Theory* (Gould 2002c) are explicitly devoted to mass extinction and the role it plays in shaping the history of life. That's thirty-six pages in a book that Steve Gould worked on for twenty years, a book that detailed how Darwin's theory "has been transformed, along his [Darwin's] original lines, into something far different, far richer, and far more adequate to guide our understanding of nature" (2002c, 24), a book so unabashedly a product of Steve's prodigiously intelligent, thoughtful, and idiosyncratic self that it will likely remain one of those unread tomes that populate people's shelves.[1] In fact, the section on mass extinction in *Structure* is so short that it could even be viewed as his greatest extended footnote. It seems almost paradoxical—so little space for what has become the focus of so much intense interest, debate, and controversy in the fields of paleontology and evolutionary biology. The casual reader might therefore equate the amount of space devoted to mass extinction in *Structure* with mass extinction's importance in Steve's evolutionary worldview (i.e., small, thus unimportant). That, however, would be wrong—the brevity of this section belies the import Steve attributed to mass extinctions, not only as the operator effecting the boldest-scale patterns in life's history but also as the ultimate example of the quirky contingency recorded in that history and the ultimate foil for natural selection as the primary agent of evolutionary pattern at the scale of geologic time.

The Structure of Science B-16

Science B-16 ("The History of the Earth and Life"), Steve's long-time signature course at Harvard, introduced hundreds, maybe thousands, of Harvard undergraduates to geology, Earth history, the nature of science, and Steve's take on the history of life (see also Ross, this volume). The changing syllabus and the content of lectures reflected the development of his thoughts and were often a bellwether of what would later turn up in an essay or book. *Time's Arrow, Time's Cycle* (1987c), for example, Steve's book on the discovery of geologic time and the nature of history, grew out of the lectures on geologic time that he developed for the course, rather than the other way around. In the late 1980s and early 1990s B-16 classes, Steve approached mass extinction from two different but intertwined perspectives: pattern, and process. Pattern focused on the changes in diversity history wrought by the event. Process included things like testing periodicity or exploration of mechanism—like extraterrestrial impact. Though interdependent, it was pattern that mattered most in these lectures, because the patterns dictated by mass extinctions meshed with his beliefs in the nonprogessive and hierarchical nature of evolution and the primacy of unpredictable contingency. What Steve was really interested in was how these mass extinction events affected (or effected, as the case might be) the direction of evolution and how that interfaced with Darwinian theory.

Steve emphasized this interest in the structure of the portion of B-16 devoted to mass extinction. Instead of starting the lecture sequence with either of the two classic slides—a titanic meteor plummeting toward the earth or the Gary Larson Far Side cartoon "What really killed the dinosaurs," which depicts delinquent dinos smoking cigarettes, Steve started with the question of Darwin's views on mass extinction. In *Origin of Species*, Darwin argued that what appeared to be simultaneous extinction of fossil organisms at particular horizons were in fact artifacts resulting from the extreme imperfections of the geologic record. If the record were better, the argument continued, such "events" would be shown to be either completely spurious or, at most, times of slightly elevated rates of extinction. That the abysmal quality of the fossil record masks the true pattern of gradual extinction and origination was Darwin's interpretation, but not the only one consistent with the data. It could instead be true that the patterns of apparent stasis and abrupt simultaneous extinction displayed in the rock record are the correct interpretation. Why then, Steve asked the class, would Darwin argue for the former and so strongly against the existence of mass extinction, particularly when there is nothing about evolution or the mechanics of natural selection incompatible with either interpretation?

The answer fed back into one of Steve's leitmotifs—the fallacy of progress in evolution. Darwin, Steve said, was committed to a worldview in which progress toward increasing complexity and competitive ability, toward fundamentally "better" organisms, was innate. But Darwin also knew that natural selection per se does not guarantee progress; it works by differential survivorship and reproduction with respect to the local environment. As the environment, including biotic and abiotic factors, changes, so will the selective pressures—and not necessarily in any particular direction. To reconcile the inherently nonprogressive nature of natural selection with the belief that organisms had become "better" over time, Darwin felt that the dominant cause of extinction was failure in biotic competition. If that were true, the fossil record should record a sequence of better organisms superseding inferior organisms throughout life's history. In this view, increasing complexity and progress are inherent and natural selection, over the fullness of geological time, will, by steady accumulation, produce the entire pattern of macroevolutionary diversity. Steve explained to the class, however, that mass extinctions "throw a wrench into the comfortable, progressive works here" because they may abrogate natural selection as the cause of extinction.

But, Steve asked, does mass extinction automatically preclude progress? To answer this question he presented three models of what determines which lineages survive and which lineages perish during mass extinction events: (1) "Turning Up the Gain" (TUG), (2) "Random," and (3) "Different Rules." Each of these models, he said, has different implications for the nature of life's history.

In the "TUG" model, we are increasing the intensity of extinction, and there's no denial we are outside the rates of "normal" evolution; mass extinctions, however, simply supercharge the causes in ordinary evolutionary processes, rather than reflect processes qualitatively outside natural selection. Adaptation is boosted because times are really tough, "turning up" the ordinary stresses. As environmental pressures increase, so do selection pressures, thereby increasing the rate at which competition, lowering of reproduction, or increasing mortality drives species to extinction. Species that would eventually go extinct anyway, because they are competitively inferior, just drop out sooner. Species that would survive under ordinary levels of stress survive. Though the rates at which it takes place are greatly accelerated, natural selection, in this model, remains the primary agent of macroevolutionary (at the level of lineages and clades) pattern. Whatever groups survive to re-diversify would have done so whether there had been a mass extinction or not and the unfolding of the history of life remains perhaps as Darwin envisioned.

The "Random" model is the other end of the spectrum of models of survivorship and extinction during mass extinction episodes. Instead of the steady hand of selection weeding out the losers due to inferior characteristics, there is no relationship whatsoever between survivorship and any morphological or physiological characteristic in this scenario. Here the probability of extinction of any particular lineage during a mass extinction is truly random (cf. Raup 1991), regardless of what its morphological adaptations to normal survivorship are, or anything else about that lineage—it's simply and chillingly a throw of the dice. Victims were just in the wrong place at the wrong time. This model of mass extinction abrogates any macroevolutionary patterning by "pure" Darwinian mechanisms. If this model is the dominant mode, then each mass extinction event is a kind of slate-cleaning reset. Unlike the TUG model, surviving clades may be the worst suited for whatever comes next, the best, or something in between.

Finally, in the "Different Rules" model, the adaptations that have given your lineage (to make it personal) your edge in the competitive Darwinian world are suddenly worthless, because of an abrupt change in environmental conditions. For example, you may be the ultimate in fish evolution, but if the water dries up, that won't matter—you're gone. In this scenario, survivorship is not conferred by being a better Darwinian organism, but instead by some quirk, some characteristic or characteristics evolved for reasons unrelated to utility in this extinction event or for no particular reason at all. Different Rules differs from the Random model in that while it doesn't matter whether you're advantaged or not under normal situations, your survival is not random—it hinges on some feature you possess. Steve liked to cite Kitchell et al. (1986), who argue that diatoms survived the postulated K-T impact-induced darkness scenario (Alvarez et al. 1980) because they have dormancy mechanisms evolved in response to the long night of high-latitude winters or to intervals of silica-poor seawater produced by their own exuberant reproduction. The linchpin of this interpretation is that diatoms did not evolve dormancy mechanisms to survive the effects of an impact, but for other reasons. Different rules for survival means a different patterning agency for macroevolutionary diversity history.

Presentation is an important part of argument, and Steve usually presented these three scenarios in this order to the class as a rhetorical device, giving the Different Rules model the last and weightiest position and the TUG model the noteworthy but weaker position at the front. It is perhaps obvious, but this order also matched the relative weight he assigned to each model in evolutionary theory. Steve always took pains to state that there is no denying that turning up the ordinary stresses in evolution does occur and regularly praised Vermeij's (1987) *Evolution and Escalation* as the best and most thoughtful demonstration. Nevertheless, it was the Different Rules model that he emphasized as his preferred dominant mode for mass extinction.

Many paleontologists would now agree that it is a dominant mode, but championing it was important to Steve because it dovetailed with a second guiding leitmotif, the primacy of historical contingency in life's history. If mass extinctions are dominated by Different Rules survivorship, then mass extinctions become the primary patterning agent in redirecting, re-canalizing, and nullifying evolutionary direction; they impose the largest-scale contingencies on the pattern of life's history. Different Rules also allows historical explanation of (or at least allows us to try to explain) why lineages survive or don't, something that the Random model does not.

The Random model always ended up in the middle and weakest rhetorical position. Steve allowed that it could be true and felt that it probably applied in clades with few lineages (because the fewer the lineages in your clade, the more likely that you might be snuffed for no particular reason at all, other than you were near the nadir of your clade's particular diversity at the time of the extinction). Steve liked to point out that the random nature of atomic decay and the ahistorical nature of atoms were precisely what allows us to accurately date historical events, but he was least interested in a random model of mass extinction because, if extinction were truly random, there would be no history to be discovered. A dominantly ahistorical mechanism for mass extinction would have been as anathema for Steve as mass extinction itself was for Darwin. We all have our biases. According to Steve, Darwin's was progress. Steve's was historicity.

The Alvarez Impact

Mass extinctions fit perfectly into Steve's belief in contingency as long as they were "catastrophic", that is, they represented events of such short duration (geologically speaking) and ubiquity or greatness of effect that the normal working of Darwinian adaptation was incapable of causing (or avoiding) a lineage's fate. Steve *was* interested in extinction mechanism, but this interest was always linked to how different mechanisms affected the duration, intensity, and uniqueness of any extinction event.

As Taylor (2004) noted, research from two directions kick-started the relatively dormant research area of extinction into one at the forefront of the field during the 1980s: (1) work by Raup and Sepkoski (e.g., Raup and Sepkoski 1982, 1984, 1986; Sepkoski and Raup 1986) on extinction rates and periodicity, based on Jack Sepkoski's Phanerozoic invertebrate familial and generic range compilations,[2] and (2) the publication of the Alvarez hypothesis (Alvarez et al. 1980) of terminal Cretaceous extraterrestrial impact. For Steve, both these lines of inquiry supported his assertion that mass extinctions were bigger, more frequent, quicker, and "more different" than previously thought (e.g., 1985f), bolstering the position of mass extinction as the ultimate contingent event.

In particular, the impact theory appealed to him for several reasons. First, it was attractive because of its external (extraterrestrial), rather than internal (sea-level, for example), induction. In B-16 lectures he frequently associated, rightly or wrongly, external causes with catastrophic results, while internal causes he implied were more likely to produce effects longer in gestation and that appeared more gradually. (Obviously he knew that either kind of event could produce either kind of outcome; he just leaned toward one.) Second, the proposed "kill mechanism" of global darkness shutting off primary productivity that the Alvarez hypothesis implied perfectly dovetailed with a Different Rules model of causation. Third and finally, unlike so many previous hypotheses about mass extinctions, this one was testable; it made lots of predictions amenable to rigorous analysis. Steve always emphasized to students that science is about posing answerable questions; untestable questions, in contrast, though often interesting, do no fall within the purview of science. For example, Steve felt that much of the conflict surrounding evolution and creation resulted from a basic misunderstanding about the kinds of things science can and cannot answer, sometimes on both sides.

Steve became a powerful advocate of the Alvarez hypothesis because it was external, catastrophic, non-random, and testable. He also derived a certain amount of pleasure from what seemed to him a great irony. The Alvarez hypothesis had initially seemed to provide a general theory of mass extinctions based on repeated bolide impacts (e.g., Alvarez et al. 1984), and Steve himself appeared to embrace that possibility (e.g., 1985f). As the 1980s passed, however, and the evidence for impacts at the other major extinctions was not realized (but see Becker et al. 2001; Ellwood et al. 2003; Olsen et al. 2002), it eventually seemed that instead of producing a general theory, the Alvarez team had instead discovered the ultimate example of contingency—the explanation for a single, wonderful, extinction, the K-T. Perhaps he was only making the best of how the data seemed to point, but Steve seemed to tell this story with a kind of quiet glee. He remained, however, always grateful for and respectful of the insight, work, and contribution that the Alvarezes and their colleagues had made.[3]

In retrospect it may be that both opinions are correct. Workers are now identifying combinations of internal environmental factors and external events that may together result in mass extinction. Bottjer (2004), for example, documents that events of the end-Triassic extinction may be linked to a combination of extended environmental stress coupled with a trigger mechanism.

Tiers of Time

At the end of the mass extinction unit in B-16, Steve reviewed his documentation of Darwin's methodological bias—an inherent extrapolationism tied up with Lyell's uniformities of rate and state (sensu 1965b, 1967d). For Darwin, Steve would say, the world was smooth enough that extrapolation of small processes operating over time allowed and explained the great events in life's history. Steve then would pose the following question: "But, what if time is tiered?" In other words, what if effects at different timescales operate in different ways? Rather than picturing a smooth, rolling (and implicitly genteel) temporal landscape where Darwinian evolution has the same set of effects everywhere, Steve asked us to imagine a rugged, stepped, and possibly forbidding[4] landscape where other processes, though rarely encountered, effect the boldest changes on life's history.

It fascinates me that although contingency was one Steve's primary themes in person, as a speaker, and as a writer, his treatment of the most powerful, far-reaching, and "pure" examples of contingency in the history of life—mass extinctions—feels almost peripheral. He himself never did any primary research on the topic; he says as much in *Structure*. Maybe he just thought it was obvious. I wonder if it was because his real love was working "with" Darwin, that is, discovering the interplay between selection and contingency, not against him.

Epilogue: One Individual in Darwin's World

Steve felt a strong personal connection to Charles Darwin. He used the last line of *The Origin of Species* as the title for his *Natural History* column. He modeled the structure of *Structure* on that book. More than once he talked about touching Darwin "through only two or three intermediaries" (1993l, 223): Colbert to Osborn to Darwin, or Colbert to Osborn to Huxley to Darwin, depending on whether you counted Osborn's story of once shaking hands with the great man himself. More importantly, Darwin was not far from the surface in any technical discussion as Steve turned ideas over in his mind and probed how each related back Darwin's perspectives or predictions. Even Steve's cherished iconoclasm resonated in a small way, for him, I think, with the effects of the publication of *The Origin.* I write these comparisons not to antagonize those who will feel they're an extension of hubris, nor to end with an effusive panegyric, but to say that every instance of connection is telling, because Steve fully saw himself as extending Darwin's work beyond the limits imposed by the scientific knowledge and cultural understanding of his (Darwin's) day. Adding to our understanding of how evolution works was Steve's way of celebrating that he was part of Darwin's world.

Acknowledgments

Nan Crystal Arens' thoughtful insight and spirited discussion greatly improved this essay. Linda Ivany and David Schindel shared their thoughts on Steve and mass extinctions, and Warren Allmon kindly sleuthed where Steve put his Darwin number in writing. The opinions and interpretations herein are based on my personal notes and memories; any errors are my own.

Steve once told me one should remain loyal to one's advisor— not in the sycophantic way, but in the sense of maintaining respect as years pass and ideas ebb and flow. I see what he meant and he was right.

References

Alvarez, L. W., W. Alvarez, F. Asaro, and H. V. Michel. 1980. Extraterrestrial cause for the Cretaceous-Tertiary extinction: Experimental results and theoretical interpretations. *Science* 208: 1095–1108.

Alvarez, W., E. G. Kauffman, F. Surlyk, L. W. Alvarez, F. Asaro, and H. V. Michel. 1984. Impact theory of mass extinctions and the invertebrate fossil record. *Science* 223: 1135.

Becker, L., R. J. Poreda, A. G. Hunt, T. E. Bunch, and M. R. Rampino. 2001. Impact event at the Permian-Triassic boundary: evidence from extraterrestrial noble gases in fullerenes. *Science* 291: 1530–33.

Bottjer, D. J. 2004. The beginning of the Mesozoic: 70 million years of environmental stress and extinctions. In P. D. Taylor, ed., *Extinctions in the history of life,* 99–118. Cambridge: Cambridge University Press.

Ellwood, B. B., S. L. Benoist, A. El Hassani, C. Wheeler, and R. E. Crick. 2003. Impact ejecta layer from the mid-Devonian: possible connection to global mass extinctions. *Science* 300: 1734–37.

Kitchell, J. A., D. L. Clark, and J. A M Gombos. 1986. Biological selectivity of extinction: a link between background and mass extinction. *Palaios* 1: 504–11.

Olsen, P. E., C. Koeberl, H. Huber, M. Et-Touhani, and D. V. Kent. 2002. Continental Triassic-Jurassic boundary in central Pangea: Recent progress and discussion of an Ir anomaly. In C. Koeberl and K. G. MacLeod, eds., *Catastrophic events and mass extinction: impacts and beyond*, 505–22. Boulder, CO: Geological Society of America.

Raup, D. M. 1991. *Extinction: Bad genes or bad luck?* New York: W. W. Norton.

Raup, D. M., and J. J. Sepkoski, Jr. 1982. Mass extinctions in the marine fossil record. *Science* 215: 1501–3.

———. 1984. Periodicity of extinctions in the geologic past. *Proceedings of the National Academy of Sciences, USA* 81(3): 801–5.

———. 1986. Periodic extinction of families and genera. *Science* 231: 833–36.

Sepkoski, J. J., Jr., and D. M. Raup. 1986. Periodicity in marine extinction events. In K. Elliott David, ed., *Dynamics of extinction*, 1–36. New York: John Wiley and Sons.

Taylor, P. D. 2004. Extinction and the fossil record. In P. D. Taylor, ed., *Extinctions in the history of life*, 1–34. Cambridge: Cambridge University Press.

Vermeij, G. J. 1987. *Evolution and escalation: an ecological history of life.* Princeton, NJ: Princeton University Press.

7

Stephen Jay Gould—What Does It Mean To Be a Radical?[1]

Richard C. Lewontin and Richard Levins

The public intellectual and political life of Steve Gould was extraordinary, if not unique. First, he was an evolutionary biologist and historian of science whose intellectual work had a major impact on our views of the process of evolution. Second, he was, by far, the most widely known and influential expositor of science who has ever written for a lay public. Third, he was a consistent political activist in support of socialism and in opposition to all forms of colonialism and oppression. The figure he most closely resembled in these respects was the British biologist of the 1930s, J. B. S. Haldane, a founder of the modern genetical theory of evolution, a wonderful essayist on science for the general public, and an idiosyncratic Marxist and columnist for the *Daily Worker* who finally split with the Communist Party over its demand that scientific claims follow Party doctrine.

What characterizes Steve Gould's work is its consistent radicalism. The word *radical* has come to be synonymous with *extreme* in everyday usage: *Monthly Review* is a *radical* journal to the readers of the *Progressive*, Steve Gould underwent *radical* surgery when tumors were removed from his brain; and a *radical* is someone who is out in left (or right) field. But a brief excursion into the *Oxford English Dictionary* reminds us that the root of the word radical is, in fact, *radix*, the Latin word for *root*. To be radical is to consider things from their very root, to go back to square one, to try to reconstitute one's actions and ideas by building them from first principles. The impulse to be radical is the impulse to ask, "How do I know that?" and, "Why am I following this course rather than another?" Steve Gould had that radical impulse and he followed it where it counted.

First, Steve was a radical in his science. His best-known contribution to evolutionary biology was the theory of punctuated equilibrium that he developed with his colleague Niles Eldredge. The standard theory of the change in the shape of organisms over evolutionary time is that it occurs constantly, slowly, and gradually with more or less equal changes happening in equal time intervals. This seems to be the view that Darwin had, although almost anything can be read from Darwin's nineteenth-century prose. Modern genetics has shown that any heritable change in development that is at all likely to survive will cause only a slight change in the organism, that such mutations occur at a fairly constant rate over long time periods, and that the force of natural selection for such small changes is also of small magnitude. These facts all point to a more or less constant and slow change in species over long periods.

When one looks at the fossil record, however, observed changes are much more irregular. There are more or less abrupt changes in shape between fossils that succeed each other in geological time with not much evidence for the supposed gradual intermediates between them. The usual explanation is that fossils are relatively rare, and we are only seeing occasional snapshots of the actual progression of organisms. This is a perfectly coherent theory, but Eldredge and Gould went back to square one and questioned whether the rate of change under natural selection was really as constant as everybody assumed. By examining a few fossil series in which there was a much more complete temporal record than is usual, they found evidence of long periods of virtually no change punctuated by short periods during which most of the change in shape appeared to occur. They generalized this finding into a theory that evolution occurs in fits and starts, and provided several possible explanations, including that much of evolution occurred after sudden major changes in environment. Steve Gould went even farther in his emphasis on the importance of major irregular events in the history of life. He placed great importance on sudden mass extinction of species after collisions of large comets with Earth and the subsequent repopulation of the living world from a restricted pool of surviving species. The temptation to see some simple connection between Steve's theory of episodic evolution and his adherence to Marx's theory of historical stages should be resisted. The connection is much deeper. It lies in his radicalism.

Another aspect of Gould's radicalism in science was in the form of his general approach to evolutionary explanation. Most biologists concerned with the history of life and its present geographical and ecological distribution assume that natural selection is the cause of all features of living and extinct organisms and that the task of the biologist, insofar as it is to provide explanations, is to come up with a reasonable story of why any particular feature of a species was favored by natural selection. If, when the human species lost most of its body hair in evolving from its apelike ancestor, it still held on to eyebrows, then eyebrows must be good things. A great emphasis of Steve's scientific writing was to reject this simplistic Panglossian adaptationism, and to go back to the variety of fundamental biological processes in the search for the causes of evolutionary change. He argued that evolution was a result of random as well as selective forces and that characteristics may be the physical byproducts of selection for other traits. He also argued strongly for the historical contingency of evolutionary change. Something may be selected for some reason at one time and then for an entirely different reason at another time, so that the end product is the result of the whole history of an evolutionary line and cannot be accounted for by its present adaptive significance. Thus, for instance, humans are the way we are because land vertebrates reduced many fin patterns to four limbs, mammals' hearts happen to lean to the left while birds' hearts lean to the right, the bones of the inner ear were part of the jaw of our reptilian ancestors, and it just happened to get dry in east Africa at a crucial time in our evolutionary history. Therefore, if intelligent life should ever visit us from elsewhere in the universe, we should not expect them to have a human shape, suffer from sexist hierarchy, or have a command deck on their space ship.

Gould also emphasized the importance of developmental relations between different parts of an organism. A famous case was his study of the Irish Elk, a very large extinct deer with enormous antlers, much greater in proportion to the animal's size than is seen in modern deer. The invented adaptationist story was that male deer antlers are under constant natural selection to increase in size because males use them in combat when they compete for access to females. The Irish Elk pushed the evolution of this form of machismo too far and their antlers became so unwieldy that they could not carry on the normal business of life and so became extinct. What Steve showed was that for deer in general, species with larger body size have antlers that are more than proportionately larger, a consequence of a differential growth rate of body size and antler size during development. In fact, Irish Elk had antlers of exactly the size one would predict from their body size and no special story of natural selection is required.

None of Gould's arguments about the complexity of evolution overthrows Darwin. There are no new paradigms, but perfectly respectable "normal science" that adds richness to Darwin's original scheme. They typify his radical rule for explanation: always go back to basic biological processes and see where that takes you.

Steve Gould's greatest fame was not as a biologist but as an explicator of science for a lay public, in lectures, essays, and books. The relation between scientific knowledge and social action is a problematic one. Scientific knowledge is an esoteric knowledge, possessed and understood by a small elite, yet the use and control of that knowledge by private and public powers is of great social consequence to all. How is there to be even a semblance of a democratic state when vital knowledge is in the hands of a self-interested few? The glib answer offered is that there are instruments of the popularization of science, chiefly science journalism and the popular writings of scientists, which create an informed public. But that popularization is itself usually an instrument of obfuscation and the pressing of elite agendas.

Science journalists suffer from a double disability: First, no matter how well educated, intelligent, and well-motivated, they must, in the end, trust what scientists tell them. Even a biologist must trust what a physicist says about quantum mechanics. A large fraction of science reporting begins with a press conference or release produced by a scientific institution. "Scientists at the Blackleg Institute announced today the discovery of the gene for susceptibility to repetitive motion injury." Second, the media for which science reporters work put immense pressure on them to write dramatic accounts. Where is the editor who will allot precious column inches to an article about science whose message is that it is all very complicated, that no predictions can be made, that there are serious experimental difficulties in the way of finding the truth of the matter, and that we may never know the answer? Third, the esoteric nature of scientific knowledge places almost insuperable rhetorical barriers between even the most knowledgeable journalist and the reader. It is not generally realized that a transparent explanation in terms accessible to the lay reader requires the deepest possible knowledge of the matter on the part of the writer.

Scientists, and their biographers, who write books for a lay public are usually concerned to press uncritically the romance of the intellectual life, the wonders of their science, and to propagandize for yet greater support of their work. Where is the heart so hardened that it cannot be captivated by Stephen Hawking and his intellectual enterprise? Even when the intention is simply to inform a lay public about a body of scientific knowledge, the complications of the actual state of understanding are so great that the pressure to tell a simple and appealing story is irresistible.

Steve Gould was an exception. His three hundred essays on scientific questions, published in his monthly column in *Natural History*, many of which were widely distributed in book form, combined a truthful and subtle explication of scientific findings and problems, with a technique of exposition that neither condescended to his readers nor oversimplified the science. He told the complex truth in a way that his lay readers could understand, while enlivening his prose with references to baseball, choral music, and church architecture. Of course, when we consider writing for a popular audience, we have to be clear about what we mean by *popular*. The Uruguayan writer Eduardo Galeano asked what we mean by writing for "the people" when most of our people are illiterate. In the North there is less formal illiteracy, but Gould wrote for a highly educated, even if nonspecialist, audience for whom choral music and church architecture provided more meaningful metaphors than the scientific ideas themselves.

Most of the subjects Steve dealt with were meant to be illustrative precisely of the complexity and diversity of the processes and products of evolution. Despite the immense diversity of matters on which he wrote there was, underneath, a unifying theme: that the complexity of the living world cannot be treated as a manifestation of some grand general principle, but that each case must be understood by examining it from the ground up and as the realization of one out of many material paths of causation.

In his political life Steve was part of the general movement of the left. He was active in the anti-Vietnam War movement, in the work of Science for the People, and of the New York Marxist School. He identified himself as a Marxist but, like Darwinism, it is never quite certain what that identification implies. Despite our close comradeship in many things over many years, we never had a discussion of Marx's theory of history or of political economy. More to the point, however, by insisting on his adherence to a Marxist viewpoint, he took the opportunity offered to him by his immense fame and legitimacy as a public intellectual to make a broad public think again about the validity of a Marxist analysis.

At the level of actual political struggles, his most important activities were in the fight against creationism and in the campaign to destroy the legitimacy of biological determinism including sociobiology and racism. He argued before the Arkansas State Legislature that differences among evolutionists or unsolved evolutionary problems do not undermine the demonstration of evolution as an organizing principle for understanding life. He was one of the authors of the original manifesto challenging the claim of sociobiology that there is an evolutionarily derived and hardwired human nature that guarantees the perpetuation of war, racism, the inequality of the sexes, and entrepreneurial capitalism. He continued throughout his career to attack this ideology and show the shallowness of its supposed roots in genetics and evolution. His most significant contribution to the delegitimation of biological determinism, however, was his widely read exposure of the racism and dishonesty of prominent scientists, *The Mismeasure of Man*. Here again, Gould showed the value of going back to square one.

Not content simply to show the evident class prejudice and racism expressed by American, English, and European biologists, anthropologists, and psychologists prior to the Second World War, he actually examined the primary data on which they based their claims of the larger brains and superior minds of northern Europeans. In every case the samples had been deliberately biased, or the data misrepresented, or even invented, or the conclusions misstated. The consistently fraudulent data on IQ produced by Cyril Burt had already been exposed by Leo Kamin, but this might have been dismissed as unique pathology in an otherwise healthy body of inquiry. The evidence produced by Steve Gould of pervasive data cooking by an array of prominent investigators made it clear that Burt was not aberrant, but typical. It is widely agreed that ideological commitments may have an unconscious effect on the directions and conclusions of scientists. But generalized deliberate fraud in the interests of a social agenda? What more radical attack on the institutions of "objective" science could one imagine?

Being a radical in the sense that informs this memorial is not easy because it involves a constant questioning of the bases of claims and actions, not only of others, but also of our own. No one, not even Steve Gould, could claim to succeed in being consistently radical, but, as Rabbi Tarfon wrote, "It is not incumbent on us to succeed, but neither are we free to refrain from the struggle."

8

Evolutionary Theory
and the Social Uses
of Biology[1]

Philip Kitcher

Introduction

Stephen Jay Gould was well known to many people for many different reasons: as a prominent paleontologist and evolutionary theorist, as a historian of the earth and life sciences, but perhaps most of all as a brilliant essayist who was able to make all sorts of difficult ideas accessible. From his earliest columns in *Natural History,* there was a clear desire to inform, to clear up widespread confusions, and to make the sciences he loved come alive for his readers. Steve's enthusiasms led him to undertake taxing evangelical missions; he was tireless in exposing the fallacies and distortions of "Creation Science," for example.[2] But I want to start with a different set of social concerns, those which stemmed from his deep sympathy with the disadvantaged and his resolute opposition to ventures in biology that were misleadingly portrayed either by scientists or by their public interpreters as buttressing inegalitarian views.

So, quite early in his career, Steve wrote columns in *Natural History*—collected in the first volumes of his essays—on muddled thinking about race, on brain size and intelligence, on human sociobiology, and on the wrongs that eugenics had done to specific people. Some of this work was drawn together in his prize-winning book *The Mismeasure of Man* (1981), which probed the uses and abuses of nineteenth-century craniometry, biological theories of criminality, and twentieth-century intelligence testing. In some instances his investigations drew him away from the areas of science in which he was an active contributor. Human sociobiology, however, was directly linked to evolutionary theory, and, in criticizing it, Steve was led to elaborate some of the themes that were central to his revisionist views about the evolutionary process.

The early criticisms are quite gentle and quite specific. In "Biological Potentiality vs. Biological Determinism" (1976n), Steve admitted admiration for much of the material in E. O. Wilson's *Sociobiology: The New Synthesis* (Wilson 1975). He objected to the sweeping claims of that book's final chapter on the grounds announced in the title of his article. Wilson had advocated a crude genetic determinism, and, in reviewing a wide range of historical episodes, Steve had become convinced both that determinist claims are frequently adopted prematurely and uncritically, and that their social impact is extremely damaging. He ended his essay by noting that, although "Wilson's aims are admirable," his "strategy is dangerous," vitiated by a commitment to determinism.

Between 1975 and 1980, Steve's reactions to the use of biology in support of inegalitarian doctrines became both more wide-ranging and more entangled with his vision of a more expansive evolutionary theory. He had, of course, already created a considerable stir among theoretical evolutionists with the publication in 1972 of the first paper on punctuated equilibrium (or at least the first paper to use that term), jointly authored with Niles Eldredge, to whom Steve, with characteristic openness, has always given primary credit for the central idea (1972e). Punctuated equilibrium was originally explained within the context of the then-orthodox Mayrian view of speciation, and it focused on the geometry of phylogenetic trees. As the 1970s proceeded, however, the emphasis on stasis as the dominant mode of a species' existence was linked to Steve's interest in developmental constraints (the major theme of his academic—or, as he sometimes unfairly called it, "unreadable"—book *Ontogeny and Phylogeny* [1977e]); by the early 1980s, he was also embedding the original account of punctuated equilibrium within an expanded, hierarchical view of selection, one that would, he claimed, generate a more complete evolutionary theory than the orthodox legacy from Darwin (1982g).[3] Delighted creationists loved to misquote him, and popular magazines ran headlines on the demise of Darwinism; Steve persistently, and lucidly, exposed the mistakes.

Yet, as I've suggested, his emphasis on an expansion of Darwinian evolutionary theory shifted the ways in which he viewed the social uses, and particularly the abuses, of biology. He came to regard the unsavory ventures, particularly in sociobiology, as an elaboration of a crude and blinkered evolutionary viewpoint, one trapped in a narrow adaptationism and insensitive to the subtle multiplicities of the modes of selection. The charge was leveled forcefully in one of his most famous papers, "The Spandrels of San Marco," jointly authored with Richard Lewontin (1979k), in which the pop sociobiologist David Barash was lambasted in the context of a more general argument for liberalizing evolutionary theory. By the mid-1980s, Steve was firmly committed to the exposure of "cardboard Darwinism" (1986j), a vulgar practice that he came to attribute to Richard Dawkins in particular, and to writers like Helena Cronin and Daniel Dennett whom he often conceived (unfairly, I think) as Dawkinsian satellites. The hardening of the antithesis continued from then until the end of his career.

I've reviewed a small part of Steve's work because I think that the prominence of the controversy with Dawkins is likely to distort our understanding of the issues and of the value of what Steve accomplished. In effect, evolutionary theorists, anthropologists, psychologists, and philosophers seem to be invited to buy one of two—and only two—packages. Either you are against pop sociobiology and its equally vulgar descendants, in which case you are a revisionist who believes in expanding Darwinism, or you are an orthodox evolutionist who must accept, perhaps regretfully, the in-your-face, like-it-or-lump-it, news from the pop sociobio front. I don't want either of these packages, and I think our options are more extensive. In what follows, I'll try to explain why. The result will be an appreciation of Steve that doesn't correspond at all points to his own self-conception—but I hope it will be an appreciation, nonetheless.

Local critique

How should we think of evolutionary theory and its applications to controversial issues? One possible answer is to suggest that the theory consists of a set of principles, more or less general truths, that are supplemented by detailed claims about particular contexts and used to derive conclusions that apply in those contexts. So conceived, the project of expanding evolutionary theory would take the form of adding new principles and/or replacing existing principles by more inclusive versions. One means of doing so would be to focus on some particular concept deployed in the standard formulations of evolutionary theory and to propose that it's unnecessarily restrictive. For example, one might think that our current notions of "selection" or "adaptation" are too narrow, that they need to be replaced by more general concepts.

There are familiar reasons for worrying about this approach to evolutionary theory. For more than two decades, philosophers have argued that the attempts to axiomatize the theory trivialize it (Lloyd 1983; Kitcher 2003, chap. 9). An alternative, and I think more fruitful, perspective is to view the theory as a collection of models (Lloyd 1988; Thompson 1988) or strategies for answering questions about the history of living things, sometimes locally and in the short term, sometimes more globally and with attention to a large temporal range (Kitcher 1993, chap. 2). To expand evolutionary theory, according to this conception, would be to increase the arsenal of the working evolutionist, by showing how to devise new kinds of models or question-answering strategies. Paradigms of expansion are provided in major works of the modern synthesis, particularly those of Fisher, Wright, Dobzhansky, and Simpson, and, more recently in the development of evolutionary game theory by Maynard Smith, in the theory of kin selection (among other things) by Hamilton, and the revival of group selection by D. S. Wilson and Elliott Sober (Hamilton 1996; Maynard Smith 1982; Sober and Wilson 1998; for elaboration of my perspective on the work of Hamilton and Maynard Smith, see chap. 3 of Kitcher 1985).

Distinguishing these perspectives on evolutionary theory is important for reconstructing the controversies in applying evolutionary considerations to social issues. Frequently, pop sociobiologists and evolutionary psychologists portray themselves as elaborating the consequences of Darwinian evolutionary theory; these consequences may be disturbing or regrettable, but the only way of avoiding them is to become a creationist (Alexander 1987). In effect, these manifestoes for pop sociobiology and evolutionary psychology conceive evolutionary theory as a set of general principles that bold new researchers are resolutely applying to the study of human behavior. The actual situation is quite different. The low-budget ventures of pop sociobiology and their debased recapitulations in the work of David Buss, Randy Thornhill, and Craig Palmer offer loose sketches of evolutionary models in support of conclusions about xenophobia, jealousy, sexual desire, and rape (Buss 1994; Thornhill and Palmer 2000; for critique see Vickers and Kitcher 2003). When those models are closely scrutinized, they are seen to depend on dubious assumptions; alternative models are available, models that generate very different conclusions. So, from the perspective of the second approach to evolutionary theory, there's an easy way to avoid the claims of pop sociobiology without embracing creationism. We should accept the usefulness of the tools that Darwinian evolutionary theory provides, but insist that they are carefully and rigorously deployed. Just as there's no general argument for pop sociobiology as a whole, so too there's no single refutation, no "stake-in-the-heart move" (Oyama 1985), but rather a need to scrutinize each of the uses of evolutionary strategies that the pop sociobiologists make.

To be sure, a detailed review of pop sociobiology or pop evolutionary psychology may display recurrent patterns—a yen for avoiding any precision in presenting evolutionary models or a predilection for using only a minute fragment of the strategies available. So, faced with a new claim about the evolutionary basis of familial homicide (say) one may suspect that one of the typical misuses has been perpetrated. But that has to be shown in the case at hand, by focusing on the details of the evolutionary explanation that has been offered.

It's worth having a name for this way of exposing the abuses of evolutionary theory, and I'll call it the practice of *local critique*. Local critique is time consuming, and the enemy often seems hydra-headed, prompting a yearning for something more global, a "stake-in-the-heart" move (Oyama 1985; Kitcher 2000). Of course, if there were some deep, systematic error in contemporary evolutionary theory, some respect in which it was too limited, those who recognized the need for expanding it might be able to settle issues more efficiently. I trace a route from the most exacting and time-consuming version of local critique to a much more ambitious project. Consider the following four possibilities: (1) The misuses of evolutionary theory occur because of faulty deployment of the available evolutionary strategies and there is no pattern to the mistakes; (2) The misuses of evolutionary theory occur because of faulty deployment of the available evolutionary strategies and there are systematic biases that generate these mistakes (neglect, for example, of certain kinds of available models); (3) The misuses of evolutionary theory occur because the arsenal of available strategies is too limited (that arsenal needs expanding); (4) The misuses of evolutionary theory occur because the principles employed by the misusers are mistaken (they are limited by containing overly narrow concepts).

First, local critique is the best we can do. Second, Steve was often extremely effective in local critique, and, in particular, he offered important defenses of (2). Third, Steve hoped to argue for more ambitious claims, sometimes for (3), sometimes for (4), and his attempts to offer an expansion of evolutionary theory were not successful. I'll try to defend these claims and to do so in a way that brings out what I take to be Steve's important accomplishments.

Two routes from San Marco

Let's start with a famous and controversial paper (1979k), one that has inspired much discussion among people from many disciplines. What exactly is the significance of the spandrels? Gould and Lewontin begin by characterizing an "adaptationist program," an approach that atomizes organisms into traits and then subjects them to optimality analysis. There's a relatively straightforward way to develop the critique of this program, namely by noting that for a trait to be given an explanation by appeal to natural selection it has to be the case that its genetic basis is such that among the suite of characteristics to which that basis gives rise (in the typical environments) variation in the forms of the focal trait is the dominant contribution to variation in reproductive success, and, furthermore, that the genetic variation available in the population has to allow for the attainment of the optimum. When biologists simply assume that the effects of a single characteristic on reproductive success can be considered in isolation from the impact of other traits to which the pertinent genetic basis gives rise, they are frequently making quite unwarranted assumptions; and, as a plethora of well-known examples–the simplest of which is heterozygote superiority—demonstrates, there is no guarantee that fixation of whatever genotype is associated with the optimal phenotype is to be expected (see Templeton 1982, and chap. 7 of Kitcher 1985).

Yet if that is the principal point of the criticism, a diagnosis of a systematic error that might be uncovered in recurrent pieces of local critique, then there's an obvious reply. This is old news. Evolutionary biologists were already aware that the operation of selection is constrained by the ways in which a suite of phenotypic traits is bound together in ontogenesis, and any immersion in standard population genetics ought to teach the moral about optimization. Gould and Lewontin saw the dismissive reply coming, and they attempted to forestall it.

> [S]ome evolutionists will protest that we are caricaturing their view of adaptation. After all, do they not admit genetic drift, allometry, and a variety of reasons for nonadaptive evolution. They do, to be sure, but we make a different point. In natural history, all possible things happen sometimes; you generally do not support your favored phenomenon by declaring rivals impossible in theory. Rather, you acknowledge the rival but circumscribe its domain of action so narrowly that it cannot have any importance in the affairs of nature. Then, you often congratulate yourself for being such an undogmatic and ecumenical chap. We maintain that alternatives to selection for best overall design have generally been relegated to unimportance by this mode of argument. (1979k)

As they went on to note, what is preached as possible on the holy days is often dismissed from consideration in the workaday world.

In terms of the view of evolutionary theory as providing a collection of question-answering strategies, it's easy to see what this version of the critique of adaptationism amounts to: certain strategies that ought to be considered in devising explanations are being forgotten in evolutionary practice and that means that evolutionary theory has effectively been narrowed. It's not old news to remind practitioners of this and to demonstrate by citing prominent examples how pervasive the neglect of nonadaptive approaches is. Much of the ensuing discussion does precisely that, and does it brilliantly—Gould and Lewontin pointed out the peculiarity of seeking a selective account of the reduced front limbs of *Tyrannosaurus,* and the sloppiness of David Barash's tale of anti-cuckoldry behavior in mountain bluebirds. Yet the essay doesn't restrict itself to exposing such follies. It proceeds to offer a rein-terpretation of Darwin's own commitments to adaptationism, to offer a "partial typology" of alternatives and to propose that some developments of evolutionary theory in continental Europe indi-cate possibilities for enriching the standard framework.

These later passages tend to undermine the significance of the line of argument I've been praising, for they suggest that the invo-cation of spandrels is worth nothing until some new expansion of the class of evolutionary strategies has been provided. Many readers, even readers as different as Ernst Mayr (1983) and Daniel Dennett (1995, 267–82), have reacted to the latter sections of the paper by maintaining (i) that insofar as Darwin offered genuine alternatives to natural selection they are already embraced by contemporary evolutionary theory, (ii) that the partial typology of alternatives are recognized as theoretical possibilities within stan-dard evolutionary theory, and (iii) that insofar as the appeal to *Baupläne* resists incorporation into the standard framework, it's completely unintelligible. In essence, the last three sections of the essay shift the claim from stage (2) of my four-stage progression to stage (3) or stage (4). Those readers who conclude that no new model has been provided and that no new evolutionary principle has been articulated then take these failures to be crucial. The genuine achievements at stage (2) are overlooked, and the essay is written off as much ado about nothing.

I suggest that the right route from "Spandrels" is to backtrack and embed the essay firmly in the project of local critique. That was not, however, the route that Steve chose. In the large book he bequeathed to evolutionary theorists and to historians and philoso-phers of biology (2002c), he provided an extended commentary on the significance of spandrels. Some passages can readily be assimi-lated to the more modest project of local critique, especially impor-tant in connection with evolutionary accounts of human behavior.

> A failure to appreciate the central role of spandrels, and the general importance of nonadaptation in the origin of evolutionary novelties, has often operated as the principal impediment in efforts to construct a proper evolutionary theory for the biological basis of universal traits in *Homo sapiens*—or what our vernacular calls "human nature." (2002c, 1264)

Two slight amendments: perhaps not "the principal impediment" but surely an important one; and not an "evolutionary theory" directed at human nature but bits and pieces of evolutionary explanation. With those amendments, I'm tempted to insist that aspiring evolutionary psychologists meditate on the sentence every day before beginning their work.

But Steve wanted to go much farther and regard the appreciation of the importance of spandrels as an important part of an expanded evolutionary theory. He writes:

> The expansion of spandrels under a hierarchical theory of selection establishes the most interesting and intricate union between the two central themes of this book—the defense of hierarchical selection (as an extension and alteration of Darwin's single-level organismal theory) on the first leg of the tripod of essential components in Darwinian logic; and the centrality of structural constraint (with non-adaptively originating spandrels as a primary constituent) for a rebalancing of relevant themes, and as a correction to the overly functionalist mechanics of selection on the second branch of the tripod (or branch of the tree—see figs. 1–4). (2002c, 1267)

Setting aside details (details which are, I think, difficult even for a reader of the whole book to identify clearly), two important things seem to go on in this complicated sentence. First, Steve rises to the challenge of saying what differentiates his appreciation of spandrels from the recognition of constraint within standard evolutionary theory (the "old news" challenge) by declaring that his account of "hierarchical selection" expands the role of spandrels; second, he appears to be conceiving evolutionary theory not as a collection of strategies or models but rather as a set of principles, of which he is providing more general versions.

I believe that Steve should be given credit for an important expansion of the evolutionary theory that descended from the modern synthesis: together with Niles Eldredge he developed a strategy for extrapolating from the fossil record to reconstruct phylogenetic trees, presenting a revolutionary geometry that views speciation events as relatively fast (in geological terms) and stasis as the predominant mode of a species' life (1972e; 1977c). His defense of the applicability of this strategy (given in chap. 9 of [2002c]) is, so far as I can judge, extraordinarily thorough and cogent. It is the product of immense knowledge of details of the records of various lineages, details that are used to explore the credentials of rival phylogenetic hypotheses. If the "Spandrels" essay criticized evolutionary practice for failing to consider the full range of possible explanations and neglecting to use data to decide among alternatives, then Steve's defense of the broad applicability of punctuational models can be seen as an object lesson in how to avoid being vulnerable to that criticism. From the standpoint of the original essay of 1972 (1972e), he should surely have seen this extended vindication as a tremendous achievement.

From the late 1970s on, however, he was advertising more. The reform of phylogenetic geometry was to be only the prelude to an expansion of our understanding of the mechanisms of evolution. Partly inspired by the philosophical thesis that species are individuals (a thesis proposed by Michael Ghiselin [1974] and David Hull [1978]), Steve has been arguing that species are "units of selection," and that we need a "hierarchical" account of selection that accommodates these "higher-level" processes. As *The Structure of Evolutionary Theory* candidly admits (2002c, 29), his attempts to articulate an account of species selection have sometimes been misguided and confused; characteristically, he acknowledges the contributions of students, co-authors, and friends who tried to put him right. The main task of the book is to present the version of the hierarchical theory at which he has finally arrived.

Expansion deflated

The ambitions of this project have often been derided by Steve's detractors, by biologists and philosophers who have claimed that he was saying nothing new or that his proposals were thoroughly confused (Dawkins 1988; Dennett 1995, chap. 10). In my judgment, the critics have not seen the issues completely clearly, and they have overlooked the real importance of some of Steve's ideas about macroevolution, in particular his well-defended claims about the geometry of phylogenetic trees. Yet, uncharitable as they have often been, I think they have recognized two important points. The first is that theses about the metaphysical status of species are unlikely to issue in novel claims about evolutionary processes. The second, and more significant, is that despite the vast amount of ink lavished upon the idea of "higher-order" selection, there has been no obvious addition to the class of evolutionary strategies. Two other great evolutionary theorists, Hamilton and Maynard Smith, were able to demonstrate in brief, precise essays the power of a novel method of conceiving evolutionary phenomena. The writings on species selection don't yield anything remotely comparable.

Perhaps this is unfair. For there have been, for almost two decades now, central exemplars to illustrate how species selection is supposed to work. The most widely cited of these is what Steve identifies as the "classic example" (2002c, 660) of Tertiary gastropods: some species form large populations; others break up into a large number of small populations. The latter characteristic—a characteristic of species, not of individuals—raises the speciation rate, allowing for the opportunity of a broader range of characteristics among descendants and enhanced possibility of surviving major environmental shifts (2002c, 709–10). So far as one can tell, the planktotrophic gastropods (those that form large populations) aren't individually either fitter or less fit that the nonplanktotrophic gastropods (those that form small populations), but there are important differences at the level of species range, duration and long-term consequences. Species of planktotrophic gastropods occupy larger ranges and endure longer; nonplanktotrophic gastropods seem able to pass through evolutionary bottlenecks.

A standard negative response to the citation of this example is to suggest that it is a single instance, and one that can be written off as an evolutionary curiosity. This seems to me incorrect. As Steve himself points out, evolutionary investigations geared to the standard framework aren't likely to uncover a multitude of examples of higher-level selection; and, in any case, the gastropod example plainly has importance for our understanding of large-scale evolutionary phenomena. I think that the real diffi-culty is more fundamental, that unlike the expansions of our evolutionary arsenal I've taken as paradigmatic—the contribu-tions of Hamilton and Maynard Smith—this case can't serve as an exemplar of a new evolutionary strategy, one that was not previ-ously accommodated.

To explain why I believe this, it will help to start with a simpler example, one that Steve uses to motivate his ideas about higher-order selection. In this hypothetical example, designed strictly for illustrative purposes, we're to imagine a "wondrously optimal fish, a marvel of hydrodynamic perfection," living in a pond. Darwinian individual selection has shaped the gills so that these fish thrive in well-aerated water, and the competition has been sufficiently fierce to allow virtually no variability in gill architecture. In the same pond lives another species of fish, the "middling fish," with less wonderful gills, a species that manages a marginal existence on the muddy fringes. Steve writes:

> Organismic selection favors the optimal fish, a proud creature who
> has lorded it over all brethren, especially the middling fish, for ages
> untold. But now the pond dries up, and only a few shallow, muddy
> pools remain. The optimal fish becomes extinct. The middling
> species persists because a few of its members can survive in the
> muddy residua. (2002c, 666)

Steve goes on to claim that we can't explain the persistence of the middling species in terms of individual selection alone.

> The middling species survived *qua* species because the gills varied
> among its parts (organisms) ... the middling species prevailed by
> species selection on variability—for this greater variability imparted
> an emergent fitness to the interaction of the species with the
> changed environment.

Is this right?

The obvious first reaction to this scenario is to note that we don't have a detailed model of just how the fitness values work. That's crucial, because a simple-minded model would divide the pond into two environmental types—Clear and Muddy. Optimal fish have high fitness in Clear and zero fitness in Muddy; middling fish have significantly lower fitness in Clear and slightly more than zero fitness in Muddy. The initial state is one in which there are vast patches of Clear and small patches of Muddy. The final state is one in which the entire environment is Super-Muddy. If Super-Muddy is lethal for all the middling fish, then no evolutionary explanation of their persistence, however many levels of selection we allow, looks promising. If Super-Muddy isn't lethal, then we can expect individual selection to shape the characteristics of the middling species after the cataclysm, and after all the optimal fish have vanished.

Let's take this a step further. Let G be some representative genotype for the optimal fish at those chromosomal regions where there are differences with the middling fish, and let G^* be a corresponding genotype among the middling fish. (It would be more realistic to consider ranges of species-distinctive genotypes, but there's no harm in simplifying.) There's an obvious way to approach the pertinent fitnesses. Start by assuming that the only difference consists in probabilities of survivorship to sexual maturity; fecundity, ability to attract mates, and so forth are constant across genotypes and across environments. Given Steve's account, it seems reasonable to suppose that the rates of survivorship might be as follows:

Table 8.1

	Clear	Muddy	Super-Muddy
G	1	0	0
G^*	0.5	0.1	0.01

If we now formalize fitnesses in each environment in the usual way, we shall have the following relationships.

Table 8.2

	Clear	Muddy	Super-Muddy
G	1	0	0
G^*	0.5	1	1

Assuming that the populations are reasonably large and that the fecundity rates are high, it is not hard to show that the effect of selection across many generations will be to set the frequency of G at virtually 100 percent in Clear, and the frequency of G^* at virtually 100 percent in Muddy.

Now add a further assumption. There's a small probability that, in each generation, the environment will switch from 95 percent Clear and 5 percent Muddy to 100 percent Super-Muddy (with a significant probability of subsequent reversion). It is not hard to show that, under these conditions, the expected outcome, after a long enough run of generations, is that *whatever the environment then found* G^* will be completely prevalent. Moreover, this will hold even if the probability of the environmental switch is extremely small. (In fact, the crucial assumptions for the scenario are: (a) that there be no significant probability of mutations that will generate G from G^*, and (b) that the fecundity rate is sufficiently high that, even with the low rate of survivorship of G^* in Super-Muddy, the fish population will not decline to zero.)

There's nothing particularly novel here—and that's the point. Resistance to the hypothetical example rests on the suspicion that the scenario can adequately be treated by deploying the familiar tools of unexpanded, nonhierarchical, evolutionary theory. Does the same apply to the much-cited example of the gastropods?

I think it does. Again let G, G^* be representative genotypes at the chromosomal regions where the planktotrophic and nonplanktotrophic species are differentiated. In the actual environment, I'll assume that the fitnesses of G and G^* are the same. We now consider a spectrum of environments such that each member of the spectrum has some tiny probability of replacing the current environment. Thus, in considering the long-run representation of each genotype, we must focus on the probabilities that particular environments come to obtain and on the fitnesses that the genotypes would have in such environments. Because the nonplanktotrophs break up into smaller breeding populations, the range of genomes in which G^* is embedded is going to be larger—probably considerably larger—than the range of genomes in which G is embedded. There's going to be a sizable subclass of the spectrum of environments in which the fitness of G^* is nonzero while the fitness of G is zero, and the converse won't hold. Given a sufficiently long time-span, we can predict that there's a high probability that one of the environments present will come to obtain, and thus that G* will become prevalent.

Perhaps, however, we can save species selection by amending the example slightly, supposing that the outcome is not one in which the frequencies of G and G* are different but one in which they are differently distributed among species: suppose, for concreteness, that there are just as many of each genotype at the time at which we take stock, but the Gs are all found in a single species, while the G*s are discovered in ten species.[4] Plainly the kinds of probabilistic argument I have outlined will not explain this outcome. Yet a slight modification will do the trick. Instead of focusing on the one-generation probabilities of survivorship (and the standard fitnesses that derive from them), we can consider the probabilities that, after m generations, a genotype will have bearers in n different species. Specifying these probabilities we can model the evolutionary process in a way that will enable us to derive descriptions of distributions.

This last suggestion brings out a general feature of the approach I am recommending. In effect, I'm considering probabilistic models that focus on individual genotypes but that expand the kinds of probabilities normally considered, effectively generalizing the notion of fitness to consider effects across many generations and other effects besides replication. But the expansion envisaged hardly counts as revolutionary, for it is continuous with the types of reasoning and modeling that are part and parcel of everyday neo-Darwinism.

I anticipate a standard response. The model I've sketched for the example of the gastropods is simply a bookkeeping maneuver, and it hides the crucial causal details of the case. As Steve rightly claims "natural selection is a causal process" (2002c, 665), and he concludes from this that we can't view genic (or genotypic) models as adequate. But what exactly are the causal details behind the process of selection that ends with the victory of nonplanktotrophic gastropods? These gastropods come to be prevalent because they form small breeding populations, so that there's a higher incidence of speciation events and a higher chance that a descendant species will acquire characteristics that allow for life in a radically different environment. That's one way to give the causal description.

Here's another. Because they have G* the individual organisms disperse in particular ways that increase the chances that the G* genotype will be associated in later generations with a broad variety of genetic backgrounds, and hence increase the probability that G* will be embedded in a genome that can interact with a radically different environment to give rise to organisms that can survive, reproduce and transmit G*. To be concrete, and perhaps absurdly simple, suppose that the difference between G* and G is expressed in different forms of five enzymes, and that the G* versions of these enzymes provoke movements at various stages in the gastropod's life that lead to small, discrete populations, that those discrete populations increase the rate of speciation, that the higher rate of speciation increases the chance of survival in a radically different environment, and that, in consequence, the nonplanktotrophs win and the G* genotype spreads. Here we have a causal story. There is no particular reason to think that this causal story has to be analyzed in a particular fashion for the purposes of talking about selection; no particular segmentation of it is privileged. Thus, once we have the model (which I've only outlined) and the whole causal story, it's a matter of convention whether we say that selection is at the "level" of the species, at the "level" of the organism, or at the "level" of the genotype (or even the gene). Consistent with honoring the causal story we can declare that the non-planktotrophic gastropods, or their genotypes, are, in a certain sense, individually fitter than their planktotrophic rivals, for their genotypes have higher probability of being represented in future generations *as a result of the activity of the enzymes which those genotypes produce.*

I'm indulging in what Steve called "Necker cubing," an error he was inclined to see as less dire than that of proclaiming the gene as the one and only unit of selection—he called it "a kindly delusion" (2002c, 656); the "error" descends from Richard Dawkins (1982), and was fully embraced in Sterelny and Kitcher (1988). But I think that Steve, like many other philosophers and biologists, makes the charge that this is an error because he is held captive by a picture. Darwin gave us a metaphor, the image of natural selection. Now the breeder, interested in a particular property of the flower or the pigeon, does select for a particular trait. Nature doesn't. Where there are causes of differential reproduction we rightly focus on them in providing explanations in terms of natural selection, but the ways in which the causal chains are to be segmented is entirely up to us. If we choose, then we can attend to the most proximate causal factors, and this will sometimes incline us to talk of higher levels of selection; but we can always press the causal analysis further back, and, if we do so, we'll identify different units. All that matters is that the models we construct be adequate to the phenomena, generating the right predictions and doing so in a way that fits the causal process that produces the outcomes.

Conclusion

In aiming to provide a hierarchical expansion of evolutionary theory, Steve neither provided his colleagues with a new class of precise mathematical models (as both Hamilton and Maynard Smith did) nor did he draw attention to an important type of evolutionary phenomenon that can only be modeled in a novel way. That, I believe, is why his claims about the expansion of evolutionary theory were often dismissed, sometimes caustically, by those who admire the work of Hamilton and Maynard Smith. In consequence, Steve's enormous range of insights and accomplishments are in danger of being undervalued. The original thesis of punctuated equilibrium, supported through nearly thirty years of patient research and reasoning—research and reasoning summarized in chapter 9 of *The Structure of Evolutionary Theory* (2002c)—is one of the principal contributions to evolutionary theory in the past decades.

Beyond that were Steve's insightful exercises in local critique. As I've already suggested, the "Spandrels" provides a general diagnosis of ills in evolutionary practice, supplementing it with attention to individual examples of abuse. Steve loved to quote the aphorism that God (or sometimes the devil) resides in the details. Whichever version one chooses, he was both gourmand and gourmet for details, and, throughout his career, he offered innumerable new perspectives on particular phenomena, biological and nonbiological. His work demonstrated clearly that we don't need an expanded evolutionary theory to expose the misuses of biology. What's required is to probe with a sensitivity to the exact character of the claims and the evidence provided, to press the details. And in other, less polemical, contexts too, Steve's delight in detail led people to understand issues that they might have thought beyond them, to promote public appreciation of science among a broad population of readers. He showed how it is possible to make the sciences accessible without vulgarizing them, and at a time when we need greater public involvement in scientific research, his accomplishment was extraordinary. It would be deeply wrong, I believe, either to ignore, or to denigrate, that aspect of his work. In fact, I am not sure what we shall do without him.

References

Alexander, R. 1987. *The biology of moral systems.* New York: De Gruyter.

Buss, D. 1994. *The evolution of desire.* New York: Basic Books.

Dawkins, R. 1982. *The extended phenotype.* San Francisco: Freeman.

———. 1988. *The blind watchmaker.* London: Longmans.

Dennett, D. 1995. *Darwin's dangerous idea.* New York: Simon & Schuster.

Ghiselin, M. 1974. A radical solution to the species problem. *Systematic Zoology* 23: 536–44.

Hamilton, W. D. 1996. *Narrow roads of gene land.* New York: Freeman.

Herrnstein, R. J., and C. Murray. 1994. *The bell curve.* New York: Free Press.

Hull, D. 1978. A matter of individuality. *Philosophy of Science* 45: 335–60.

Kitcher, P. 1985. *Vaulting ambition: Sociobiology and the quest for human nature.* Cambridge, MA: MIT Press.

———. 1993. *The advancement of science.* New York: Oxford University Press.

———. 2000. Battling the undead: how (and how not) to resist genetic determinism (most readily accessible as chap. 13 of Kitcher 2003).

———. 2003. *In Mendel's mirror.* New York: Oxford University Press.

Lloyd, E. A. 1983. The nature of Darwin's support for the theory of natural selection. *Philosophy of Science* 50: 112–29.

———. 1988. *The structure and confirmation of evolutionary theory.* Westport, CT: Greenwood.

Maynard Smith, J. 1982. *Evolution and the theory of games.* Cambridge: Cambridge University Press.

Mayr, E. 1983. How to carry out the adaptationist program. *American Naturalist* 121: 324–33.

Michael, J. S. 1988. A new look at Morton's craniological research. *Current Anthropology* 29: 349–54.

Oyama, S. 1985. *The ontogeny of information.* New York: Cambridge University Press.

Sesardic, N. 2000. Philosophy of science that ignores science: Race IQ and heritability. *Philosophy of Science* 67: 580–602.

———. 2003. Review of Naomi Zack philosophy of science and race. *Philosophy of Science* 70: 447–49.

Sober, E., and D. S. Wilson. 1998. *Unto others.* Cambridge, MA: Harvard University Press.

Sterelny, K., and P. Kitcher. 1988. The return of the gene. *Journal of Philosophy* 85: 335–358.

Templeton, A. 1982. Adaptation and the integration of evolutionary forces. In R. Milkman, ed., *Perspectives on evolution,* 15–31. Sunderland, MA: Sinauer Associates.

Thompson, P. 1988. *The structure of biological theories.* Albany, NY: SUNY Press.

Thornhill, R., and C. Palmer. 2000. *A natural history of rape.* Cambridge, MA: MIT Press.

Vickers, A. L., and P. Kitcher. 2003. Pop sociobiology reborn: The evolutionary psychology of sex and violence. In Cheryl Travis, ed., *Evolution, gender, and rape.* Cambridge, MA: MIT Press.

Wilson, E. O. 1975. *Sociobiology: The new synthesis.* Cambridge, MA: Harvard University Press.

9

Stephen Jay Gould's Evolving, Hierarchical Thoughts on Stasis

Bruce S. Lieberman

Introduction

Steve Gould made so many lasting contributions to paleontology and evolutionary biology that it is hard to identify any one as most significant. It is clear, however, that his work developing punctuated equilibrium with Niles Eldredge (e.g., Eldredge, 1971; 1972e; 1977c) will rank as one of the most important. Punctuated equilibrium posits that speciation takes place relatively rapidly compared to the total duration of the species and occurs in small, isolated populations usually along the margin of the species range. Because of the relatively sudden nature of speciation and because it happens in small, isolated populations, its occurrence in the fossil record will be difficult to observe. Punctuated equilibrium also posits that throughout most of their evolutionary history species are stable, displaying what was termed morphological stasis; stasis may be punctuated equilibrium's "most important contribution to evolutionary science" (2002c, 874). The development of punctuated equilibrium represented an important theoretical breakthrough by Eldredge and Gould made possible because each author had collected abundant data from the fossil record, and also each author had a thorough understanding of evolutionary theory. The latter was particularly important.

Eldredge and Gould (1972e) incorporated elements of the work of Mayr (1963), especially his ideas on allopatric speciation. They also incorporated elements of Simpson's (1944) work, especially his ideas on the nature of evolution in the fossil record and his recognition that sometimes evolution in the fossil record was glacial and other times rapid. Simpson (1944), however, focused primarily on the sudden appearance of higher taxa in the fossil record and argued that this was due not to gaps in the fossil record, as Darwin (1859, 1872) and subsequent generations of evolutionary biologists argued, but due rather to times of rapid evolution. Eldredge and Gould (1972e), by contrast, focused primarily on the sudden appearance of new species in the fossil record and the relevance of this to evolutionary theory. While indirectly Darwin (1859, 1872) and even Simpson (1944) (directly) argued that this sudden appearance of new species was merely a preservational bias or gap in the fossil record, Eldredge and Gould (1972e) contended that it was a result of the allopatric model of speciation, and that small, isolated populations tend to evolve much more rapidly than large panmictic ones.

Eldredge and Gould (1972e) also added various insights on the nature of the fossil record and species, and drew on the data they had collected (e.g., Eldredge 1971; Gould 1969i) about closely related subspecies complexes in the fossil record and how these evolved through time. They blended these various elements and insights into what amounted to something new: a theory—in some ways Darwinian at base because it, of course, implied common descent but reliant on profound variation in evolutionary rates and identifying species as an entity of paramount evolutionary importance. The two latter precepts were in direct contrast to Darwin (1859, 1872), who had largely emphasized the uniformity of evolutionary rates and who had in effect categorized species as an arbitrary milestone (see Mayr 1982, 1988) within a continuum of evolutionary change.

In some ways Gould and Eldredge's work and its emphasis on the value of the fossil record for evolutionary theory was antithetical to the work of Darwin, who in some ways denigrated the significance of the fossil record. Yet clearly neither Gould nor Eldredge thought that Darwin was completely wrong. The suggested revisions of Darwin's ideas were not only well founded but done in a respectful and tasteful way and thus part of a general vision that in order for science to progress ideas and theories needed to be revised. Steve's perspective was for expansion, not replacement of Darwinism (1982g; Futuyma 2002). Part of the reason Steve may have received so much criticism for being an anti-Darwinian is because the current scientific generation has witnessed the apotheosis of Darwin, a phenomenon partly explicable as a reaction to creationism. In fact, Steve had tremendous respect for Darwin as a person and a scientist. For example, he often described Darwin as his personal hero in his *Natural History* essays, and indeed the title of these essays, "This View of Life," was derived from the last sentence of Darwin's (1859) *Origin of Species*. Moreover, part of the theoretical impetus for the genesis of punctuated equilibrium, the work of Mayr and Simpson, should have suggested that Eldredge and Gould were squarely in line with the traditions of evolutionary biology, from Darwin on down to the founders of the evolutionary synthesis: Dobzhansky, Mayr, and Simpson. It was probably no coincidence that Eldredge and Gould developed their ideas at the same institution that hosted the founders of the evolutionary synthesis during the critical parts of their careers: the American Museum of Natural History (AMNH).

Stasis and Punctuated Equilibrium

How common is stasis?

Two provocative aspects of punctuated equilibrium, in terms of the new ground it broke, the reaction it engendered, and the criticism it received, had to do with the extent to which stasis within individual species lineages was the rule and what caused stasis. The former issue has been dealt within in a series of papers that were reviewed in Stanley and Yang (1987), Lieberman et al. (1995), Eldredge et al. (2005) and the references therein. Within sexual species, stasis appears to be the dominant mode, but important exceptions have been documented, especially by Geary (1990), notably a student of Steve's. Eldredge and Gould (1972e; Gould and Eldredge 1977c, 1993j; Eldredge 1985a; Gould 2002c) emphasized that stasis as an aspect of punctuated equilibrium could not be refuted or corroborated with a few cherry-picked examples. Instead, the relevant scientific question that needed to be addressed was one of relative frequency; the answer to this question could emerge only from a large number of analyses.

Gould and Eldredge's views on punctuated equilibrium and the prevalence of stasis were never really dogmatic. It was clear that Eldredge and Gould (1972e; Gould and Eldredge 1977c) regarded punctuated equilibrium as the dominant, though not exclusive, evolutionary mode, just as Darwin (1859, 1872) regarded uniform gradual change as the dominant, though not exclusive, mode. For example, Darwin's (1859, 1872) iconography of evolution from the lone figure in his book did show a few lineages that were stable over long periods of time out of many that were gradually evolving. Similarly, in Eldredge and Gould's (1972e) hypothetical figure of anticipated species morphology through time, all species are shown to be stable. Probably part of the reason for this iconography is that Darwin and Eldredge and Gould sought to distinguish their work from that of earlier conceptions, and thus they were bound to emphasize their viewpoint more, as is standard practice in science (Kuhn 1962; Hull 1988).

Eldredge and Gould (1972e) and Gould (1989d, 1996d) also tried to make the case that punctuated equilibrium described a general pattern of change in a variety of systems, not just evolutionary biology. It was the very predilection for making the connection between such seemingly disparate subjects as baseball and biology that was part of Steve's genius, but it also raised the hackles of other scientists who were predisposed to disagree with his ideas on evolutionary theory. Notably the discussions of Steve's penchant for Marxism in Eldredge and Gould's (1972e) paper probably created in some an almost visceral reaction (see further discussion of this point in Allmon, this volume).

I have come to the conclusion that stasis is probably the rule, though exceptions also exist, and even in stable lineages oscillation does occur. This is based on: my reading of the literature (see Lieberman et al. 1995; Eldredge et al. 2005); morphometric measurements of more than one thousand specimens of two species of Middle Devonian brachiopods, including statistical analyses of patterns of change in these measurements during the species' roughly five million year history, which showed overall stasis (see Lieberman et al. 1994, 1995); and similar studies (still unpublished) on the morphology of the Middle Devonian trilobites *Dipleura dekayi* and *Greenops boothi* that were again based on statistical analyses of measurements from hundreds of specimens that also showed stasis (the former during undergraduate research at the AMNH, the latter as a graduate student at the AMNH and Columbia University working along with undergraduate researcher Courtney Reich).

There was and is a strong scientific selection pressure to identify patterns of gradual change in fossil lineages (Gould 1972e, 1977c). This is for no other reason than that examining large numbers of samples, taking thousands of measurements, and applying various statistical analyses seems like a Sisyphean effort if the result is to document little or no change; such an interpretation is mollified, however, if one accepts the statement that emerged from Gould and Eldredge (1977c) like a mantra, "Stasis is data." Based on the effort involved, such studies as that of Stanley and Yang (1987), which documented stasis in a large number of species, are particularly impressive. In spite of all of these data showing stasis, some evolutionary biologists continue to subscribe to a view of species as evanescent where species morphology is continually changing. This again points out how correct Eldredge and Gould (1972e) were when they argued that outlook and training partly influence the patterns that scientists see.

Defining stasis

One aspect that deserves consideration is how Gould (and Eldredge) sought to define stasis. They always emphasized that tests of stasis must rely on statistical treatments of large quantities of data, not generalities (1977c). Furthermore, because stasis did not require lineages to be obdurate (Lieberman and Dudgeon 1996), some evolutionary change through time was permissible and indeed expected, as long as that change was oscillatory and produced no net statistically discernible shift (1977c, 2002c). Notably, this is also the definition of stasis in Lande (1986). This is why Sheldon's (1993) operational definition of stasis, which assumes even oscillatory change is a partial refutation of the stasis model, is not an accurate characterization of stasis under punctuated equilibrium. Bookstein's (1987) interesting ideas on testing stasis using techniques from the statistics of random walks are also noteworthy in this regard. In effect, he argued that recovering a pattern of morphological change compatible with a random walk would refute punctuated equilibrium's predicted pattern of stasis. Given that Eldredge and Gould (1972e; Gould and Eldredge 1977c; Eldredge 1989; Gould 2002c) never argued that stasis must be monolithic, a variety of morphological random walks through time could in fact be compatible with stasis, while some, of course, would not be. Again, although Bookstein's (1987) approach was innovative, his concept of stasis and the null hypothesis he defined for stasis, in effect allowing no change through time, was overly strict and not in line with what the authors of punctuated equilibrium intended.

The ontology of species and its relevance to identifying stasis

Another crucial aspect of punctuated equilibrium and issues of stasis has to do with the ontology of species and their epistemology in the fossil record (Eldredge 1982, 1985b). The Darwinian ontology of species as arbitrary waypoints in an evolutionary stream differed fundamentally from the vision of species as real, stable entities suggested by Eldredge and Gould (1972e) and laid out in detail by Eldredge (1979, 1982, 1985a, b), Gould (1980c, 1982g, f, 1990e, 2002c), and Vrba (1980, 1984). Eldredge and Gould (1972e) needed to reorient their readers to their revised ontology of species. This is why punctuated equilibrium is more a theory about the nature of species than the nature of speciation, and it was this aspect of the theory that was particularly novel (Lieberman, 1995). (Gould [1982g], Eldredge [1979], Vrba [1980, 1984], and Vrba and Eldredge [1984] in fact recognized that this aspect of punctuated equilibrium also had important implications for species selection.)

Noteworthy here is the vision of stasis and change that emerges from scientists whose theoretical outlook and species ontology differed from that of Eldredge and Gould; for example, Phillip Gingerich. Gingerich (1976) published several diagrams presenting data he had collected on molar tooth morphology from Cenozoic mammals occurring in the Bighorn basin of western North America. Gingerich (1976, 1983) concluded that his data showed evidence of gradual change, and he divided many of his species up into gradually diverging anagenetic lineages. Some of these lineages seemed to show oscillatory change and doubled back on earlier lineages such that Gould and Eldredge (1977c) and Eldredge (1989) argued that there were several stable species that changed in an oscillatory manner. Part of the difference in interpretation has to do with different views on the ontology and epistemology of species. Gingerich, because of his species ontology, epistemology, and theoretical orientation, might predict to see gradually anagenetically diverging lineages existing as continua that could be broken up into different species, while Eldredge and Gould, because of their species ontology, epistemology, and theoretical orientation, might predict to see stable species that show some reversible oscillations. Eldredge and Gould's paper (1972e), in fact, was one of the first scientific papers that I know of (though not the first scholarly paper, as they acknowledge), that stated that theoretical outlook and orientation partly influenced a scientist's conclusions, and Eldredge and Gould deserve credit for their intellectual honesty in this regard.

Gould on the Causes of Stasis

Early mechanisms for stasis, developmental constraint

There are several mechanisms that might explain why stasis is such a prominent pattern, and in effect this is the opposite side of the coin of the issues considered by Ross and Allmon's (1990) important paleontological perspective on the causes of evolution. Eldredge et al. (2005) recently comprehensively reviewed the mechanisms for stasis, and therefore this will not be the focus here (see also Williamson 1987); instead, the emphasis will be explicitly on Steve's views on the causes of stasis and how and why his views changed through time. Steve's ideas on the mechanisms of stasis span thirty years of published work from 1972 to 2002. The original mechanism for stasis emphasized by Eldredge and Gould (1972e) focused on lower-level entities in the genealogical hierarchy and invoked some form of constraint relating to genetic homeostasis and the stability inherent in individual development. Van Valen (1982) also endorsed this mechanism, and Steve (1980c) concentrated more specifically on developmental constraint, arguing that change within species might be restricted to relatively few morphological pathways. This emphasis on self regulation and organismal development was clearly an attractive idea to Steve given his interest in development and ontogeny (e.g., 1973i, 1977e). Since 1972, however, numerous examples have emerged in which that developmental constraint is absent. This includes the recognition of the substantial morphological changes that occurred during the domestication of plants and animals by humans (Williamson 1987); here Steve (2002c, 880) argued that constraint "does not play the strong role that I initially advocated." Rather, he argued that such constraints were more important at higher taxonomic levels. This was probably related to a fundamental shift in Gould's theoretical interests regarding organismal development. Early on he was mostly, though not solely, interested in patterns of morphological change within closely related lineages and their ontogenetic expression: for example, his work on allometry (1966b) and then ontogeny and phylogeny (1973i, 1977e). Later, however, his interests in development came to focus more on how it affected the origin of bauplans, first with his interests in the Burgess Shale (1989d) and later homeobox genes (1997m, 2002c; see also Dorit, this volume).

Stabilizing selection and stasis

Stabilizing selection is a mechanism that has frequently been cited as contributing to stasis (e.g., Charlesworth et al. 1982; Van Valen 1982; Williamson 1987; Cheetham et al. 1994) although Eldredge and Gould (1972e) did not mention it. There are in fact various problems inherent with invoking stabilizing selection as a mechanism for stasis. For example, such a mechanism requires an invariant environment over long periods of time, and such conditions are known not to prevail (see Williamson 1987; Eldredge et al. 2005). Also, stabilizing selection was originally framed for the case of single populations and did not consider what would happen to species containing several populations; extrapolating between patterns and processes in the single population case and the patterns and processes in a species containing several populations is not straightforward (Lieberman et al. 1995; Lieberman and Dudgeon 1996, 2002c; Eldredge et al. 2005). In addition to the theoretical problems, the mechanism is also weakened by a lack of empirical support from paleontological studies. For example, in a test of stasis and its causes in Middle Devonian brachiopods, Lieberman et al. (1994, 1995; Lieberman and Dudgeon 1996) found no evidence that stabilizing selection played a role in mediating stasis

Stasis and the nature of subdivided populations

Instead of stabilizing selection, Lieberman et al. (1995) and Lieberman and Dudgeon (1996) described a mechanism that focused on the organization of species into several demes, each of which occur in distinct environments (see also Eldredge et al. 2005). In effect, stasis emerges as each deme within the species adapts or randomly drifts individualistically to a particular habitat in a subsection of the entire species' geographic range; the result within the species will tend to be stasis, because as different parts of the species are diverging independently the sum of the changes will tend to cancel each other out and lead to no net change. The more environments a species is distributed in at any one time, the less change it will show through time (Lieberman et al. 1995; Lieberman and Dudgeon 1996, 2002c). Steve referred to this as a mechanism mediating stasis that was related to the nature of subdivided populations (2002c). The mechanism also predicts that a species broken up into a few or even one deme distributed in a few or even one environment will be more likely to diverge through time. The latter prediction makes sense in light of the allopatric model of speciation because it is the isolated, single population that tends to diverge.

Likely one reason this mechanism of stasis appealed to Steve was that it is a macroevolutionary explanation that focused on levels in the genealogical hierarchy (sens. Eldredge and Salthe 1984 and Eldredge 1986) above the level of the individual organism (2002c, 881). Steve argued that this mechanism explained stasis in a way that is not reducible to strict Darwinian, organismally based natural selection (2002c, 884).

Given the important role that Steve saw for stasis in evolutionary theory (2002c, 874), it was probably crucial that there might be some macroevolutionary explanation for it. (This was given added significance now that developmental constraint did not seem to be a prominent explanation for stasis.) This also meshed with Steve's view that to understand evolution there needed to be some understanding of the uniqueness and common features of all the hierarchical levels in the biological world (2002c, 830). In effect, this matches the contention (Feynman 1965; Lieberman et al. 1993) that advancing our understanding in science requires expanding our knowledge of the connections between various hierarchies. Further, Steve probably thought it was advantageous that stabilizing selection could not be the sole or even an important mechanisms for stasis (2002c, 874) since this mechanism was simply in line with standard population genetic thinking and largely ignored hierarchical inputs.

Steve's ultimate emphasis on hierarchies reflected his deep commitment to macroevolutionary theory. This interest in macroevolution was one of the links between all of his different intellectual and research endeavors.

Other mechanisms for stasis

The mechanism relating to the subdivided nature of populations differs from Sheldon's (1996) *plus ça change* model in several respects. Sheldon's (1996) model suggests that in a varying environment species morphology will tend to remain stable, whereas by contrast environmental stability leads to change in species morphology. The model was developed to explain something that has troubled scientists since Darwin's day: how such species as the wooly mammoth could persist unchanged through the dramatic environmental swings it experienced during the glacial period (Falconer 1868: see discussion by 2002c). Lieberman *et al.* (1995) and Lieberman and Dudgeon (1996) differed with the *plus ça change* model in that they argued it was important to distinguish between the species level and the population level; what matters is the environmental context of the populations within the species. This is the level where the adaptations to the various habitats a species experiences occur, not at the species level. By focusing solely on the environment the whole species experiences, not the various populations of the species, the plus ça change model misses an important aspect of the patterns and processes of stasis.

Worthy of note regarding causes of stasis is Steve's (2002c, 801–2) discussion of Futuyma's work (1987). Futuyma (1987) argued that the spatial locations of habitats tend to shift through time as environments change. Because of this, local populations may merge and interbreed or even go extinct, making much of the geographic differentiation of populations ephemeral. Speciation is important because reproductive isolation and speciation guard evolutionary change and make it permanent. Thus, Futuyma's idea (1987) is more a theory about how the formation of species leads to change rather than why species are stable through time. He described how change is injected into the phylogenetic stream. If one were to try to visualize Futuyma's (1987) theory as being about stasis, however, as Futuyma (2002) viewed it, then his theory makes a set of predictions about stasis and its causes. Specifically, his theory predicts that as environments change through time, as they tend to do, habitats shift. Populations adapted to these habitats will merge and homogenize, and this gene flow will promote stasis (Futuyma 2002). This in fact partly matches the predictions of the plus ça change model, if one were to extend Sheldon's (1996) focus to the population level, as during times of environmental and habitat change separate populations might merge more frequently. Futuyma (1987) (see also discussion in Gould 2002c, 801–2) also suggested, however, that the local extinction of populations that accompanies habitat shifts might make much of the geographic differentiation that occurs within species ephemeral. In one respect this is clearly true. In another respect, however, based on the work of Lieberman et al. (1995), Lieberman and Dudgeon (1996), and the discussion in Gould (2002c), one might predict that a species with fewer populations present in fewer habitats should have greater potential for future evolutionary change. Thus, actually sometimes the phenomenon described by Futuyma (1987) could accelerate change rather than promote stasis.

Although Steve did not single out Futuyma (1987) as presenting a general mechanism for stasis, one thing that likely appealed to him about the mechanism was that it is hierarchical to the extent that it focused on populations existing within species and the effect this has on patterns of stasis and change within species lineages through time.

Conclusions

One of the most groundbreaking theoretical advances in evolutionary biology during the last thirty years was Eldredge and Gould's (1972e) development of punctuated equilibrium. This, with the recognition that species are stable for long periods of time, contributed to a hierarchical expansion in evolutionary biology, as scientists came to focus more on the patterns and processes in entities above the level of the individual organism (Lieberman 1995). For instance, ideas on group selection did not originate with Eldredge and Gould (1972e), but these concepts and the related concept of species selection were given added significance by the fact that species are not ephemeral (Lieberman 1995; Lieberman and Vrba 1995, 2005). Stasis also means that the fossil record is the only place to observe the long term dynamics of species; being evolutionarily significant entities, this has real relevance to evolutionary biology. Thus, as mentioned already, there were many key contributions that Steve made to evolutionary theory, but twin themes that frequently emerged from his research and writings were the significance the fossil record and a hierarchical way of thinking had for evolutionary theory.

Paleobiology as a discipline benefited immensely from Steve's contributions, because they advanced the prestige of the field and spurred new insights. Sometimes biologists were (and are) more reluctant to acknowledge Steve's contributions, yet one of the leaders in the field of evolutionary biology acknowledged that "evolutionary biology has an immensely broader perspective than [before 1972 and punctuated equilibrium and now recognizes] stasis, constraints, multiple levels of selection, differential clade diversification, and historical contingency as valid principles worthy of research; and that Gould played a leading role in bringing about these changes" (Futuyma, 2002, 663). It is depressing that Steve will not be around to witness, comment on, and contribute to the changes in evolutionary biology and paleontology in the coming decades. Still, as scientists crawl along the road toward some emergent truth, we recognize that Stephen Jay Gould, because of the impetus of his research and vision, spurred a quantum leap forward down that road.

Acknowledgments

I am grateful to W. Allmon and R. Ross for inviting me to participate in this volume; to R. Kaesler for comments on earlier versions of this paper; and to N. Eldredge, J. Thompson, P. Brakefield, S. Gavrilets, D. Jablonski, J. Jackson, R. Lenski, M. McPeek, and W. Miller III, all members of the working group on ecological processes and evolutionary rates at the National Center for Ecological Analysis and Synthesis, University of California, Santa Barbara (NSF DEB-94–21535), for discussions of stasis. This research was supported by NSF-EAR-0106885, a Self Faculty Award, NASA Astrobiology NNG04GM41G, the Yale Institute for Biospheric Studies, and the Yale Peabody Museum of Natural History.

References

Bookstein, F. L. 1987. Random walk and the existence of evolutionary rates. *Paleobiology* 13: 446–64.

Charlesworth, B. R., R. Lande, and M. Slatkin. 1982. A neo-Darwinian commentary on macroevolution. *Evolution* 36: 474–98.

Cheetham, A. H., J. B. C. Jackson, and L.-A. C. Hayek. 1994. Quantitative genetics of bryozoan phenotypic evolution. II. Analysis of selection and random change in fossil species using reconstructed genetic parameters. *Evolution* 48: 360–75.

Darwin, C. 1859. *On the origin of species* (reprint 1st ed.). Cambridge, MA: Harvard University Press (1964).

———. 1872. On *the origin of species by means of natural selection; or the preservation of favored races in the struggle for life* (reprint 6th ed.). New York: New American Library.

Eldredge, N. 1971. The allopatric model and phylogeny in Paleozoic invertebrates. *Evolution* 25: 156–67.

———. 1979. Alternative approaches to evolutionary theory. *Bulletin of the Carnegie Museum of Natural History* 13: 7–19.

———. 1982. Phenomenological levels and evolutionary rates. *Systematic Zoology* 31: 338–47.

———. 1985a. *Time frames.* New York: Simon and Schuster.

———. 1985b. *Unfinished synthesis.* New York: Oxford University Press.

———. 1986. Information, economics, and evolution. *Annual Review of Ecology and Systematics* 17: 351–69.

———. 1989. *Macroevolutionary dynamics.* New York: McGraw-Hill.

Eldredge, N., and S. N. Salthe. 1984. Hierarchy and evolution. *Oxford Surveys in Evolutionary Biology* 1: 184–208.

Eldredge, N., J. N. Thompson, P. M. Brakefield, S. Gavrilets, D. Jablonski, J. B. C. Jackson, R. E. Lenski, B. S. Lieberman. M. A. McPeek, and

W. Miller III. 2005. The dynamics of evolutionary stasis. In E. S. Vrba and N. Eldredge, eds., *Macroevolution. Diversity, disparity, contingency. Essays in honor of Stephen Jay Gould*, 133–45. Supplement to *Paleobiology* 31(2).

Falconer, H. 1868. *Palaeontological memoirs and notes*. C. Murchison, ed. 2 vols. London: Robert Hardwicke.

Feynman, R. 1965. *The character of physical law*. London: BBC.

Futuyma, D. J. 1987. On the role of species in anagenesis. *American Naturalist* 130: 465–73.

———. 2002. Stephen Jay Gould à la recherché du temps perdu. *Science* 296: 661–63.

Geary, D. H. 1990. Patterns of evolutionary tempo and mode in the radiation of *Melanopsis* (Gastropoda; Melanopsidae). *Paleobiology* 16: 492–511.

Gingerich, P. D. 1976. Paleontology and phylogeny: patterns of evolution at the species level in early Tertiary mammals. *American Journal of Science* 276: 1–28.

———. 1983. Rates of evolution: effects of time and temporal scaling. *Science* 222: 159–61.

Hull, David L. 1988. *Science as a process*. Chicago: University of Chicago Press.

Kuhn, T. S. 1962. The *structure of scientific revolutions*. Chicago: University of Chicago Press.

Lande, R. 1986. The dynamics of peak shifts and the pattern of morphological evolution. *Paleobiology* 12: 343–54.

Lieberman, B. S. 1995. Phylogenetic trends and speciation: analyzing macroevolutionary processes and levels of selection. In D. H. Erwin and R. L. Anstey, eds., *New approaches to speciation in the fossil record*, 316–37. New York: Columbia University Press.

Lieberman, B. S., and S. Dudgeon. 1996. An evaluation of stabilizing selection as a mechanism for stasis. *Palaeogeography, Palaeoclimatology, Palaeoecology* 127: 229–38.

Lieberman, B. S., and E. S. Vrba. 1995. Hierarchy theory, selection, and sorting. *BioScience* 45: 394–99.

———. 2005. Stephen Jay Gould on species selection: 30 years of insight. In E. S. Vrba and N. Eldredge, eds., *Macroevolution. Diversity, disparity, contingency. Essays in honor of Stephen Jay Gould*, 113–121. Supplement to *Paleobiology* 31(2).

Lieberman, B. S., W. D. Allmon, and N. Eldredge. 1993. Levels of selection and macroevolutionary patterns in the turritellid gastropods. *Paleobiology* 19: 205–15.

Lieberman, B. S., C. E. Brett, and N. Eldredge. 1994. Patterns and processes of stasis in two species lineages of brachiopods from the Middle Devonian of New York State. *American Museum of Natural History Novitates*, 3114: 1–23.

————. 1995. A study of stasis and change in two species lineages from the Middle Devonian of New York state. *Paleobiology* 21: 15–27.

Mayr, E. 1963. *Animal species and evolution.* Cambridge, MA: Belknap/ Harvard University Press.

————. 1982. *The growth of biological thought.* Cambridge, MA: Belknap/ Harvard University Press.

————. 1988. *Toward a new philosophy of biology.* Cambridge, MA: Belknap/ Harvard University Press.

Ross, R., and W. D. Allmon, eds. 1990. *Causes of evolution. A paleontological perspective.* Chicago: University of Chicago Press.

Sheldon, P. 1993. Making sense of microevolutionary patterns. In D. R. Lees and D. Edwards, eds., *Evolutionary patterns and processes,* 19–31. Linnaean Society Symposium vol. 14.

————. 1996. Plus ça change—a model for stasis and evolution in different environments. *Palaeogeography, Palaeoclimatology, Palaeoecology* 127: 209–27.

Simpson, G. G. 1944. *Tempo and mode in evolution.* New York: Columbia University Press.

Stanley, S. M., and X. Yang. 1987. Approximate evolutionary stasis for bivalve morphology over millions of years: A multivariate, multilineage study. *Paleobiology* 13: 113–39.

Van Valen, L. M. 1982. Integration of species: stasis and biogeography. *Evolutionary Theory* 6: 99–112.

Vrba, E. S. 1980. Evolution, species and fossils: how does life evolve? *South African Journal of Science* 76: 61–84.

————. 1984. What is species selection? *Systematic Zoology* 33: 318–28.

Vrba, E. S., and N. Eldredge. 1984. Individuals, hierarchies and processes: towards a more complete evolutionary theory. *Paleobiology* 10: 146–71.

Williamson, P. G. 1987. Selection or constraint?: A proposal on the mechanism for stasis. In K. S. W. Campbell and M. F. Day, eds., *Rates of evolution,* 129–42. Boston: Allen and Unwin.

10

Stephen Jay Gould

The Scientist as Educator

Robert M. Ross

In Steve Gould's reply after receiving the National Association of Geoscience Teacher's Neil Miner Award for teaching in 1984, he started his article: "I have never understood how people can read newspapers and drink coffee, never gazing out of the window, as they fly over the folded Appalachians. We must make our students unable ever again to commit such a sin against nature's beauty and their own intellects" (1984g). This, in fact, was the same example that Steve often used with his teaching assistants to explain what he wanted undergraduate students to get out of his popular Harvard course. Twenty years later, long after they'd forgotten all the factual details of the course, he hoped they would be more likely to look out their plane window with wonder and inquire about the origin of the sights below. This, at least he so claimed, summarized his most basic view of science education for the nonscientist done well.

Stephen Gould received the Neil Miner Award not for his writing, for which he received so many other awards and with which we are all so familiar, but for his undergraduate teaching. Because teaching is by its nature so much more local, few outside Harvard were exposed directly to Steve's teaching and therefore most missed his more coherent approach to teaching an entire body of knowledge spread over the course of a semester.

Many authors have commented on Gould's scientific legacy and his life as a popular writer, but few have examined his life as an educator, particularly in the classroom. For example, it is notable that in Michael Shermer's (2002) excellent review of Gould's work, the only reference to Gould's writings on education is one mention of "teaching" as one of the topics within the bin titled "interdisciplinary" (one of several that Shermer used to categorize all of Gould's writings). Because few have examined Gould as a teacher, many seem not to have recogized his long service in this aspect of his career. Goldberg (1997) writes, interestingly, "Strictly speaking, Stephen Jay Gould is not an educator. He's a paleontologist, Harvard professor, and arguably America's finest and most commercially successful serious science writer...his career in education has been devoted to highly motivated graduate students and undergraduates."

The present chapter is a brief attempt, personal and anecdotal, to recount Gould's approach to education, undergraduate education in particular. It is informed in part from my own experience as a graduate student "teaching fellow" (Harvard's term for a teaching assistant) for three years in his large undergraduate class, from input from a number of people who knew him and saw him teach from the late 1950s on, and of course from his writings.

Like many of us, Steve was a teacher in many different contexts: he mentored several dozen PhD students; he taught several undergraduate courses, from one of Harvard University's most popular "core" classes for nonscience majors to advanced classes, and he ran a weekly "brown bag" seminar on current research. I will concentrate more on Steve's popular core class "History of Life" (Science B-16) than other areas of teaching, first, because I am most familiar with it, but more substantively, because it is the one course he taught consistently throughout his career and because it reflected many of the essential intellectual themes that also pervade his scientific interests in paleontology and in his popular writing.

1. Science B-16

Science B-16, the "History of the Earth and its Life" (later short-ened to the "History of Life") was to his teaching what his stretch of 300 essays in *Natural History* were to his writing: over twenty-five years of consistent devotion. Steve is well known to have made every lecture every year, even through two bouts of cancer, as if to emulate the hitting streak of DiMaggio that he so admired (Gould 1988h). When he arrived at Harvard in the late 1960s he began team teaching in what was more of an introductory geology class. After about ten years he began his own course, "Nat Sci 10," which evolved into Science B-16, a course that became renowned at Harvard and that he taught for more than twenty years.

Nat Sci 10 and Science B-16 took about 100 to 300 students per year (peaking in the early 1980s when his course was most popular) on a trip through the intellectual world of Stephen Gould, perhaps the most integrated look one could get of Steve's mind, his last book (2002c) notwithstanding. A good account of this course by Steve himself is available in his published reply for the Neil Miner Award (Gould 1984g). In the course, using the style so characteristic of his essays, Steve told stories to get at bigger ideas, including planetary geology, the significance of size and shape, contingency, rates of change, and understanding the history of ideas to put into perspective our own current understandings. He wrote (Gould 1984g) that he stuck to five principles, which would be well applied to any large class for nonscience majors. Paraphrased, they were to:

1. convey the subject's excitement by focusing on those general theories that alter a layman's perception of the world such as plate tectonics and evolution;
2. use a history approach to emphasize the social human side of science;
3. emphasize the different styles of science and recognize multifaceted ways of knowing;
4. convey the practice and human side of science by assigning carefully selected research publications for reading (these went into a fat "source book" for the course);
5. increase personal contact with the students by carefully choosing section leaders to run what he called "parallel courses."

Steve insisted that he receive enough teaching fellows each semester to allow them to run discussion-based sections of ten to twenty students. These sections, though containing some standard hands-on labs on rocks and fossils that reinforced or supported his lectures, were mainly intended to be mini-courses run independently by the teaching fellows. Each teaching fellow chose their own section topics according to their interests and background, and in principle undergraduate participants chose their section according to the topics being offered.

Nat Sci 10 included three field trips: to Cape Cod, to Essex Co., Massachusetts, and an overnight camping field trip to the Connecticut River Valley. Former teaching fellow Patricia Kelley remembers, "That was quite a trip, with multiple buses and making all these kids do campfire cooking and sleep in tents! We TAs taught the students actual physical geology and Steve talked about whatever was on his mind (Thomas Burnet or whatever—I would often see these topics appear as *Natural History* columns). He did some geology too—I remember one demonstration where he gleefully tore a strand of asbestos off a hand sample, as if he were defying death (interesting foresight there)" (pers. comm. 2006). Eventually this fieldwork was standardized as one full day geological field trip along the North Shore of Boston with all 300 students, many buses, and lots of donuts.

With such a large number of assistants, Steve could also afford to insist that exams were problem-based with essay-format answers. Steve and the teaching fellows created the tests collaboratively. Gould encouraged unusual or even bizarre questions that would make students think and apply essential concepts. Patricia Kelley (pers. comm.) remembers creating a question based on course content and being asked by Steve, "Where's the anomaly in that?" In January 1986, one of the exam questions asked what we might expect to see on Miranda, a moon of Uranus, as the Voyager 2 was taking flyby pictures that very day (see Schneiderman, this volume). To deal with such a large number of written answer exams, Steve and his TAs together spent an entire eight- to twelve-hour day grading and discussing the completed exams (Steve bought the pizza).

At best Steve Gould was a natural "showman"; he taught with energy and force. He was honest and straight-talking, and therefore provocative, even if not intentionally so. He had a reputation for being well prepared for lectures, and his lecture topics and usually his individual lectures, like his essays, had a beginning and an end, with a take-home message.

The course was consistently very highly rated by students. Comments from *Harvard CUE Guides* (a student-edited review of Harvard course evaluations produced by Harvard's Committee on Undergraduate Education) reflect that Gould was an outstanding lecturer, and many of the comments run parallel to the sorts of comments one generally reads about his writing.

> [Students] note that Stephen J. Gould is one of the most knowledgeable and outstanding lecturers they have encountered at Harvard. Gould's lectures are reportedly lively and interesting and make the subject comprehensible to virtually everyone listening. His knowledge of the material is unquestioned, and while he may seem distant and unapproachable at times, several respondents note that he is actually quite willing to help with questions. [1980]

> Many students say that, because of the charm and brilliance of Professor Stephen Jay Gould and his section leaders, Science B-16 is the best course they have had at Harvard.

> Students describe Professor Gould as witty, brilliant, and articulate, a showman who knows how to raise the interest of non-scientists and to make students think about what they are learning. [1983]

> Nearly all respondents pronounce Professor Stephen Jay Gould a brilliant and dynamic lecturer; his wit and lucidity particularly delight enrollees. [1986]

> Respondents are enthralled with Professor Stephen J. Gould's presentations, describing them as engrossing, lucid, and brilliant. [1989]

His lectures were, however, also frequently cited for pomp and bias:

> He is criticized for being a bit pompous and egotistical; this turns a good number of students off, and some speculate that it even affects the way he deals with controversial issues in the course. While he freely admits his biases to the class and does deal with opposing views, many feel he could do even more to explain the opposition. [1980]

> Some criticize him for being fairly dogmatic because of his comic discussions of opposing viewpoints. But others consider his treatment of alternative views fair and entertaining. [1983]

While a large number of those surveyed accuse Professor Gould of arrogance and pomposity, a few claim that he takes pains to be accessible. A minority of respondents complain that his presentation of the material is highly biased. [1986]

A few of those commenting, however, label him supercilious, arguing that he is both inaccessible outside of class and intolerant of questions during lecture. [1989]

His distractibility in the classroom was also a long-running theme:

... [a] small number report that he is easily distracted from his theme by students' speech and movement in the lecture hall. [1986]

Over two fifths of surveyed enrollees complain that he is an easily distracted or arrogant lecturer. Polled students specifically cite his frequent interruptions when students arrive late or leave early. [1992]

As many have suggested about his writings, his lectures seem to have lost their focus over the years:

...a large minority applaud his engrossing lectures. One sixth, however, bemoan their discursive nature. [1995]

Just over one-fifth marvel at the brilliance that he brings to his presentations. However, an equal number gripe that lectures often digress. [1998]

Those polled describe Professor Stephen J. Gould as an excellent instructor (25%), calling him interesting (34%) and knowledgeable (25%). Others state that he can be boring (22%), disorganized (22%), and digressive (22%). [2001]

Student ratings of Gould's effectiveness as a lecturer declined over the course of the two decades that he taught B-16, even as the average rating for Harvard professors increased slightly (figure 10.1). Having been one of the most popular lecturer's on campus in the early 1980s, by the late 1990s he was rated below average.

The workload and competitiveness of Science B-16 were not particularly noteworthy, being generally average or even slightly below. The grade distribution was fairly low, however, and the material was challenging. Students considered the course content engaging, even if they didn't always "get it." Many students took

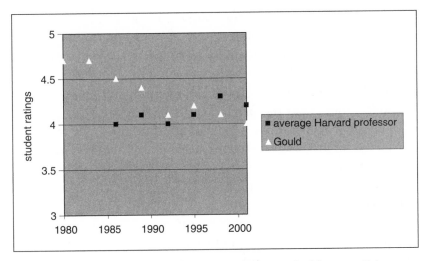

Figure 10.1. Student ratings of Gould as an instructor for his course Science B-16, The History of the Earth and of Life, on a scale of 1 to 5 at three-year intervals from 1980 to 2001.

the class because of Steve's celebrity and perhaps did not make the effort they might have to grasp what he was trying to convey. Many or even most students had some difficulty trying to figure out what was expected of them, being accustomed to memorizing large quantities of material as opposed to listening for a central concept that could be applied to other problems such as on exams.

Of course, like his essays, the very nature of his eclectic interests and dedication to preserving the complexity of the story also sacrificed clarity and comprehensibility in some cases. Teaching fellows in fact were frequently of the opinion that the ultimate success of the course depended on their absorbing Steve's views especially well and then "translating" them to their section students. Steve was well known for not dumbing down his essays, and he similarly did not skip challenging topics in his lectures. Teaching fellows occasionally discovered that he assumed some concepts such as plate tectonics and the relative position of major events in geologic time to be common knowledge, overestimating enormously what students actually knew before coming into the class. This quality of being a bit out of touch with the background of his audiences may be in part what prompted him to approach topics at a high level without cluttering his teaching with detailed content reviews.

The content of Science B-16 changed through the years, as he noted in an interview (Brown 2000): "I've taught this course on the history of Earth and life pretty much since I started; of course the content changes enormously." The course syllabus, however, reveals that although the science changed through time (in fact, Gould was sometimes reporting on science that he and colleagues were in the process of creating), a glance at syllabi from 1986 and 2001 (see table 10.1) show that the major concepts and topics remained more or less the same. (As of August 2008, material for the 2001 B-16 course, including syllabus and sourcebook references, could still be found online at the "SJG Archive" (www.sjg archive.org), a compilation of various Gould documents.)

Opinions vary even among his former teaching fellows regarding the effectiveness of his undergraduate teaching. A safe conclusion is that probably most undergraduate students absorbed at least some concepts, but few came away with all. It was, ironically, perhaps the graduate students serving as teaching fellows for several years who benefited most, because one grasped Steve's worldviews more clearly through repeated exposure and discussion. It would be fascinating to be able to test Steve's long-term teaching success with respect to his own criterion—how many of his students really are more likely to look out the plane window with interest in the origin or the landforms below?

It might have been his obvious dedication to teaching that had the greatest impact on his teaching assistants. Former graduate student Linda Ivany (Ivany 2002) wrote, "One thing that comes up over and over again when I talk with students of Steve was his unfailing dedication to teaching, particularly with respect to the large undergraduate survey course that he taught. Steve was so unlike many other research professors, who often treat teaching like a chore and put into it only the minimum required effort. Steve taught us to value and aspire to excellence in our own teaching. In all the years I was at Harvard, I don't recall him ever missing a lecture, and he was just as diligent about his office hours. His current students...tell me that in the last semester of teaching his big History of Life course, he missed two classes because of brain surgery and chemotherapy. Good excuses, you'd think, for canceling class. But he insisted on making up those lectures at the end of the semester, and he gave the last one only a week and a half before he died."

Table 10.1

Science B-16: The History of Life Course Syllabus, Spring 2001

I. Styles of Science, Modes of History, and Theories of the Earth

February 1
Questioning the New Millenium: Why we can't predict the future but can (in principle at least) explain the past.

February 6
The nature of science and the nature of history.

February 8
Contingency, and Laplace's demon at the battle of Gettysburg. The kinds of questions that science can and cannot answer.

February 13
Deep time as geology's greatest revolutionary concept. Absolute and relative dating for the earth's history.

February 15
Inherit the Wind.

February 20
The scientific revolution and the early history of paleontology.

February 22
The arrows and cycles of time. Thomas Burnet's late 17th century *Sacred Theory of the Earth.*

February 27
James Hutton's "world machine." "Time is, to nature, endless and as nothing."

March 1
Charles Lyell and the principle of uniformity. The power of cultural expectations in theories about the nature of things.

II. Evolutionary Theories and Fallacies

March 6
Charles Darwin's revolution in thought.

continued

Table 10.1 Continued

March 8
The factual basis of evolution.

March 13
Creationism as an American sociocultural phenomenon

March 15
Where adaptation and physical modeling work: size and shape from planetary surfaces to human brains to the architecture of the cathedral.

March 20
Critique of adaptationism and "evolutionary psychology": sandals and spandrels.

March 22
Midterm exam

April 3
Critique of gradualism and the theory of punctuated equilibrium.

April 5
Full House: critique of progress and the perennial Age of Bacteria, or why no one hits .400 in baseball anymore.

April 10
Wonderful Life: critique of determinism and the fractality of contingency from the origin of animals to patterns of human history.

III. Evolution and the Patterns in the History of Life

April 12
The origin and early history of life: problems of drawing conclusions from one experiment and thoughts on the grandest of all unanswerable (for now) questions: intelligent life (or life at all) on other worlds.

April 18
Using the fossil record and the developmental genetics of modern organisms to understand the relationship and early history of animal phyla.

April 20
The Earth's first two multicellular faunas: Ediacara and the Cambrian Explosion.

continued

Table 10.1   Continued

April 24
Mass extinctions: Are they catastrophic? How do they pattern the
history of life: expeditors of progress or the joker in the deck?

April 26
Contingent patterns in the evolution of vertebrates.

May 1
Human origins and modern racial variation: equality as a contingent
fact of history.

May 3
Why contingent human history will not allow us to forecast and indeter-
minate future. Reasons for optimism despite these deepest uncertainties.

2. General views on undergraduate education

Gould was well aware that large research-based universities such as
Harvard are not ideal educational centers for undergraduates, and
he made occasional reference to the excellent education he felt he
had received at Antioch, a small liberal arts college. In a passage
in *Wonderful Life* (1989d, 279), he writes:

> Several years ago, Harvard University, in an uncharacteristic act of
> educational innovation, broke conceptual ground by organizing
> the sciences according to procedural style rather than conven-
> tional discipline within the core curriculum. We did not make the
> usual twofold division into physical versus biological, but recog-
> nized the two styles just discussed—the experimental-predictive
> and the historical. We designated each category by a letter rather
> than a name. Guess which division because Science A, and which
> Science B? My course on the history of earth and life is called
> Science B-16.

In this passage he manages to say something interesting about
approaches to science and how it is taught, while taking a jab at
both the Harvard educational system and at undervaluation of the
historical sciences.

Farther on in *Wonderful Life*, Gould notes the impact upon the emphasis of the mentor-graduate student relationship, and what it means for productivity and legacy, at research-driven universities:

> ...this system is largely responsible for the sorry state of under-graduate teaching at many major research universities. A student belongs to the lineage of his graduate adviser, not the teachers of his undergraduate courses. For researchers ever-conscious of their reputation, there is no edge whatever in teaching undergraduate courses. You can do it only for love or responsibility. Your graduate students are your extensions; your undergraduate students are ciphers in your fame. I wish that this could change, but I don't even know what to suggest.

It is interesting that he recognized and felt badly about the state of undergraduate education, and perhaps even his role in it, given his own reputation as an outstanding teacher.

3. The essayist as teacher

Steve was once quoted in an interview as saying that "Anything, even the conceptually most complex material, can be written for general audiences without any dumbing down. Of course you have to explain things carefully. This goes back to Galileo, who wrote his great books as dialogues in Italian, not as treatises in Latin. And to Darwin, who wrote *The Origin of Species* for general readers. I think a lot of people pick up Darwin's book and assume it must be a popular version of some technical monograph, but there is no technical monograph. That's what he wrote. So what I'm doing is part of a great humanistic tradition" (Krasny 1997).

Steve's writing and lecturing, and even sometimes discussion, were notably similar to each other, and any review of Steve's strengths as a teacher cannot be considered independently from his development as an essayist. According to one long-term friend and fellow writer (P. Long, pers. comm.), Steve's writing bore close resemblance to his thought processes, and thus to the way he lectured, at least in part because he revised little after his initial drafts.

To quote one short biography, "Gould's mission as a writer of accessible essays and books aimed at a broad literate public is not overtly pedagogic. In this sense he was not a spokesman for science or a teacher for the masses" (Lowood 1998). He did not refer to trying to reach anyone [P. Long, pers. comm]. From Long's perspective Steve's essay career with *Natural History* magazine was a defining event, a historical contingency in his own life, an enormous opportunity without which his career might have been differently shaped. In particular, the monthly column provided a certain intellectual freedom to follow eclectic interests within the disciplined context of deadline and, at least early in the series, length constraints.

Communicating via both essays and lecturing provided inspiration for each. Some lectures and classes drew from or were based entirely around the content of his essays. On the other hand, some of his teaching early in his career later showed up in his *Natural History* columns. It is interesting to contemplate to what extent, if any, his essays were informed from the direct response of audiences to his teaching.

4. Steve and trends in science education

Although Steve was not directly involved in education outside the university, except through lectures and essays, it is evident that he cared deeply about it, at least in part through his involvement in debates about evolution in the public schools (see also Allmon, this volume).

In an interview, he said, "I think [children's interest in science is] there already. That's why I take heart. It's true that the level of scientific knowledge among adults is very low, but that's not because there isn't a natural interest. I think most kids are fascinated by the natural world. I've often said that if you could quantify the mental power involved in all the dinosaur names correctly known and correctly spelled by five-year-old kids in America, you could move any mountain on earth" (Krasny 1997).

Over the course of Steve's career, several trends and events in national K-16 science education took place. In part this involved new curricula, including Earth science curricula in the 1960s that featured problem-based learning (the Earth Science Curriculum Project; ESCP 1973). There was increased interest in constructivism and the importance of inquiry-based learning, which was codified in landmark works such as *Science For All Americans* (AAAS 1989) and *Benchmarks for Science Literacy* (AAAS 1993) from the American Association for the Advancement of Science, and the *National Science Education Standards* (NRC 1996) from the National Research Council of the National Academy of Sciences. These documents established Earth science as one of the core sciences with physics, chemistry, and biology. The American Geophysical Union published *Shaping the Future* on post-secondary Earth science education in 1996 (Ireton et al. 1996) and NSF began the Geoscience Education funding program in 1998 (NSF 1998).

He was by and large not plugged into the science education community, but Steve's teaching philosophy and emphasis were consistent with some of the recommendations coming out of these reforms, such as emphasizing science as a process and as an accessible subject to all. And he was not uninvolved as a science education advocate. Perhaps most conspicuously, he worked on keeping evolution in the public schools (e.g., 1999g). Patricia Kelley wrote, "his efforts were tireless to support the teaching of evolution in public schools and to clarify the relationship between science and religion" (Kelley 2002). One story relates that in 1981, when he appeared in Arkansas as one of six expert witnesses in a lawsuit challenging the constitutionality of that state's new law requiring teachers to cover both evolution and creationism in their biology classes, several Arkansas teachers also testified. Steve remembered that one high school teacher, asked what he would do if the law was upheld, "looked up and said, in his calm and dignified voice: 'It would be my tendency not to comply. I am not a revolutionary or a martyr, but I have a responsibility to my students, and I cannot forego them.'" Recalling the incident, Steve added a benediction: "God bless the teachers of this world" (Linder 2004).

But he did not apparently look outside of his own long-standing approaches to education, or concern himself with developments in educational research. Most of the people spoken to for this paper agree that Steve "did what he did," and changed little over the course of his university career. Any similarities to ongoing trends in education were his own parallel efforts to make science more engaging. On the other hand, some his graduate students who were heavily influenced by Steve in their own approaches to education (including the editors of this volume, e.g., Allmon and Ross 2004; Kelley and Burks 2004) now are involved in precollege and public education and are, more generally, integrated into the community of individuals seeking to improve science education.

Recently, an NSF-funded conference on reform in K-16 Earth and space education promoted the idea of an education "Revolution" (Barstow and Geary 2002) that emphasizes Earth as a system, inquiry-based teaching, and strategic use of Internet and visualization technology. Some of us already employ these strategies in our own classes, including figuring out how to better engage students in large lecture format classes. Steve's teaching was broadly consistent with at least two of these three basic aspects of the mission of the revolution, emphasizing the nature and process of science over facts, and emphasizing systems and models, but in practice the means to this revolution were outside of his apparent area of interest or activity. In particular, little in his teaching was connected to recent opportunities for data analysis and visualization. Though he was no stranger to inquiry or discussion, and he created opportunities for discussion in his sections, most of his own teaching was lecturing.

Mentoring of graduate students

Steve's style of mentoring may be less distinctive than who he was as a lecturer and essayist, but it was consistent with some of the qualities that were distinctive of him as an educator: loyalty and dedication, with high student expectations. Steve gave an unusually detailed account of graduate student mentoring in *Wonderful Life* (1989d, 139–40):

> In some fields, particularly those with large and expensive laboratories dedicated to the solution of definite problems, you must abandon all thought of independence, and work upon an assigned topic for a dissertation (choice in research is a luxury of later postdoctoral appointments). In more genial and individualistic fields like paleontology, you are usually given fair latitude in choosing a topic, and may emerge with a project uniquely your own. But in any case you are an apprentice, and you are under your mentor's thumb—more securely than at any time since the early years of

primarily school.... If you work well together, and your mentor's ties to the profession are secure, you will get your degree and, by virtue of his influence and your proven accomplishments, your first decent job.

It's a strange system with much to criticize, but it works in its own odd way. At some point, you just can't proceed any further with courses and books; you have to hang around someone who is doing research well. (And you need to be on hand, and ready to assimilate, all the time, every day; you can't just show up on Thursday afternoon at two for a lesson in separating parts from counterparts.) ... when it works.... I cannot imagine a better training.

It is interesting to reflect on the degree of idealism in this passage relative to the reality of Steve's mentoring style. No one went to work with Steve to learn the protocols of field work or the details of their taxonomic group of choice. Certainly none of us were instructed on separating parts from counterparts. But at least up until his later years, however, he was unfailingly available to meet if requested, to discuss the general issues surrounding one's work. The passage continues:

Many students don't understand the system. They apply to a school because it has a general reputation or resides in a city they like. Wrong, dead wrong. You apply to work with a particular person. As in the old apprenticeship system of the guilds, mentor and student are bound by mutual obligations; this is no one-way street. Mentors must, above all, find and provide financial support for students. (Intellectual guidance is, of course, more fundamental, but this part of the game is a pleasure. The real crunch is the search for funding. Many leading professors spend at least half their time raising grant support for students.) What do mentors get in return? This reciprocation is more subtle, and often not understood outside our guild. The answer, strange as this may sound, is fealty in the genealogical sense.

The work of graduate students is part of a mentor's reputation forever, because we trace intellectual lineages in this manner. I was Norman Newell's student, and everything that I ever do, as long as I live, will be read as his legacy.... I happily accept this tradition and swear allegiance to it—and not for motives of abstract approbation but because, again as with the old apprenticeship system, I get my turn to profit in the next generation.... The greatest benefit is an exciting lab atmosphere for the moment—but I am not insensible to the custom that their future successes shall be read, in however small a part, as mine also.

Many, perhaps most, of Steve's former graduate students can identify with the general sense of this passage, in that, though Steve was tugged in many directions, he was able to be loyal to those students with whom he felt a commitment in part by turning away those he felt were asking of his time without his prior consent. (He especially bristled at those who took his time because of his celebrity status. After one encounter he was heard to exclaim, "I'm not a [expletive deleted] tourist attraction!") He placed much trust in his students, surely in part because he had to, given his time constraints, empowering his students to choose and run their own research programs, to run their own course within his course, and to make decisions about the facility and seminars. He was long committed to a weekly brown-bag seminar, even during his illness, in which many of us actively debated with him on the issues of the time, but in which he would also often pontificate on his views, sometimes all-too-familiar, sometimes insightful tweaks on familiar themes, and occasionally surprising us.

Former student Linda Ivany expressed some of these thoughts about Steve's treatment of his graduate students (Ivany 2002): "One of the things I've always admired about Steve is that his support for his students was unfaltering. He never made any of us feel intellectually inferior; he always treated us as equals.... In fact, as one of his students recently mentioned to me, that was sometimes one of the hardest things to deal with—trying to live up to his high opinion of our intelligence and ability. But then his support and confidence in us was a great motivator as well—we knew he expected a lot of us, and we didn't want to let him down, so we worked hard."

Take away messages

I asked former student Dana Geary what she'd learned about teaching from Steve. "Never, ever, ever eat an apple while lecturing," she quipped. More substantively, following is a brief summary of a few qualities of Steve as an educator, and as a teacher in particular, to which we can aspire. Many of these have already been adopted consciously or not by twenty-plus years of graduate student teaching assistants (now out teaching their own students) whose teaching was influenced by his approach and thinking, and many others such as colleagues and undergraduate students who were influenced at least indirectly.

- Steve had high expectations of his students and his reading audiences. Few resented that and most of us rose to the challenge.
- Steve had an intense loyalty to those for whom he felt he had accepted a responsibility, although this meant being guarded of his time with those he did not.
- Steve provided a model case for making the effort to understand the origin of our own scientific beliefs and to pass on that context.
- Steve had a remarkable dedication and perseverance to carry out his teaching responsibilities that he carried with him right to the end of his life. In a way, this was an extension of his loyalty.
- Steve brought a part of himself and his experiences to teaching. While some considered this pretentious, it also gave a dynamic and personal feel to his teaching.
- Steve helped us to think differently, to think more broadly and to look ourselves for cross-cutting general principles. Of course, that is a result in part of the way he did science, but it also informs our teaching.
- Steve operated in several spheres, from the general public to undergraduate non-science majors to advanced students. He educated through lecturing, writing, discussing, and entrusting his section leaders. And he demonstrated the potential for increasing the breadth of our scholarship and teaching as an integrated part of sharing what we love.

Acknowledgments

This manuscript benefited from numerous helpful conversations with colleagues, students, and friends of Steve Gould. Among those that helped in conversations and correspondence are Phoebe Cohen, Mark Goldberg, P. Long, Andrew Knoll, Roger

Thomas, Patricia Kelley, Emily CoBabe, Alexandra Moore, Dana Geary, Linda Ivany, and Warren Allmon. Kelley, Moore, and Allmon provided helpful comments on the manuscript.

References

Allmon, W. D., and R. M. Ross. 2004. Earth system science and the new Museum of the Earth. *American Paleontologist* 12(1): 9–12.

American Association for the Advancement of Science. 1989. *Science for all Americans.* Washington, DC: American Association for the Advancement of Science.

———. (1993). *Benchmarks for science literacy.* New York: Oxford University Press.

Barstow, D., and E. Geary, eds. 2002. *Blueprint for change: Report from the national conference on the revolution in earth and space science education.* Cambridge, MA: TERC.

Brown, D. 2001. Stephen Jay Gould, from brachiopods to baseball. *Powells.com Interviews* (www.powells.com/authors/gould.html, accessed February 25, 2008).

ESCP (Earth Science Curriculum Project). 1973. *Investigating the Earth.* rev. ed. Boston: Houghton Mifflin.

Goldberg, M. F. 1997. An interview with Stephen J. Gould: Joltin' Joe and the pursuit of excellence. *Phi Delta Kappan,* 395.

Ireton, F., C. Manduca, and D. Mogk, eds. 1996. *Shaping the future of undergraduate Earth science education: Innovation and change using an Earth system approach.* Washington, DC: AGU (www.agu.org/sci_soc/spheres/, accessed February 25, 2008).

Ivany, L. 2002. Memories of Steve Gould. Transcribed talk at Paleontological Society luncheon. Geological Society of America, Oct. 28, 2002 (www.paleosoc.org/Lindaivany.html, accessed February 25, 2008).

Kelley, P. H. 2002. Remembering Stephen Jay Gould. *Geotimes,* May 2002 (www.geotimes.org/may02/WebExtra0522.html, accessed February 25, 2008).

Kelley, P. H., and R. J. Burks. 2004. The importance of teaching earth science in the public schools. *GSA Today* 14(8): 29.

Krasny, M. 1997. Stephen Jay Gould (interview). *Mother Jones,* January/February 2007, 60–63 (www.motherjones.com/commentary/columns/1997/01/outspoken.html, accessed February 25, 2008).

Linder, D. 2004. Stephen Jay Gould. In *Exploring constitutional conflicts: The evolution controversy* (www.law.umkc.edu/faculty/projects/ftrials/conlaw/gouldsj.html, accessed February 25, 2008).

Lowood, H. 1998. Stephen Jay Gould. Stanford Presidential Lectures in the Humanities and the Arts (http://prelectur.stanford.edu/lecturers/gould/, accessed February 25, 2008).

National Research Council, (1996). *National science education standards.* Washington, DC: National Academy of Science.

NSF 1998. Awards to Facilitate Geoscience Education: Announcement of Opportunity. NSF 97–174. (www.nsf.gov/pubs/1997/nsf97174/nsf97174.pdf, accessed February 25, 2008).

Shermer, M. B. 2002. This view of science: Stephen Jay Gould as historian of science and scientific historian, popular scientist and scientific popularizer. *Social Studies of Science* 32(4): 489–525.

~ 11 ~

Stephen Jay Gould

Remembering a Geologist

Jill S. Schneiderman

I was a teaching assistant (Harvard calls them "teaching fellows") from 1983 to 1987 in Steve Gould's renowned introductory course at Harvard, "The History of Earth and Its Life" (see also the chapter in this volume by my sometimes fellow Rob Ross, one of my colleagues in this endeavor). Some of this time coincided with Steve's difficult first fight with cancer. Despite his illness, Steve never missed a class or his weekly meeting with his teaching fellows—a fact of which he was justly proud.

I felt during those years that Steve was very engaged with our group of teaching fellows. Perhaps what I felt as engagement was his response to the vulnerability that one might feel during a period of grave illness. Or maybe, as a student of metamorphic petrology at the time, I wasn't as awed by him as some of the pale-ontology students might have been, and consequently Steve and I had an easy relationship. It was only 1983, he was only forty-two, I hadn't taken paleontology as an undergraduate and had not yet come to appreciate the significance of his work. So to me, at the time, Steve was just a geology professor with whom I inter-acted with ease. Nonetheless, I experienced Steve as a professor who seemed to be quite invested in the thinking of those students whom he had entrusted as leaders of what he called, the "parallel courses," more commonly thought of as discussion sections, to his large lecture course.

I'll forever be indebted to University of Wisconsin paleontologist Dana Geary, Steve's head teaching fellow at the time, for
suggesting me to Steve as a teaching fellow for the course. At my
first meeting with him, how little I appreciated that I was encountering a person who would affect my intellectual development and
my career in fundamental, meaningful, and sustaining ways from
that day forward.

I choose to remember Steve as a geologist. In fact, when his
obituary in the *New York Times* referred to Steve at "one of the
most influential evolutionary biologists of the 20th century" and
never mentioned that he was a geologist, I felt compelled to write
a cranky letter to the editor (unpublished) to complain! Other
obituaries and tributes that followed his death similarly focused
on Steve's work on the history of life and evolution as a process.
Although some writers referred to him as a paleontologist, none
called him a geologist. His professorships in zoology and biology
at Harvard and New York University notwithstanding, Steve *was*
indeed a geologist, and he would have never eschewed the label.
In fact, I know that Steve would want geologists to raise their voices
and claim him as one of us, after the fashion in which he claimed
Darwin for geology based on his work on coral atolls.

Unsurprisingly, I perceived Steve strongly as a geologist. He
wanted me as a teaching fellow, not because I knew about fossils but
because I knew about minerals and rocks. Steve required all undergraduates in his course to learn about minerals, rocks, and geological maps, and take part in a day-long field trip to examine volcanic
and sedimentary rocks along the New England coast, so we "hard
rockers" were important for the course at least in that regard. Steve
allied himself with what he lovingly called "the dummy science" (he
lamented the dearth of scientific awards and recognition bestowed
upon geology's practitioners), considered the Geological Society
of America (GSA) his organization (one might be tempted to call
him "a card-carrying member of the GSA"), and asserted geology's
critical importance for being located, in his words, "at the fulcrum
of historical and ahistorical science."[1]

Apropos of this distinction, Steve seemed each year during his
course to relish the lectures in which he articulated the systematic structures of silicate minerals. As anyone who has read his
essays and books knows, Steve perhaps loved nothing more than to

reflect on the process and nature of science, and on the activities of scientists. He asserted that there was no such thing as *the* scientific method and went to great lengths to demonstrate for his students examples of different styles of doing science some appropriate to historical science and others appropriate to ahistorical science. He quoted Wordsworth's poem "The Tables Turned" to convey the derision with which reductionism—the stereotypical methodology of ahistorical science—has sometimes been received.

> Sweet is the lore which Nature brings;
> Our meddling intellect
> Mis-shapes the beauteous forms of things:—
> We murder to dissect.[2]

But Steve celebrated the success of Cartesian thinking applied to the endeavor of understanding the silica tetrahedron and silicate mineralogy. In his words it was "reductionism in its most elegant form," "reductionism triumphant" for the arrangement of silica tetrahedra, whether isolated or linked into rings, chains, or frameworks, all came down to the size and charge of ions.[3] He seemed to love the fact that the submicroscopic level had implications for the macroscopic level, that out of this schema fell the physical properties of minerals such as specific gravity and cleavage, the sequence of crystallization of minerals from cooling magma, the order of weathering of minerals at the earth's surface, to name just a few mineralogical phenomena made understandable by breaking down whole systems into constituent parts. With a gleeful grin he savored the moment each semester when he got the chance to peel along its plane of cleavage a window-sized sheet of muscovite. Who above the age of seven would derive such joy from this action but a geologist?

Perhaps what also affected my perception that Steve was fundamentally a geologist, albeit one whose work had critical implications for evolutionary biology, but a geologist first and foremost, was the fact that during part of my tenure as a teaching fellow for his course, he was passionately at work on his book, *Time's Arrow, Time's Cycle: Myth and Metaphor in the Discovery of Geological Time* (1987c). As a result, during those years, he immersed us in the writings of Nicolas Steno, Thomas Burnet, James Hutton, and Charles

Lyell. I relished hearing his thoughts on neptunists and plutonists, the igneous nature of basalts, and the momentous significance of geological unconformities.

With regard to the idea of punctuated equilibrium, the stability of species in the fossil record was, in Steve's words, the "proper expression of geological time."[4] As one might have expected, Steve devoted a substantial portion of his intellectual efforts to elucidating the idea of geological time and its history. It's this appreciation of geological time among other aspects of the science that made Steve a geologist. Indeed, Steve referred to geological time as "his discipline's greatest contribution to human thought," and wrote "I love geological time—a wondrous and expansive notion that sets the foundation of my chosen profession" (1990s; 1993l, 48). And it was not only the formulation of relative geological time that he loved. Steve enjoyed being able to render the scientific history of our understanding of the earth's age beginning with Lord Kelvin's 1866 paper "The Doctrine of Uniformity in Geology Briefly Refuted" and ending with the discovery of radiometric dating. As Steve wrote, "However elegant his calculations, they were based on a false premise [the assumption that the earth's current heat is a residue of its original molten state and not a quantity constantly renewed] and Kelvin's argument collapsed with the discovery of radioactivity early in our century" (1983w; 1985y, 128). Among the lessons Steve would have his geologist colleagues learn from that episode in the history of science, was "Geologists should have trusted their own intuitions from the start and not bowed before the false lure of physics" (1983w; 1985y, 128).

In her essay "Geology: The Bifocal Science," Sue Kieffer refers to geologists as the "elders of the planet" for our ability "to see the planet in three spatial dimensions" as well as across the fourth dimension of time. She writes, "Just as the person who looks through bifocal lenses can see both far and near objects, our science allows us to look at the evolution of our planet at many scales of space and throughout all of geological and astronomical time" (Kieffer 2004). Another mark of Steve's "geologicalness," if you will, was his appreciation of space and scale. In one of my favorites among his essays, "The Great Scablands Debate" (1978o), Steve tells the tale of the vindication of J Harlen Bretz, whose catalog of evidence of a catastrophic flood in southeastern Washington initially failed to

convince the geological community about the origin of the chan-
neled scabland's features. Steve wrote:

> But when the first good aerial photographs of the scablands were
> taken, geologists noticed that several areas on the coulee floors
> are covered with giant stream bed ripples, up to 22 feet high and
> 425 feet long. Bretz, like an ant on a Yale bladderball, had been
> working on the wrong scale. He had been walking over the ripples
> for decades but had been too close to see them.... Observations can
> only be made at appropriate scales. (1978o; 1980h, 199)

This appreciation of another cornerstone of geology—spatial variation over distance and time—may have also motivated his writings
on continental drift, plate tectonics, and planetary phenomena.

In the early 1980s I'd listened to Steve in lecture wrestle with
the question, "was plate tectonics a universal physics?" While doing
so, he reveled in images of planets and their satellites in space,
marveled at the space program's accomplishment of melding
knowledge and wonder. He was taken by the idea that

> simple rules of size and composition would set planetary
> surfaces....Small bodies...(would) experience no internal forces
> of volcanism and plate tectonics and no external forces of atmo-
> spheric erosion. In consequence, small planets and moons should
> be pristine worlds studded with ancient impact craters neither
> eroded nor recycled during billions of years. Large bodies, on the
> other hand, maintain atmospheres and internal heat machines.
> Their early craters should be obliterated, and their surfaces, like
> our earth's, should bear the marks of continuous, gentler action.
> (1989t; 1991a, 507)

But Voyager images of Io, had led him to question his 1977
assertion that "the difference (between planetary surfaces) arises
from a disarmingly simple fact—size itself, and nothing else"
(1991a, 491–492).

In January 1986, when the Voyager spacecraft was scheduled
to flyby Uranus and its moons, Steve and his teaching chose to
include on the upcoming final exam a question related to the
flyby (see also Ross, this volume). The date of the exam had been
set for the same morning that Voyager would be relaying photos

of Uranian moons back to earth.[5] The question asked students on what basis one might predict the level of activity on the surface of Miranda, the smallest of the planet's major moons. "Miranda's countenance," as Steve put it, with its subsequently revealed fractured and reaggregated terrain, undid whatever conviction Steve had left in the size hypothesis. But to Steve's thinking, the discovery about Miranda's intense surface activity, as well as the later discovery of activity on Neptune's moon Triton, led to the realization that, as he wrote, "Planets are physical bodies that require historical explanations.... [they]) are not a repetitive suite, formed under a few simple laws of nature. They are individual bodies with complex histories." What a vindication of sorts, for a geologist who had worked throughout his career to show in his words, "that the sciences of history may be different from, but surely not worse than, the sciences of simpler physical objects... large and adequate theories usually need to forage for insights in both physics and history" (1991a, 497–98). Steve joked in writing, that he wished his publisher would lose all unsold copies of *The Panda's Thumb* in which he made his prediction about planetary surfaces.[6] But if they did, we'd also lose the record of his prescience regarding plate tectonic theory. Steve wrote, "I am not distressed by the crusading zeal of plate tectonics.... My intuition... tells me that it is basically true" (1977o; 1977f, 167).

I experienced Steve as a geologist, not only through his lectures and writings, but in the field. After I earned my PhD and went to teach at Pomona College, he visited on a few occasions. He examined the San Onofre breccia with interest for its historical importance as evidence of tectonic activity recognized by A. O. Woodford in 1925. And with enthusiasm for his late 1980s pudginess and the return to health that it signaled, he clambered with me in unselfconsciously ungainly fashion over Precambrian gneisses and Cretaceous granites in Joshua Tree National Park. Steve loved the desert (see 2002f), and I found this fact somewhat odd. Since we were both Jewish kids from Queens with alma maters of P.S. somethings, I figured he'd be more inclined to the lush green landscapes of the Catskills—like I was.

Our common cultural background drew me to Steve. "Jewish geologist" is practically an oxymoron. Steve was fond of recounting his relatives' response to his plan to become a paleontologist:

"That's a profession for a Jewish boy?" (1977e, ix). And I can hear him laughing at my characterization of our field as "the goyish science." Steve was a port in a storm for me. There were few women graduate students in geology at Harvard when I studied there, and no women occupied tenure-track faculty positions in the Department of Geological Sciences until my last year. During my time at Harvard, any women around were likely to be shiksas. I bonded with Steve as a New York Jew.

Steve averred that he was uninterested in the politics of science. Yet he used science to affect politics, in particular to work for social justice. He wrote and lectured about the limitations of biological theories of race and sex. I learned from him that geology could be put to work in the service of social justice, something that I try to do today in my own work on the distribution of environmental risk (Schneiderman and Sharp 2003). In a walk together over the Roebling Suspension Bridge from Cincinnati to Covington, Kentucky, and back during the 1992 annual GSA meeting, Steve encouraged me to pursue a GSA Congressional Science Fellowship to put my geological knowledge to work on policy issues that mattered to me. (I did.)

The years since Steve died have offered up many opportunities to miss this twentieth-century polymath. At such moments as the address to the NBER Conference on Diversifying the Science and Engineering Workforce on January 14, 2005, Harvard's president Lawrence Summers attributed the minority status of women in science to "different availability of aptitude at the high end," I feel acutely Steve's absence, for I know he would have helpful remarks to make about the perils of biological determinism and essentialism. If Steve were still alive, he would not be surprised that evolution is once again under siege across this country; he always asserted that any moment in social history is simply a pendulum's slice of a social spectrum. But he would confront in writing individuals such as Rabbi David Eidensohn who, in an article published by Concerned Women for America (2005) wrote, "Evolution is not a scientific process, said Mr. Gould. 'If the Tree of Life was planted anew,' said he, 'life would not form as we know it.' Gould stood evolution on its head.... The fossil record opposes Darwinian biological evolutionary process." Steve would have told us that Rabbi Eidensohn's half-quotes and misquotes are typical creationist tactics. He would weigh in once again on the renewed debate about evolution in the classroom in eloquent and convincing terms. The December 2004 tsunami in Indonesia surely would have drawn his insightful commentary about the scale and timing of geological events and their human toll. In contrast, Steve would have delighted in the revelations of NASA's Mars exploration rovers, Opportunity and Spirit, in their pursuit of geological clues to a once-watery Martian environment. And of course he would have thrilled to the ecstatic finish of the Boston Red Sox against the New York Yankees in the 2004 World Series.

I'm currently a geology professor at Vassar College just up the river from the Palisades sill, rock I imagine Steve gazed at knowingly from the galleries of the Cloisters. At every Vassar convocation, we sing *Gaudeamus igitur*, a piece regarded as the oldest student song and the embodiment of the free and easy student life. Its origin comes from a Latin manuscript dated 1287, the music from 1794. Choral music was one of Steve's well-known loves, along with baseball and architecture. The first verse of *Gaudeamus igitur* seems to me a fitting tribute to a scientist who I will always remember as a geologist:

> Let us therefore rejoice, while we are young:
> After the joys of youth,
> After the troubles of old age
> The earth will have us.

References

Eidensohn, D. 2005. Why was evolutionist Stephen Jay Gould a "hero" of an orthodox Jew? Concerned Women for America, Culture and Family Institute, May 24, 2005, www.cultureandfamily.org/articledisplay.asp?id=656&department=CFI&categoryid=cfreport

Kieffer, S. W. 2000. Geology: The bifocal science. In J. S. Schneiderman, ed., *The earth around us: Maintaining a livable planet*, 2–17. New York: W. H. Freeman.

Schneiderman, J. S., and V. A. Sharpe. 2003. Geology and environmental justice: An example from Hawaii. In J. S. Schneiderman, ed., *The earth around us: Maintaining a livable planet*, 368–385. New York: Westview Press.

∾ 12 ∾

Gould's Odyssey

Form May Follow Function, or Former Function, and All Species Are Equal (Especially Bacteria), but History Is Trumps

R. D. K. Thomas

Few scientists in our time have had such fervent admirers and such virulent detractors as did Steve Gould. In part, this reflects contrasting elements of his strong personality. In part, it reflects his intellectual style, which was by turns magisterial, warmly humane, richly synthetic, and aggressively combative. It also reflects the fact that, for all its legitimate claims to intellectual rigor, his scientific work was intensely personal. The nature of his engagement with paleontology which he dearly loved, with evolutionary theory which he aspired to remake in his own mold, and with life, which he treasured all the more having almost lost it, is admirably illustrated by the trajectory of his thinking on the classic theme of form and function.

Steve approached the analysis of form initially through its role in the functions of the particular extinct and living animals that he studied in detail. Increasingly over the course of his career, he came to treat function in general as an essential counterpoint to other factors that influence the course of evolution, on which he focused his attention. His mature preference was to see individual organisms, evolving lineages, and the great evolutionary faunas of the Paleozoic, Mesozoic and Cenozoic all as unpredictable, far-from-equilibrium systems, characterized by unique, disjunctive, temporarily stable states, although he did not define them in these terms. This worldview was bound, sooner or later, to bring him into conflict with proponents of equilibrium models in the analysis of form, behavior, and the dynamics of evolving populations.

In his earlier work, Steve's functional interpretations of organic form were largely those of a conventional Darwinist. Studying the worm snail, *Vermicularia spirata*, he set out "to test the hypothesis that rapid upgrowth is the adaptive reason for uncoiling in *V. spirata* and, further, that attachment to the [delicately branching coral] *Oculina* favors such growth" (1969g, 437). He showed that the degree of uncoiling of the shell of *V. spirata* is indeed correlated with variation in the growth form of *Oculina* and other potential substrates. The straightest worm snail shell that he observed, however, was only 48 percent uncoiled, far from the hypothetical straight, tubular shell that represents the optimum design for upward growth (see fig. 12.1). Steve recognized that this devia-

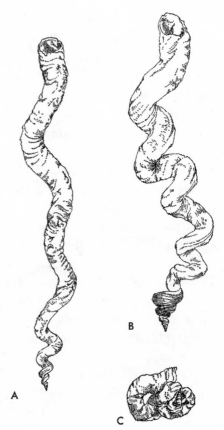

Figure 12.1. Shells of the worm snail, *Vermicularia spirata*, showing varying degrees of uncoiling. Early whorls form a tightly coiled, high-spired shell, comparable to that of typical members of the Turritellidae. Ratios of actual height to the height of a regularly coiled shell of comparable whorl length are (A) 3.2, (B) 2.0, and (C) 0.65. (From 1969g; reprinted with permission of the *Bulletin of Marine Science*)

B

A

C

tion from a paradigm based on a single function was due in part to the influence of another adaptive factor, the need for attachment sites along the length of the shell. This variable, however, did not adequately explain the extent of irregular coiling of the shell that is observed. The adult form of these animals emerges by departure, early in the course of shell growth, from a tightly coiled, high-spired juvenile shell like those of the turritellids, to which *Vermicularia* is related. This inherited growth pattern continues to influence the form of the shell throughout its development. Hence, the "ecological advantages of upward growth are counteracted by the genetic constraint of spirality" (1969g, 442).

Steve's graduate and later work on the Bermudian land snail *Poecilozonites* is best known for the use Niles Eldredge and he made of it, to illustrate their novel and since-much-debated concept of punctuated equilibrium. On the basis of detailed quantitative analysis of variation in shell shape, Steve was able to distinguish populations he recognized as distinct subspecies, with thinner and more angular, paedomorphic shells than those of long-ranging lineages from which they were derived (see fig. 12.2). These novel subspecies evolved repeatedly on lime-poor soils, developed under wet climatic conditions. Summarizing this work, Steve observed that "The major temporal variations of morphology in the *Poecilozonites bermudensis zonatus* stock are adaptive in nature" (1969h, 491).

From the outset, Steve's main interest was to explore the relationship between size and shape in these shells. Allometric growth facilitates change in shape and limits potential size increase, both in individual development and in evolving lineages. This implies that the power function k, derived from measurements of any two characters that have a nonlinear, exponential relationship, should in general be smaller in larger animals. Steve showed that this was so for three species of *Poecilozonites* (see fig. 12.3; see also analogous variation within a single species, *P. cupula*, see fig. 12.2). Hence, he confirmed his expectation that "proportion changes in phylogeny may be required correlates of trends in size variation. An alteration in shape is not ipso facto an adaptation" (1966b, 1131). He went on to argue that "Lineages whose small members maintain high k values for allometric relationships should either lack large representatives or show progressive decrease in k with increasing size" (1966b, 1138). Steve had convincingly demonstrated that

Figure 12.2. Shells of the land snail, *Poecilozonites*, described by Gould (1969h). 1–5: Five distinct Pleistocene subspecies of *P. cupula*, showing a range of variation in shell shape exceeding that observed in the allometric comparison of populations from three different species by Gould (1966b). Scale, x 2.5. 6–7: Shells of *P. bermudensis zonatus*, showing distinctive color banding characteristic of populations isolated in western and eastern Bermuda, respectively. These are the root stocks from which subspecies cited as key examples of evolution in the mode of punctuated equilibrium were derived (Eldredge and Gould 1972). Scale, x 2. (From 1969h; reprinted with permission of the President and Fellows of Harvard College)

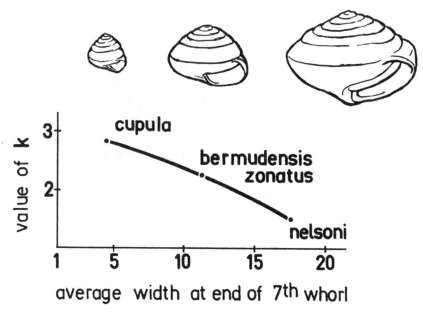

Figure 12.3. Inverse relation between k, the power function principally governing allometric growth of shell height relative to shell width, in three species of *Poecilozonites*. The greater the value of k, the more beehive-like the growing shell becomes. (Modified from figs. 3 and 4 of 1966b, and used by permission of the Paleontological Society)

this was the case in *Poecilozonites*. It is important to note, however, that there is an inherent weakness in the contention that k limits size. If members of an evolving lineage stay small, k is inferred to exert constraint. If they increase in size, the concomitant reduction in k is interpreted as adaptation to enable size increase. Both may be true, but these hypotheses require independent verification, without which they are simply "just so stories."

Steve made only one substantial foray into biomechanics. Prompted by work on shell growth patterns in the bivalve *Glycymeris* that reflect the function of the ligament and adductor muscles (Thomas 1975), he saw the opportunity to apply these ideas to a more interesting system. Steve set out to assess ontogenetic changes in shell form that might be related to swimming in scallops (see fig. 12.4). Most juvenile scallops are active swimmers; few living scallops, apart from *Amusium* and *Placopecten*, swim more than occasionally as adults. Steve began by conducting a theoretical analysis (1971f), in which he showed that a scallop's capacity to generate lift decreases with increasing size. Meanwhile, the take-off velocity

Figure 12.4. Left valve of an adult shell of *Chesapecten jeffersonius* from the Yorktown Formation (Pliocene) of Virginia. This is the one fossil species included in Steve's study of swimming in scallops. It is also said to be the first fossil illustrated and described, by the great anatomist Martin Lister, from North America. Steve would have appreciated this fortuitous connection. (From Ward and Blackwelder 1975)

required to get up off the sea bottom increases. Consequently, swimming becomes much more difficult as the animal gets larger.

On this basis, Steve predicted that shell shape and the position of the adductor muscle should change in ways that would maintain the capacity to swim for as long as possible, as the animal grows. He tested this hypothesis by measuring the dimensions of shells and muscle scars of living and fossil scallops (see fig. 12.5). In

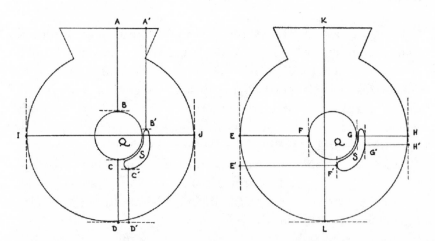

Figure 12.5. Measured variables employed in Gould's study of scallop shell form and biomechanics. The aspect ratio, which influences lift in swimming, is IJ/KL. The quick adductor muscle (Q) is mainly involved in swimming; the slow (S) or "catch" portion of the muscle holds the valves tightly shut. Gould defined the position of the adductor relative to the hinge, and hence its relative leverage, as AB/CD. (From 1971f; reprinted with permission of the Palaeontological Association)

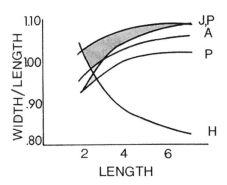

Figure 12.6. Changes in aspect ratio (IJ/KL, see fig. 12.5) of scallop shells with increasing size and among taxa. J: *Chesapecten jeffersonius*, large, thick-shelled adults, often with barnacles attached to the upper valve, certainly did not swim; P: *Placopecten magellanicus* (samples from two populations), an active, lifelong swimmer; A: *Amusium ballotti*, thin shells, reinforced by interior ribs, the most active swimming scallop; H: *Hinnites multirugosus*, cemented in coral or among rocks. The aspect ratio increases modestly, but at a decreasing rate, with increasing shell/body size in the swimmers, including juveniles of *Chesapecten*. (From 1971f; reprinted with permission of the Palaeontological Association)

more active swimmers, he found that shell width increased faster than shell length, with increasing size, as predicted (see fig. 12.6). Like wings, shells with a greater aspect ratio generate more lift. In contrast, the shell of the rock scallop, *Hinnites*, becomes more and more elongate, as it expands ventrally into open space, away from the crevice in which it is cemented.

In swimming scallops, the fast-closing, "quick" portion of the adductor muscle increases disproportionately in size. It also migrates ventrally, away from the hinge, increasing its mechanical advantage (see fig. 12.7). Thus, the animal's capacity for jet propulsion, accomplished by expelling water rapidly from the mantle cavity, is enhanced. *Hinnites*, however, defies the predictions set by these arguments for a nonswimmer. Its adductor muscle also increases disproportionately in size, and it migrates ventrally *more* than do those of the swimmers. Steve (1971f, 89) recognized the problem that these results pose for his argument: "But why should a non-swimmer show relative increase in muscle size? And does this not invalidate a claim that such relative increase therefore provides benefits to swimmers?" He circumvented these difficulties in a quite conventional way, like any right-thinking adaptationist. Numerous observations show

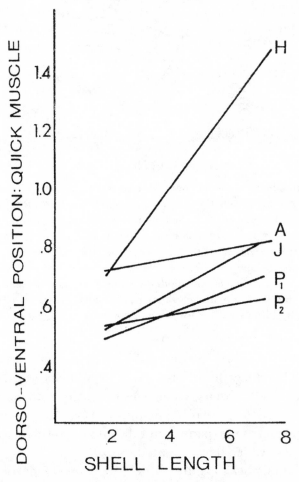

Figure 12.7. Ventral migration of the quick muscle during scallop ontogeny. This allometry facilitates continued swimming, but it has an alternative function in *Hinnites*, as explained in the text. Key as for fig. 12.6. (From 1971f; reprinted with permission of the Palaeontological Association)

that sessile bivalves require large, quickly contracting adductor muscles to expel pseudofeces and sediment, to cleanse the mantle cavity. So, the enlarged adductor muscle has a different function, unrelated to swimming, in *Hinnites*. As to the ventral migration of the adductor, this is an incidental effect, not necessarily adaptive in itself, but rather a byproduct of the elongation of the shell, which has a different adaptive purpose. Returning to his main argument, Steve concluded this study with as ringing an endorsement of functional adaptation as one might find:

"each allometric trend proceeds in a direction that provides better design for swimming and counteracts the difficulties of increasing size" (1971f, 91).

In his studies of *Poecilozonites* (1969g), Steve employed factor analysis to identify groups of characters that consistently exhibit strong covariance. Impressed by the utility of this technique, he applied it to measurements of pelycosaurs that had been made by Romer and Price (1940). In the case of cranial characters, a group of closely related variables was found to be linked to the preorbital length of the skull (see fig. 12.8). In this restudy (1967a), Steve followed Romer and Price in their interpretation of this linkage as a functional complex, associated in multiple lineages with the anterior displacement of increasingly specialized dentitions that emerged in carnivorous pelycosaurs. Analysis of postcranial data likewise yielded results showing a direct relation to function, driven in this case by allometric changes in proportion dictated by scale (see fig. 12.9). One group of variables, principally bone lengths, is related to overall body size; another is determined by body weight and the thicknesses of limb bones required to support it. Hence, Steve found in factor analysis a technique that identified two sorts of patterns in a data set. One picked up variables related by convergence—strictly speaking, it is parallel evolution, in this case—based on a paradigm of efficient jaw function. The other reflected scaling requirements of increasing size, and not on change that might otherwise have been inferred to represent novel adaptation in response to changing demands of the external environment.

In all these reports on work done early in his career, Steve was led by his interest in allometry to explore "laws of form" largely from a functional point of view, albeit with quite regular acknowledgments of the influence of scale, growth patterns, and evolutionary history. He was especially impressed by the utility of quantitative methods that could be applied to characterize, analyze, and model relations between form and function, with the potential to establish a new science of form (1970c). It is not surprising, in light of these studies, that Steve was drawn to the work of D'Arcy Thompson (1942). He appreciated Thompson's broad scholarship, his fine literary style, and his capacity for synthesis of disparate ideas, old and new (1971b). Steve particularly admired Thompson's elegant transformations, by

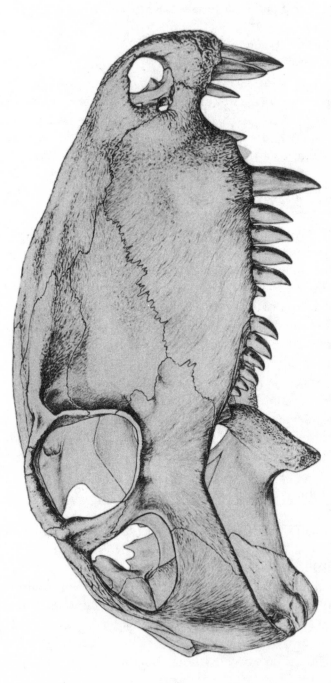

Figure 12.8. Skull and upper jaw of *Dimetrodon limbatus*. This is one of the pelycosaurs measured by Romer and Price and incorporated in Gould's (1967a) study of applications of factor analysis to group variables in functional complexes and to establish clusters of potentially related taxa. Note the anterior displacement of front teeth, creating a large diastema, in this carnivore. The set of measurements associated with this condition is identified by R-mode factor analysis as a tightly integrated complex (From Romer and Price 1940; reprinted with permission of the Geological Society of America)

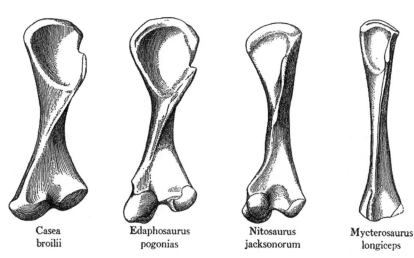

Casea	Edaphosaurus	Nitosaurus	Mycterosaurus
broilii	pogonias	jacksonorum	longiceps

Figure 12.9. Femora of four of the Permian pelycosaurs yielding data used by Gould (1967a) to document integration of variables that govern the ability to support substantial body weights on land. Scales: x 0.65, 0.25, 0.52, 1.03. (Modified from fig. 37 of Romer and Price 1940, and used by permission of the Geological Society of America)

means of which the forms of fishes, skulls, or other structures could be systematically related to one another. These were constructed graphically, by deforming rectangular grids, a technique that antici-pates the much later emergence of landmark analysis. Moreover, Steve recognized (1971b) that Thompson's emphasis on intrinsic formal properties of organic design was complementary to Darwinian evolutionary theory, as another avenue toward a more complete understanding of nature. This enthusiasm waned, however, as Steve later became increasingly unwilling to accept the determinism that is implicit in Thompson's worldview. Ultimately, he would exclaim, in his final magnum opus, *The Structure of Evolutionary Theory*, "If physical forces shape organisms directly, then their prior histories don't matter" (2002c, 1181).

Meanwhile, well schooled in the theory of allopatric speciation and cognizant of the penetrating analysis of rates of evolution provided by Simpson (1944) in *Tempo and Mode*, Eldredge and Gould (1972e) were energetically engaged in developing and then promoting the concept of "punctuated equilibrium." This led Steve to focus ever more intently on evolutionary mechanisms that had the potential to explain the actual, one-time running of the tape (his own oft-repeated metaphor), the particular history of life on Earth.

Steve's extensive multidisciplinary work on patterns of variation and dynamics of evolutionary change in the Bahamian land snail, *Cerion* was undertaken in this context. Here, allometric growth patterns expressed in shell form are invoked largely as inherited, historical constraints, and function is often downplayed in the arguments developed. This is well illustrated by Steve's elegant analysis of the growth and form of the most divergent of *Cerion* shells, the "smokestack dwarfs" and "smokestack giants" (see fig.12.10). In this study (1984j), Steve showed that smoke-stacks can evolve in two ways, each involving a simple modification of *Cerion*'s pattern of shell growth, at the extremes of its size range. In small shells, reduction in whorl width yields smokestack dwarfs if the normal number of whorls is maintained. On the other hand, adding extra whorls of normal proportion produces smokestacks at unusually large shell sizes. Smokestacks have evolved repeat-edly in both ways, but they do not occur in *Cerion* populations of average size. Steve attributed this phenomenon to channeling by developmental constraint, citing the coherence and apparent independence of sets of characters related to whorl number and

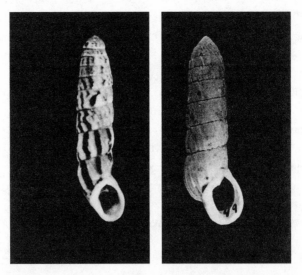

Figure 12.10. Aberrant "smokestack" shells of dwarf (left) and giant (right) species of the Bahamian land snail *Cerion*. Scaled to a common height to illustrate differences in their proportions. The height of *C. excelsior* is more than 2.5 times that of *C. pauli*. (From 1984j; reprinted with permission of the Paleontological Society)

to whorl size. Simple modifications of the growth process give rise to smokestacks at extreme sizes, whereas two different sorts of change would have to be coordinated to generate smokestacks in the midrange of shell size. Hence, constraints of geometry, in conjunction with a developmental program inherited in common by all species of *Cerion*, determine the observed pattern of evolutionary change.

Interestingly, in developing this analysis, Steve did not take up the suggestion that progenesis was involved in the evolution of smokestack dwarfs, whereas the giants appear to be neotenic. The observed growth patterns are consistent with this interpretation, which is corroborated by the fact that smokestack dwarfs consistently occur in disturbed habitats. Given the book on heterochrony (1977e) that may well stand as his most important work, it is striking that Steve did not interpret these data in this context. To do so, however, would turn the balance of his argument from constructional constraint back to natural selection, which directly determines life history strategies. Evidently, Steve preferred to explain *how*, in terms of growth process and pattern, these unusual forms have evolved in isolated populations, rather than *how* they emerged as outcomes of selection acting in local environments.

In parallel with these studies, which are based on extensive fieldwork and large sets of laboratory data, Steve devoted increasing attention to the development of evolutionary theory. The appearance of his article, with Richard Lewontin, on the now-notorious spandrels of San Marco and the supposedly Panglossian paradigm (1979k), marks a turning point in Steve's assessment of the role of adaptive function in evolution. The object of this polemic was to emphasize the roles in evolution of causal factors other than natural selection and outcomes apart from optimum function (see fig. 12.11). The target was in many respects a straw-man, as few evolutionary biologists were truly panselectionists, even in the wake of what Steve had aptly called (1983k) the "hardening" of the mid-twentieth-century "Modern Synthesis" of genetics, natural selection, and other aspects of evolutionary biology (see Mayr and Provine 1980). Perhaps on this account, or perhaps because Gould and Lewontin laced their argument with titillating examples and rhetorical flourishes, this article provoked a maelstrom of debate within and far beyond

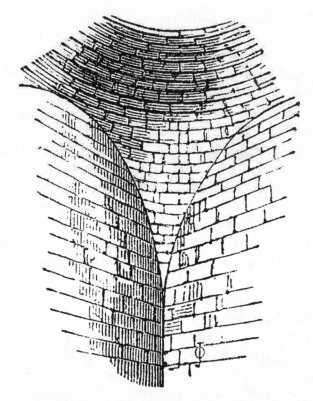

Figure 12.11. A spandrel or pendentive occurs where the surface of a dome extends down along the curves of arches that support it. In this unornamented, Byzantine dome, the function of the spandrel as a part of the ceiling is obvious. However, the shape of the spandrel is an incidental consequence of the structure of the dome. The option created by the spandrel to accommodate a painting or other ornaments, especially those that take advantage of its particular shape, is an exaptation, as noted by Gould (1997f) in later writing on this topic. (From J. H. Parker, *A Concise Glossary of the Terms Used in Gothic Architecture*, 6th ed., 1882)

evolutionary biology. In the process, strong arguments bearing on the weakness of untested "adaptive stories" and the influence of historical contingency, pleiotropy, and other developmental factors on organic form were set aside by many commentators. Despite the fact that Gould and Lewontin had marshaled Darwin's own evolutionary pluralism in their support (1979k, 597), the essay was widely construed as an anti-Darwinian diatribe. Gould and Lewontin overstated their case, but their argument has been very fruitful, if not entirely vindicated. Biologists and paleontologists are now much more generally willing to interpret their data in relation to a variety of causal factors, acting at more than one hierarchical level, than they were a generation ago (e.g., Pigliucci and Kaplan 2000; Grantham 2004). This shift is most evident in the burgeoning field of evolutionary developmental biology. The extent to which variation, and hence evolutionary innovation, are not merely constrained but facilitated by developmental pathways, is now widely appreciated (Kirschner and Gerhart 1998; Carroll et al. 2001; for an earlier, more tentative view, see Maynard Smith et al. 1985).

Seeking to establish pride of place for paleobiology as a science in which hypotheses are rigorously tested to discredit false hopes and establish general laws, Steve expressed "ambiguous feelings about standard functional morphology" (1980b, 111). His earlier enthusiasm for the emergence of "a new science of form" (1970c) had evaporated. Now, it seemed to him that assessment of function against mechanical or physiological benchmarks established little more than that most organisms are relatively well designed. He placed more value on studies that showed how form departs from functional paradigms, as a result of constructional constraints or historical contingency. Endorsing analyses of theoretical morphospace, he ventured the expectation that large realms are unexploited due to accidents of history, despite the fact that authors of most such studies had explained these omissions largely in terms of functional advantage and constructional constraint (e.g., Raup 1966; McGhee 1980).

Collaboration with Elisabeth Vrba led to recognition of a distinction between adaptation, defined more narrowly than hitherto, and the novel concept of exaptation (1982e). Adaptation occurs where a character emerges as a direct result of natural selection based on its concurrent utility, as in the divergence of beaks of Darwin's finches in response to the availability of different food supplies. Exaptation occurs when a character molded by selection to serve one purpose is co-opted to undertake a new function, as when feathers presumed to have evolved for thermoregulation were redeployed as wings, or when a character with no prior function is likewise co-opted. It seems very likely that many characters commonly regarded as adaptations in fact originated by exaptation. If this is so, evolutionary change that has been attributed by default to natural selection owes much to the fortuitous availability of states, with or without prior functions, that could be co-opted to serve new purposes. Hence, evolutionary change that is commonly assumed to have emerged by adaptation in response to natural selection in fact owes much to chance and historical contingency.

This shift in emphasis from functional adaptation to historical contingency as a preferred explanation for organic form reflects a theoretical commitment that dominated the latter half of Steve's career. He was determined to expunge every expression of the deeply rooted "idea of progress" from our account of life's fitful proliferation here on earth (1985f, 1996d). But, effective, albeit not necessarily efficient, function is essential to the survival and differential reproductive success of individuals in the evolution of local populations. Over the long haul—billions of years—organisms of increasing size and complexity have emerged, adding new levels of organization to existing communities and exploiting habitats or modes of life that were not open to their predecessors. Enhancements of function on all scales of structural organization, from molecules to ecosystems, have underwritten this expanding scope of activities. These realities stand in the way of Steve's enterprise. To circumnavigate them, he argued that selection for improved function within populations is of only local significance, and that long-term increases in complexity are a side effect of undirected change, arising from the fact that the earliest organisms were necessarily small and less complex. Steve saw long-term trends as effects of species selection (e.g., 1982g, 1985f), a process in which he supposed function and natural selection to have no direct role.

In Steve's last, exhaustively comprehensive account of his "expansion" of Darwinian evolutionary theory, constraints of form and function are not overlooked, but they constantly play second fiddle to historical contingency. As causal factors, function determines what may be adaptive in the right circumstances, and constructional constraint prescribes what the organism can or cannot grow, as continuously modified products of its development. These factors define those predictable sets of forms that *can* evolve in any given circumstance, and they explain why convergence is ubiquitous. They *cannot* fully explain, Steve insists, the "lovely puzzles" posed by the evolution of "actual organisms in real places" (2002c, 1338).

Ironically, functional design and constructional constraint together have the potential to explain how macroevolutionary trends may be established. According to Steve, the "wonderful life" on the Middle Cambrian sea floor recorded by the Burgess Shale was an evolutionary lottery (1989d). Descendants of most major players on this dark stage soon succumbed to extinction, but a few improbable survivors went forth and multiplied. They gave rise to divergent species, genera, and families, in a modest number of "lucky" phyla. But, if key elements of organic design, such as anatomical baupläne and viable skeletal constructions, are common *properties* of taxa in the clades to which they give rise, these clades are themselves units of selection. No systematic assessment of functional advantage and constructional potential of the Burgess Shale animals has been undertaken. Consequently, the differently weighted roles assigned by Steve (1989d) and Conway Morris (1998) to chance and necessity in the Cambrian radiation of higher taxa have more to do with their personal philosophies than the available data.

Steve preferred contingency to any sort of more general determinism on personal and ideological grounds. His rich and provocative evolutionary theory—at least the key punctuational and hierarchical parts of it—does not require that the effects of natural selection, speciation, or extinction must be unbiased in their directions. But Steve's humanity, his commitment to free will and personal responsibility, did require this. It gave rise to a highly personal evolutionary synthesis in which historical contingency takes the dominant role (see fig. 12.12). In study of the evolution of individual taxa and in analysis of the radiation and extinction of faunas that spanned hundreds of millions of years, Steve sought above all to explain those individual and collective properties of living organisms that witness their particular historical origins. History, to know what actually happened, when and where it occurred, Steve argued (1986a) citing Darwin's "long argument" as his model, is all that really matters. This is the substance of evolutionary relationships. Form does not follow function, it is merely mediated by it. In evolution, form follows prior form, recurrently constituting T. S. Eliot's "present moment of the past."

Figure 12.12. The (punctuated) Ascent of Stephen Jay Gould, or Portrait of the Evolutionist as a Provocateur. "A tiny little twig on the enormously arborescent bush of life," no more significant than uncountable bacteria? Surely, he doth protest too much! (Cartoon by Tony Auth, reproduced by permission of the author and the *Philadelphia Inquirer*)

References

Carroll, S. B., J. K. Grenier, and S. D. Weatherbee. 2001. *From DNA to diversity: Molecular genetics and the evolution of animal design.* Malden, MA: Blackwell Science.

Conway Morris, S. 1998. *The crucible of creation: The Burgess Shale and the rise of animals.* New York: Oxford University Press.

Grantham, T. A. 2004. Constraints and spandrels in Gould's *Structure of evolutionary theory. Biology and Philosophy* 19: 29–43.

Kirschner, M., and J. Gerhart. 1998. Evolvability. *Proceedings of the National Academy of Sciences, USA* 95: 8420–27.

Maynard Smith, J., R. Burian, S. Kauffman, P. Alberch, J. Campbell, B. Goodwin, R. Lande, D. Raup, and L. Wolpert. 1985. Developmental constraint and evolution. *Quarterly Review of Biology* 60: 265–87.

Mayr, E., and W. B. Provine, eds. 1980. *The evolutionary synthesis: Perspectives on the unification of biology.* Cambridge, MA: Harvard University Press.

McGhee, G. R. 1980. Shell form in the biconvex articulate Brachiopoda: a geometric analysis. *Paleobiology* 6: 57–76.

Pigliucci, M., and J. Kaplan. 2000. The fall and rise of Dr. Pangloss: adaptationism and the *Spandrels* paper 20 years later. *Trends in Ecology and Evolution* 15: 66–70.

Raup, D. M. 1966. Geometric analysis of shell coiling: general problems. *Journal of Palaeontology* 40: 1178–90.

Romer, A. S., and L. W. Price. 1940. Review of the Pelycosauria. *Geological Society of America, Special Paper* 28: 1–538.

Simpson, G. G. 1944. *Tempo and mode in evolution*. New York: Columbia University Press.

Thomas, R. D. K. 1975. Functional morphology, ecology and evolutionary conservatism in the Glycymerididae (Bivalvia). *Palaeontology* 18: 217–54.

Thompson, D. W. 1942. *On growth and form*, 2nd ed. 2 vols. Cambridge: Cambridge University Press.

❧ 13 ❧

The Tree of Life

Stephen Jay Gould's Contributions to Systematics

Margaret M. Yacobucci

All of the varied subjects that Steve Gould took up during his career fall into a few coherent and interrelated themes (see fig. 13.1). Steve challenged us to see beyond the conventional view of evolution as gradual, predictable progress in a world governed entirely by selective forces. Instead, he emphasized the roles of constraint and contingency—those unique circumstances and historical moments that shape and limit the evolutionary patterns we observe. Steve advanced his view of evolution as a hierarchy of individuals and agents operating at different taxonomic, spatial, and temporal scales as a way to expand our understanding of the causes of evolution.

This chapter illustrates some of the ways in which Steve's research and writing on systematics reflect these fundamental agendas. I focus on four topics: the iconography of evolutionary trees, Steve's own systematic work on the Caribbean land snail *Cerion*, his complex views on the systematic approach of cladistics, and the reality of species as evolutionary entities.

The Tree of Life

In much of his popular writing, Steve referred to that great symbol of systematic work, the Tree of Life, or *Etz Chayim* (עץתיים) in Jewish tradition. The Tree of Life imagery has a long history of

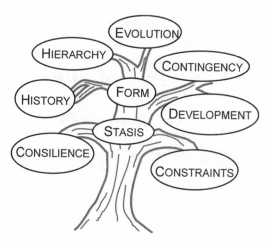

Figure 13.1. Core themes within Gould's body of work. Virtually all of his writings relate to one or more of these concepts.

meaning in theology and art, signifying the abstract principles of wisdom, strength, life, and growth (see fig. 13.2). As Charles Darwin and others realized, the Tree of Life symbol is readily appropriated to represent the evolutionary history of organisms (1993x). Often, though, the iconography of these trees contains

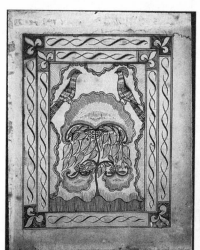

Figure 13.2. The Tree of Life is a popular subject in religious art. *Left:* Illustrated manuscript, artist unknown, Museum of Fine Arts, Boston. *Right:* Wall painting, East Turkestan, artist unknown. Among other meanings, the Tree of Life is a (feminine) symbol of wisdom: Wisdom "is a tree of life to them that lay hold upon her, and happy is every one that holdest her fast" (Prov. 3:18).

embedded messages of linear, predictable progress (1989d). Such trees are more than just an expression of nested sets; rather, the relative position of groups on these trees implicitly conveys each group's relative complexity, importance, or worth (O'Hara 1996). Consider, Steve pointed out, Ernst Haeckel's famous tree (complete with bark) representing the phylogeny of humans (see fig. 13.3). Haeckel's choice of placement of each group, and the size of the twig on which each sits, merely expresses a new form of the classic Chain of Being leading to humans, rather than providing an accurate depiction of life's diversity. Other trees display an "up-and-out" trajectory away from a few purportedly primitive ancestors, presenting the evolutionary history of such groups as the inexorable, steady, gradual, and progressive expansion of more improved descendants away from their limited, less adequate parents (see fig. 13.4).

For Steve, however, the Tree of Life illustrates and celebrates the richly complex and historically contingent evolution of life on Earth. Steve challenged traditional iconographies, pointing out the way they reinforce notions of gradualism and predictable progress. Most importantly, he offered testable alternatives, such as the decimation and diversification model he favored for the radiation of modern animals during and after the Cambrian Explosion (see fig. 13.5) (1989d).Why do evolutionary trees always have to be narrowest at the bottom and widest at the top? For that matter, why must trees show evolution as a continual process of gradual, directional change? The paleontological record suggests a very different iconography, in which change is concentrated at branching points while lineages remain in stasis for most of their duration (see fig. 13.6) (1972e). With the proposal of punctuated equilibrium, the image of the Tree of Life has been forever altered.

The Tree of Life metaphor captures well both the hierarchical taxonomic *structure* of biodiversity and the evolutionary *processes* of branching and divergence. Steve's own systematic work always confronted this tension between the static configuration of named objects and the dynamic evolutionary processes that produced them. Steve argued that one should never place a higher priority on naming entities than on searching for explanatory mechanisms. For example, he saw extensive intraspecific variability not as a taxonomic nightmare but as a "delight," providing valuable

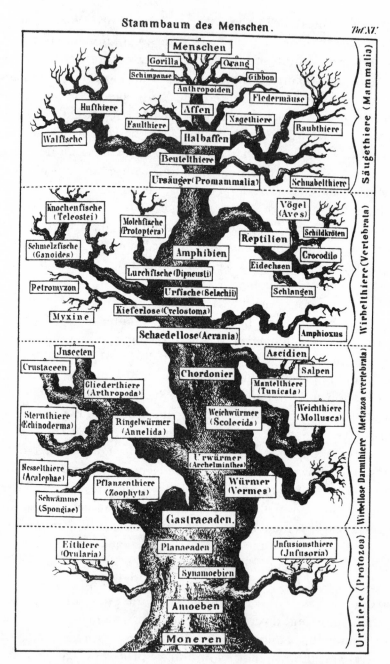

Figure 13.3. Ernst Haeckel's tree of human phylogeny (Haeckel 1874). This depiction revisits the classic Chain of Being, with diverse, successful groups like insects (*Insecten*) and hoofed mammals (*Hufthiere*) relegated to small side branches so that humans may take pride of place at the top of the great central trunk of the tree. (See fig. 20, 171, 1977e.)

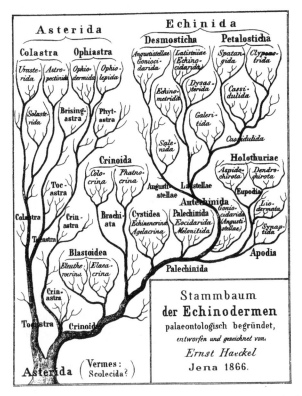

Figure 13.4. Ernst Haeckel's phylogeny of echinoderms (Haeckel 1866). Note the "up-and-out" trajectory, with a few closely spaced ancestors radiating out in steadily increasing diversity. (See fig. 4.6, 266, 1989d.)

information on how evolution works. That such variation makes the classification of species difficult is a small price to pay for such evolutionary insight.

Cerion's "Exuberant" Morphological Diversity—The Importance of History and Developmental Constraint

Steve's delight in variability was a good thing, since most of his systematic contributions, of course, involved the notoriously variable pulmonate land snails, beginning with the subgenus *P.* (*Poecilozonites*) on which he worked for his PhD dissertation (1969g; Gould 1967), and continuing quickly thereafter with the Caribbean snail *Cerion*. Steve deliberately chose this maximally variable group because he hoped he could shed new light on how

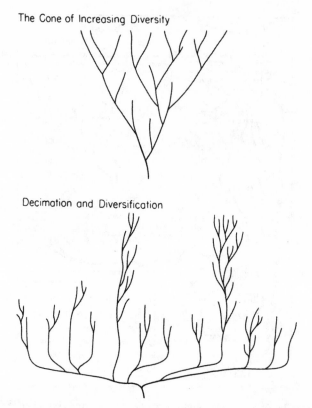

The Cone of Increasing Diversity

Decimation and Diversification

Figure 13.5. The conventional iconography of evolutionary trees (*top*) and Gould's proposed alternative (*bottom*; 1989d). In the cone of increasing diversity, a clade gradually and progressively expands both its diversity and morphological disparity. The decimation and diversification model, on the other hand, proposes that maximal disparity of form occurs early in a clade's history, and is then pared down by extinction. The survivors may later diversify, but only within narrow limits of form. (See fig. 1.17, 46, 1989d.)

form evolves to match environment. In particular, he focused on relating morphological variations to the geographic and environmental distributions of the various *Cerion* populations. Steve's goal in this systematic work was always to use observed patterns of variation to infer the evolutionary processes that produced them, or as he put it, to "extract evolutionary data from shells" (1969i, 409). It was not enough for him to reclassify or reorganize the *Cerion* faunas he encountered—his work must provide new insight into how evolutionary changes occur.

As an aside, Steve and I were talking once about the notion of different male and female intelligences, in particular the idea that

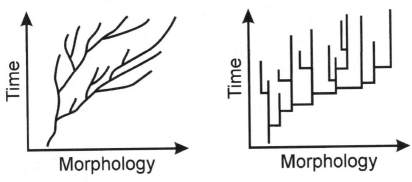

Figure 13.6. The alternative iconographies of phyletic gradualism (*left*) and punctuated equilibrium (*right*) (1972e). In the model of phyletic gradualism, evolutionary change is continual, gradual, and often directional, with the same type of change occurring both within species and during speciation events. The pattern of punctuated equilibrium locates evolutionary change at branching events, with lineages in net stasis throughout most of their stratigraphic durations, resulting in a very different tree iconography. Note that in both models, a overall trend exists toward morphologies on the right, although the cause of the trend is clearly different. Under phyletic gradualism, selection within species drives all taxa toward the right, while under punctuated equilibrium, a bias in the direction of (successful) speciation creates the pattern.

men tend to be more quantitative and rigidly logical while women tend to be more verbally fluent and narrative. Needless to say, neither of us had much use for this idea. I noted that I tended to be fairly quantitative in outlook, and Steve told me that the most quantitatively gifted and logical person he had ever known was his colleague Elisabeth Vrba. He then confessed that, under this false dichotomy, he himself clearly thought like a girl!

While few would dispute the notion that Steve tended toward the verbose and the narrative, he was certainly no slouch when it came to quantitative approaches either. All of his systematic work was actually based upon multivariate analyses, mostly factor analyses of one sort or another. He devised his own standardized multivariate morphometric method for analyzing *Cerion* morphology and patterns of covariation in shell characters (see 1984i for an overview). Indeed, it is somewhat ironic that Steve essentially used numerical taxonomy or phenetic methods, that quantitative and computerized approach developed in the 1960s, given both his general distrust of computers and his absolute commitment to making evolutionary inferences, something pheneticists traditionally disavow.

Steve spent decades methodically applying his approach to various *Cerion* populations, beginning with simple cases, then applying the principles learned from them to more complex situations. Steve's first systematic work on *Cerion* was a 1969 paper on the relatively simple case of *Cerion uva*, type species for the genus, from the islands of Aruba, Bonaire, and Curaçao (1969c). In this work, he used four shell characters to separate the species into four geographic groups and multiple, environmentally controlled, intraregional variants. (Later, Steve would revisit *Cerion uva*, using a larger set of measurements to resolve a long-standing disagreement about whether shell variations were due to distinct geographic populations or more localized patchy variations in physical conditions. What Steve found, in typical fashion, was that both sides of the debate were correct—a few geographically based groups could be recognized through multivariate analysis, but also apparent were the smaller-scale variations attributable to differences in local environments [1984i, 1995q].)

Steve continued his work on *Cerion* for the rest of his career, steadily producing journal articles and monographs throughout the 1970s, 1980s, and 1990s, moving to Caribbean islands with more and more diverse *Cerion* faunas (1969e, 1971d,i, 1974k, 1978e, 1979e, 1980k, 1984i, j, 1985g, 1986d, 1987d,e, 1988b, 1989c, 1990c, 1992j, 1996g, 1997e). Many of the lessons learned in these studies were synthesized in the 1986 monograph co-authored by longtime collaborator David Woodruff (1986d). This work focused on tiny New Providence Island in the Bahamas, where more than ninety species of fossil and living *Cerion* had been described—the most diverse and taxonomically complex case of all.

Using their now standardized protocol of performing factor analyses on nineteen shell measurements, Gould and Woodruff showed that all these variants could be separated out into only two morphs, the ribbed form *Cerion glans* and the mottled form *Cerion gubernatorium* (see fig. 13.7). These two "semispecies," their hybrids, and three extinct fossil species really comprise the entire *Cerion* fauna of the island—some serious taxonomic lumping! Steve even apologized in his taxonomic description for having to name one of the three fossil forms, *C. clenchi*, as a new species, thereby adding to the morass of *Cerion* nomenclature (see fig 13.8).

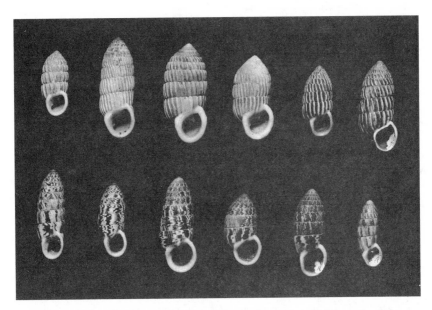

Figure 13.7. Demonstration of the pronounced variation within each of the two morphotypes of *Cerion* on New Providence Island, Bahamas. The *upper row* shows variants within the ribbed morph, *Cerion glans*; the *lower row* shows variants within the mottled morph, *Cerion gubernatorium*. The tallest shell (*upper row, second from left*) is 34.2 mm tall. See original source for detailed information on individual specimens. (See fig. 2, 396, 1986d.)

Gould and Woodruff discovered that the two morphs, *C. glans* and *C. gubernatorium*, are actually found all over the northern Bahamas. They argued that these forms do not merely indicate two separate lineages; rather, each morph has evolved repeatedly and independently in similar habitats. The mottled *C. gubernatorium* forms display more variation within populations and tend to transition smoothly between the three phases of allometric growth recognized in all *Cerion* (see fig. 13.9), while populations of the ribby *C. glans* tends to show a more abrupt, discontinuous shift during growth. Hence, the two morphs represent two easily-achieved but distinct, conserved ontogenetic pathways or "developmental systems" within the genus, defined by several independent covariance sets.

The real key to understanding the systematics of these forms, they argued, was to recognize the correlation of morphology with geographic location and environmental context. New Providence Island lies on the edge of a carbonate bank (see fig. 13.10). The

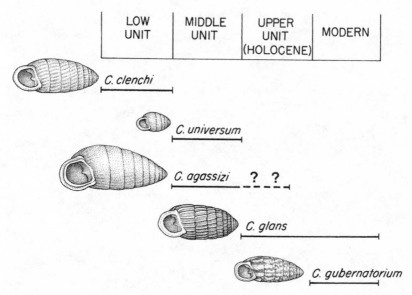

Figure 13.8. Fossil and living species of *Cerion* from New Providence Island, Bahamas. In an enormous exercise in taxonomic lumping, Gould and Woodruff (1986d) condensed seventy-one named extant species and seven fossil species into the two "semispecies," *C. glans* and *C. gubernatorium*, nine fossil species into *C. agassizi*, and three into *C. universum*. However, they had to erect a new species name for the distinctive fossil *C. clenchi*. (See fig. 37, 479, 1986d.)

ribby morphs of *Cerion* consistently live near modern coasts along bank edges, while the mottled morphs are found in the interior of the modern islands and on modern coasts that lie at ancient bank interiors. Gould and Woodruff suggested that the conventional explanation for this distribution—that the two morphs represent

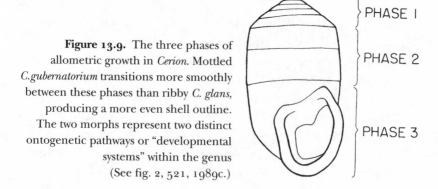

Figure 13.9. The three phases of allometric growth in *Cerion*. Mottled *C. gubernatorium* transitions more smoothly between these phases than ribby *C. glans*, producing a more even shell outline. The two morphs represent two distinct ontogenetic pathways or "developmental systems" within the genus (See fig. 2, 521, 1989c.)

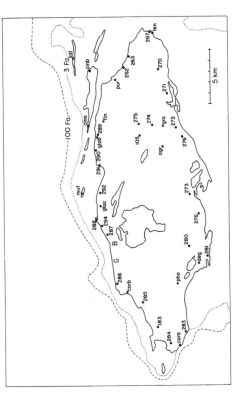

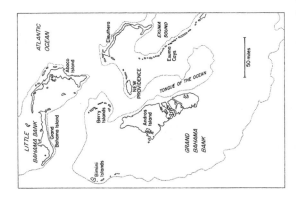

Figure 13.10. Maps of the Great Bahama Bank (*left*) and of New Providence Island (*right*; with Gould and Woodruff locality numbers). Note the location of New Providence Island relative to the bank interiors and edges—the island's south and east coasts border bank interiors; the north and west coasts border bank edges. While both ribbed and mottled morphs can be found along the island's modern coastlines, ribbed *C. glans* is distributed along bank edges; mottled *C. gubernatorium* is found both in the modern island interior and on modern coasts along bank interiors—a position that would have been island interior during the Pleistocene. Hence, *C. gubernatorium*'s modern geographic distribution may reflect an historical preference for island interiors. (*Right*: fig. 3, 397; *Left*: fig. 7, 404, 1986d.)

adaptation to the different environments found in their modern ranges—may not be right. Rather, the current distribution may actually be an historical relict of Pleistocene geography and habitats, with the ribbed morph living on Pleistocene coasts and the mottled morph in what *was* island interior during the Pleistocene. This historical explanation implies that the geographic distribution of the morphs is remarkably stable over long time spans, a controversial idea for *Cerion*.

The results from New Providence highlight the central themes of much of Steve's systematic work. A synthesis of data from morphology, ecology, geography, stratigraphy, and genetics—a "consilience of inductions," to use William Whewell's phrase (1840)—enabled him to revise the taxonomy of *Cerion* and explain elements of its evolutionary history. He highlighted the potential of quantitative techniques for improving taxonomic practice and providing new approaches to problems of variation, adaptation, and constraint. Steve's work characteristically challenged selectionist explanations for correlations of form and habitat—why forms are found where they are—emphasizing instead the roles of developmental constraint and the contingent geologic histories of populations.

A Paleontologist's View of Cladistics, or Why We Love Paraphyletic Groups

Even though cladistics is a core methodology of modern systematics, Steve has a reputation as an opponent or debunker of the field. However, his views on the matter were more complex, and he certainly never minded his students, including me, using cladistic approaches in our dissertations. Steve has also been quoted as saying that he was agnostic when it came to cladistics (see 1995o), but that's not right either—he really had quite strong and thought-out opinions on the subject.

There were some aspects of the cladistic approach, to be sure, with which Steve took issue. He never liked the way that *unique* traits were marginalized or outright ignored by cladists, who dismiss such specializations as contributing no phylogenetic signal, that is, data useful for grouping taxa by closeness of evolutionary descent. Evolution, though, is more than *just* branching order. We want to know, Steve argued in a *Natural History* piece, about the origin of pelycosaur sails and saber-tooth teeth and armadillo armor and other evolutionary quirks of particular groups (1995o). Unique traits may not help us tell who's related to whom, but they tell us a lot about the evolutionary "stories" of the groups that have them.

The condemning of paraphyletic groups as "artificial" also challenged Steve's common-sense take on classification. He was always being asked about that vexing "birds are dinosaurs so dinosaurs aren't extinct" issue. As he noted once in an interview, "the fact that a...lineage of small running dinosaurs evolved into birds *doesn't* mean that dinosaurs are still around....It's not what in a vernacular sense we *mean* when we say that dinosaurs are dead....a sparrow is not a *Tyrannosaurus*" (Brown 2001). This "birds are dinosaurs" issue was also problematic for Steve because it gives the false impression of a transformationist view—that is, that dinosaurs *turned into* birds in an anagenetic fashion, rather than birds *branching* from a small subset of dinosaurs. (This linear, transformationist view plagues popular understandings of human evolution as well, feeding into several cultural biases, including racist notions of relative "advancement" [1989d].)

While Steve acknowledged that cladistics provided "the purest of all genealogical systems for classification" (1987a, 69), he saw all sorts of pitfalls to basing classifications on cladograms, that is, strictly by branching order (1992r). To do so, he argued, is to miss many aspects of morphology and function, and the ecological roles that groups play, as highlighted in the infamous salmon-lungfish-cow example (see fig. 13.11). The grouping of salmon plus lungfish is paraphyletic, since it does not include all of the group's descendants, and is therefore "unnatural" according to strict cladists. However, to insist on classifying lungfish with cows, rather than with salmon, because they share a more recent common ancestor, is to deny the obvious common morphological and ecological traits that unite the two fish. Indeed, that we can talk about "fish" and be generally understood demonstrates the obvious similarities among such animals. Since the position and hierarchical arrangement of all the nodes on a cladogram changes as new taxa are added or existing datasets are revised, cladistic-based classifications are also thought to be more unstable than those based on the Linnaean system, which

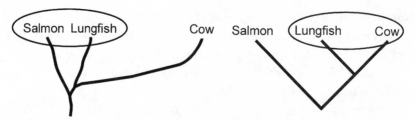

Figure 13.11. The now-infamous salmon-lungfish-cow problem. In a conventional classification, salmon and lungfish are grouped together as types of fish, while cows are separated out by a great many distinguishing morphological and ecological characteristics (*left*). Under cladistic conventions, however, the entity "salmon + lungfish" is a paraphyletic group (because it excludes some descendants) and is therefore "unnatural." It is the grouping "lungfish + cows" that forms a natural clade, or monophyletic group (*right*). Cladists distinguish between *clades*, definable units reflecting closeness of ancestry, and *grades*, such as "fish," which are impossible to define precisely, if often intuitively understood. (See Gee 1999, 141–50, for an interesting account of the origin of this classic problem.)

can hinder the usefulness of such classifications for identification and communication.

Steve also balked at the more extreme version of pattern cladism, which abandoned the central goal of phylogenetic systematics by arguing that the process of evolution was inherently unknowable and that, therefore, systematists should stick with merely documenting the branching patterns represented by cladograms (1986a). Removing the study of process, of the *causes* of branching relationships, takes away the most powerful aspect and greatest contribution of cladistics to evolutionary study—its emphasis on history.

As Steve consistently affirmed in his writings, cladistics is the best available method for reconstructing branching order, and therefore for working out historical patterns in the origin and evolution of traits (1986a, 1987a). He often cited cladistic studies in his popular essays and books, using the results to illustrate new challenges to the perception of gradual, predictable, and progressive evolutionary change. For instance, Steve cites as an example of cladism's utility the 1995 study of Daniel Blackburn, who tested gradualist and punctuated equilibrium models for the evolution of viviparity and placentation in lizards and snakes (2002c). It had always been assumed that lineages had to pass through a gradual sequence of three intermediate stages before reaching full placentotrophy. Blackburn's analyses, however, showed that in all one-hundred-plus instances in which viviparity evolves, not even once does the branching order reflect this hypothetical linear trend.

For similar reasons, Steve also loved the way the American Museum of Natural History in New York redesigned its fossil mammal halls in the mid-1990s to reflect mammals' phylogenetic branching order. This mode of display disrupts the standard iconography of a progressive, linear sequence leading to humans, placing us in a more accurate, if less self-aggrandizing, position about halfway through the mammal hall (see fig. 13.12) (1995o). Cladograms can directly challenge the "cones of increasing diversity" and ladders of progress that so often masquerade as accurate depictions of the Tree of Life.

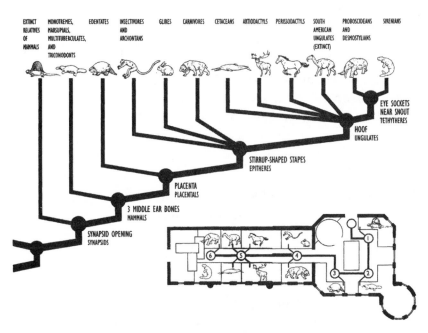

Figure 13.12. Floor plan for the American Museum of Natural History's Hall of Mammals. The displays, redesigned and opened to the public in early 1995, are innovatively arranged according to the branching order of the major mammal groups on a cladogram, rather than as a ladder of progress toward modern humans. Note humans' position at number 4. (1995o, 13; original source is the American Museum of Natural History pamphlet, *Mammals and Their Extinct Relatives: A Guide to the Lila Acheson Wallace Wing.*)

Steve frequently acknowledged the power of cladistic approaches to test for the relative frequency of punctuated equilibrium (2002c). In a paper Steve called a "breakthrough in the application of quantitative modeling to cladistic patterns of evolution" (2002c, 817), Wagner and Erwin (1995) showed that the proportion of polytomies to dichotomies in well-resolved clades will reflect the proportion of punctuational events. If an ancestral species is in stasis and produces multiple daughter species in punctuational events, its daughters will acquire only ancestral characters (plesiomorphies) and autapomorphies (derived characters unique to each species). Branching order among the daughters will then be unresolvable, producing a polytomy (see also Yacobucci 1999).

Perhaps most importantly, Steve argued that the cladist's emphasis on branching was essential for recognizing species as discrete and real evolutionary entities (1992i, 2002c). As he noted in his "big book," "[m]any...have failed to recognize that the so-called cladistic revolution in systematics rests largely upon this insistence that species (and all taxa) be defined as discrete historical individuals by branching" (2002c, 605). Steve's grand, hierarchical view of macroevolution depends on this recognition. The ontological reality of species as individuals defined by a branching point "birth" and an extinction "death" makes species eligible to be macroevolutionary players.

The Reality of Species and the Hierarchical Structure of Evolution

In an essay reflecting back on his early work on *Cerion uva* from the Dutch Leeward Islands of Aruba, Bonaire, and Curaçao, Steve gently berated himself for his motivation behind the study (1995q). He hoped, he said, to use this relatively simple, monospecific case to develop a better understanding of the infinitely more complex occurrences of *Cerion* in the Bahamas and other West Indian islands. He soon recognized his error. "Variation within a species doesn't tell you how to treat interactions between species; the phenomena are disparate and exist at different scales....Causal continuity does not unite all levels; the small does not always aggregate smoothly into the large." (1995q, 22).

Given this challenge to extrapolation, a theme oft-visited in his writing and fundamental to his hierarchical theory of evolution, how paradoxical it is that most of Steve's essays and scientific contributions center on a fine scale, detailed study that leads inexorably to much grander and more general conclusions about the history of life on Earth. For instance, in the original punctuated equilibrium paper (1972e), Steve uses his early work on subspecies of the land snail *Poecilozonites bermudensis* from the Pleistocene of Bermuda (1969g, 1970b; Gould 1967) as an example to show how the careful study of subtle details of geography, stratigraphy, and morphology can reveal surprising, indeed convention-challenging, tempos, and modes of evolution.

The empirical pattern of punctuated equilibrium, the abrupt appearance and long-term stasis of lineages as preserved in the fossil record, has been well-documented (1972e, 1977c, 1986i, 1993j, 2002c). Stasis is quite apparent, although explaining *why* it occurs has become of great theoretical interest among evolutionary biologists and paleontologists. Documenting geologically rapid speciation events is more operationally difficult, but not impossible, as Goodfriend and Gould (1996g) demonstrated in their elegant study of *Cerion* on Great Inagua, Bahamas. By dating individual shells found on a single mudflat, they were able to determine that the now-extinct *C. excelsior* hybridized with the newly migrated *C. rubicundum* within an interval of about 15,000 years, a punctuational event on a timescale usually not detectable by paleontologists. (A side note about all these hybrids and "semi-species" in *Cerion*: as Roger D. K. Thomas pointed out to me, it's rather ironic how often Steve's own "species" weren't really species, given the centrality of the reality of species to his hierarchical view of evolution!)

While the basic pattern of stasis punctuated by speciation has been recognized by paleontologists for decades, the causal mechanisms of punctuated equilibrium and its theoretical consequences for evolutionary paleobiology have stimulated seemingly endless debate (see Gould 2002c for a extended account of the controversies). Punctuated equilibrium requires no radical mechanism of speciation; rather, ordinary allopatric or sympatric speciation will necessarily produce a punctuated pattern in the fossil record. The causes of stasis, on the other hand, are less clearly understood; it is

308   MARGARET M. YACOBUCCI

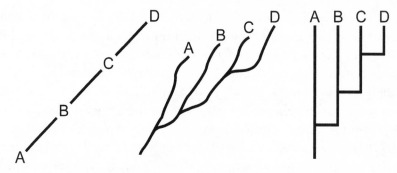

Figure 13.13. Schematic diagrams showing anagenesis (*left*), phyletic gradualism (*center*), and punctuated equilibrium (*right*). If species arise via anagenesis or phyletic gradualism, their beginning and end points are not well defined, and their character is in constant flux. Under punctuated equilibrium, in contrast, species have clearly defined speciation "births" and extinction "deaths," and enough stability during most of their stratigraphic ranges to make them discrete, definable historical individuals. The recognition of species as Darwinian individuals expands evolutionary theory into the hierarchical structure Gould found so fascinating.

likely that each of the many proposed mechanisms, such as stabilizing selection, developmental constraint, and habitat tracking, apply in specific cases.

The theoretical importance of punctuated equilibrium lies in its demonstration that species are true individuals, each with a defined birth and death, and with sufficient stability during its lifetime (see fig. 13.13; as noted earlier, the cladistic emphasis on branching highlights this view of species.) As Steve put it, "if punctuated equilibrium prevails as an empirical proposition...then species are individuals...by all vernacular criteria" (2002c, 606). Since species produce "daughters" that inherit properties from them, they (and the clades that contain them) become fully Darwinian individuals upon which selection can act. Hence, a nested set of evolutionary individuals exists, ranging from genes to cell lineages to organisms to demes to species to clades, which leads to a reformulation of evolutionary theory that emphasizes this inherently hierarchical structure (2002c).

Many evolutionary biologists have denied the importance, if not the logical possibility, of species selection, since it must necessarily (by virtue of fewer individuals and "birth" events) be weaker than organismal-level selection. However, species selection can be effective if lower levels of selection do not "go anywhere"—that is, if stasis within lineages reigns and most evolutionary change takes place at speciation events (1998b, 2002c). What was of most interest to Steve was how these different selective levels must function differently, based on differences in their essential properties and processes. For instance, Steve argued that the different mechanisms by which organisms and species actively maintain their borders of individuality will necessarily lead to the dominance of different evolutionary processes. Specifically, selection should dominate at the organismal level, while selection, drift, and drive will all be important mechanisms for change at the species level (2002c). Thus, the expansion of evolutionary theory to include other levels of selection simultaneously expands the range of evolutionary processes that play roles in shaping the long-term history of life.

Numerous other subjects within evolutionary paleobiology, from the nature of trends to the long-term structure of ecological communities, are opened up to new interpretations with the recognition of species as individuals. Permit me to let Steve have the last word: Extending our view of evolution to include species as Darwinian individuals causes us to rethink "both the pageant of life's history and the causes of stability and change in geological time" (2002c, 893). What marvelous expansions of evolutionary theory came from patient systematic studies of a few land snails!

Why Taxonomies Matter

My paleontology students anticipate with dread the section of the course on systematics. What could be duller than learning how to put names on things? It takes a bit of work to convince them that systematics is exciting, dynamic, and full of controversy, and that systematics centers on a question of greatest interest—reconstructing the evolutionary histories of life on Earth. In contrast to my apprehensive students, I have always appreciated the sense of fun and discovery Steve brought to even the most small-scale and detailed of systematic studies (see, for instance, the vermetid papers [1994e, f]), as well as his affirmation of the centrality of systematics to evolutionary theory. Darwin famously called his *Origin of Species* "one long argument" (Darwin 1859, 459). Steve identified this underlying argument as "the claim that history

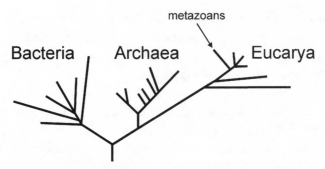

Figure 13.14. Carl Woese's radical new vision of the Tree of Life, emphasizing the deep split among the bacteria, archaea, and eukaryotes, and reducing metazoans to a trivial twig. (See Woese, Kandler, and Wheelis 1990.)

stands as the coordinating reason for relationships among organisms" (1986a, 74). Fundamentally, Darwin's insight was to recognize that the observed hierarchical organization of organisms within our taxonomies is the end-product of a long sequence of historical events. Hence, to develop a natural classification is to document the processes of evolution.

Steve spoke and wrote with great reverence about traditional systematists, from Linnaeus's determination to classify organisms on nature's terms rather than by their use to humans (1993p), to Dobzhansky and Mayr's insistence on placing systematics on an equal footing with genetics in the Modern Synthesis (2002c), to Whittington, Briggs, and Conway Morris's reinterpretation of the Burgess Shale fauna (1989d), to Carl Woese's iconoclastic presentation of life's diversity—a radically new Tree of Life (Woese et al. 1990, 1996d) (see fig. 13.14). Steve emphasized that taxonomies are always theory-driven and therefore intellectually exciting. The best systematic studies, including Steve's own, are both creative and transformative—indeed they can even be revolutionary—and further Steve's lifelong goal of better understanding the processes of evolution.

Acknowledgments

I wish to thank Warren D. Allmon, Patricia Kelley, and Robert M. Ross for inviting me to participate in the 2003 Geological Society of America Pardee Symposium, "His View of Life: Reflections on the Scientific Legacy of Stephen J. Gould," and to contribute to this volume. Tony Avruch, Linda Ivany, and Roger D. K. Thomas also provided helpful suggestions and insightful comments. And of course, I wish to posthumously thank Steve Gould, for his support and encouragement, his insights, and his wisdom.

References

Blackburn, D. G. 1995. Saltationist and punctuated equilibrium models for the evolution of viviparity and placentation. *Journal of Theoretical Biology* 174: 199–216.

Brown, D. 2001. Stephen Jay Gould, from brachiopods to baseball. *Powells .com Interviews* (www.powells.com.authors/gould.html, accessed August 1, 2008).

Darwin, Charles. 1859. *On the origin of species by means of natural selection, or preservation of favoured races in the struggle for life*. London: John Murray.

Gee, H. 1999. *In search of deep time: Beyond the fossil record to a new history of life*. New York: Free Press.

Gould, S. J. 1967. Pleistocene and recent history of the subgenus Poecilozonites (Poecilozonites) (Gastropoda: Pulmonata) in Bermuda: An evolutionary microcosm. Ph. D. dissertation, Columbia University.

Haeckel, E. 1866. *Generelle Morphologie der Organismen*, 2 vols. Berlin: G. Reiner.

———. 1874. *Anthropogenie: Keimes- und Stammes-Geschichte des Menschen*. Leipzig: W. Engelmann.

O'Hara, Robert J. 1996. Trees of history in systematics and philology. *Memorie della Società Italiana di Scienze Naturali e del Museo Civico di Storia Naturale di Milano* 27(1): 81–88.

Wagner, P. J., and D. H. Erwin. 1995. Phylogenetic patterns as tests of speciation models. In D. H. Erwin and R. L. Anstey, eds., *New approaches to speciation in the fossil record*, 87–122. New York: Columbia University Press.

Whewell, W. 1840. *The philosophy of the inductive sciences, founded upon their history*. London: John W. Parker.

Woese, C. R., O. Kandler, and M. L. Wheelis. 1990. Towards a natural system of organisms: proposal for the domains Archaea, Bacteria, and Eucarya. *Proceedings of the National Academy of Sciences USA* 87(12): 4576–79.

Yacobucci, M. M. 1999. Plasticity of developmental timing as the underlying cause of high speciation rates in ammonoids: An example from the Cenomanian Western Interior Seaway of North America. In F. Olóriz and F. J. Rodríguez-Tovar, eds., *Advancing research in living and fossil cephalopods*, 59–76. New York: Plenum Press.

∽ 14 ∽

Genetics and Development

Good as Gould

Robert L. Dorit

Steve Gould frequently reminded all of us of just how uninteresting hagiography was as history of science. He would justly rail against "Whig history"—the tendency to look back from our present vantage point and pass judgment on predecessors based on how close they had come to getting it right (right, of course, meaning "as we understand it today"). Steve wouldn't let us forget that scientists and their work need to be judged in their proper historical and intellectual context. He would also insist that the trajectory of science was influenced more than imagined by factors outside of objective dispassionate inquiry: the warts and blemishes of the profession and its practitioners matter too.

Steve had a healthy ego. As many of the essays in this volume make clear, Steve cast a long shadow in biology and paleontology over the last quarter of the twentieth century. I think it disingenuous to argue that he did not derive pleasure from the length of that shadow: ego is often a key propellant of ambition and achievement. But the most common misunderstanding about Steve confuses his ego for pretension. He was—no hagiography here—one of the least pretentious people I have known. When he would begin one of our weekly seminars with something like "As you remember from having read chapter four of '*Histoire générale*

*et particulière des anomalies de l'organisation chez l'homme et les animaux'
(Geoffroy Saint-Hilaire 1837)"* he was not trying to make us look bad,
or himself look smart. Instead, Steve's error was an intellectual
optimism that was not subject to disproof—he honestly believed
that all of us had read *Histoire générale et particulière.* Despite our
frequent good-natured ribbing, he lived in a world in which
everyone was simply dying to read these fascinating and obscure
volumes, and giving in to the urge at every opportunity. It pained
us to disabuse him of this worldview, though fortunately we never
did succeed in eroding his fundamental idealism that imagined us
all reading as voraciously and widely as he did.

Perhaps Steve's optimism is why the Museum of Comparative
Zoology was such an exciting place to be in the mid-1980s, when
so many interesting and meaningful debates were swirling around
the emerging field of macroevolution. The essays in this volume
address the many dimensions of this debate. Here, I want to
concentrate specifically on the contact zones between macroevolution—as defined by Steve—and two of the supporting girders of
modern biology: population genetics and developmental biology
(specifically what has come to be called "evo-devo"). This chapter
is explicitly not intended to be an encyclopedic review of the ways
in which these crucial intellectual vectors intersected, then and
now. Instead, this is an effort to reconstruct what seemed at the
time (and since) to be the central areas of agreement, as well as
the main points of friction. In so doing, I hope to shed some light
on the evolution of Steve's thinking about the macroevolutionary
agenda and on the role it could play in complementing the existing
edifice of the Modern Synthesis.

Population Genetics and the Emergence of Macroevolution

The publication in 1972 of the Eldredge and Gould paper (1972e)
established the framework for the notion of punctuated equilibrium. A great deal of ink has been spilled since, elaborating, and in
some cases revising the original notion of punctuated equilibrium,
dissecting its meaning, attacking its relevance, validity, and even the
motivations of its proponents (see, e.g., chapters by Allmon and
Geary in this volume). At the core of the original paper lie a series
of claims, both about pattern and about the underlying generative

processes that could account for the pattern. The first, and most modest claim, reasserts that the fossil record frequently—perhaps very frequently—fails to conform to the expectations of steady and gradual phyletic change. Instead, the pattern, according to the paper, is more often one of abrupt (at least in geological terms) appearance of forms, frequently (but not always) in conjunction with the fossil record version of speciation. New forms appeared but did not necessarily replace existing forms. The second corollary claim, still one about pattern, argues that morphological stasis is the prevailing feature of most species with a decent fossil record. No change may be monotonous, but it is not without significance, as Steve (1993n) and others (Fortey 1985; Cheetham 1986) would spend the next three decades arguing ("Stasis is data" was the *cri de coeur*). These two statements about pattern, would of course lead to some provocative claims about the underlying generative processes. Species were constrained to change little over their lifetimes, or at least little in any one direction. Only the process of speciation, with its accompanying disruption of the established genetic order could temporarily release these constraints.

Had punctuated equilibrium been a set of observations and proposals about the fossil record and its prevalent features, it would likely have attracted little attention from population genetics. Eldredge and Gould's proposal, however, must be seen in the context of the history and status of paleontology in the mid-twentieth century. Paleontology had been demoted from its role as the central source of evolutionary observations. The Modern Synthesis had largely dismissed the fossil record as too coarse and too incomplete to serve as a testing ground for the rich tapestry evolutionary hypotheses. Experimental biology and the detailed exploration of model systems (*Drosophila, E. coli,* etc.) would now serve as the source material for the interesting questions and the place to go for adjudicating among competing evolutionary hypotheses. At the center of this new evolutionary edifice, resplendent with quantitative rigor and predictive power, stood population genetics.

Steve understood the disciplinary power play that accompanied the Modern Synthesis. Not one known for being intellectually coy, he would go on to argue that punctuated equilibrium required a far more catholic approach to the evolutionary forces that shaped life on the planet. If the pattern of the fossil record was to be taken at face value, rather than rationalized away, it was not *and could not* be predicted by extrapolating from conventional evolutionary mechanisms. Stasis individuated species, and sorting and selection at higher levels now had to be included in any catalog of evolutionary forces (Stanley 1975; 1993a, 1999j). Paleontology, the one discipline capable of detecting the operation and consequences of these higher order mechanisms, now needed to reclaim its seat at the "high table."

The gauntlet had now been thrown down, and population geneticists were quick to respond. The power of population genetics theory resided in part on its ability to model a vast range of evolutionary scenarios—large and small populations, variable intensity selective effects, linear and epistatic interactions between loci, stochastic effects, and so on. In addition, population genetics, like the rest of evolutionary biology, could postulate an array of external selective forces that could propel population change in virtually any direction. Thus armed, it was relatively trivial for population genetics simulations to yield the two prominent features of punctuated equilibrium: stasis and geologically instantaneous change (Lande 1985; Newman, Cohen et al. 1985; Lande 1986; Lewin 1986; Somit and Peterson 1992).

The second antigen that triggered a strong response from population genetics concerned the genetic mechanisms that caused, or at least accompanied, speciation. The initial punctuated equilibrium papers had placed significant emphasis on the destabilizing effects of Mayrian peripheral isolate allopatry, including the breakup of "co-adapted gene complexes," the consequences of sampling and drift on the Wrightian landscape, and the accelerating effects of selection of reinforcement mechanisms. Attractive as these speciation-related genetic phenomena were as an explanation for rapid morphological change at speciation, their reality was (and remains) contested territory (Lande 1980; Barton and Charlesworth 1984; Carson and Templeton 1984; Lande 1985; Newman, Cohen et al. 1985; Lande 1986; Lewin 1986; Coyne 1994; Gavrilets and Hastings 1996; Slatkin 1996; Turelli, Barton et al. 2001; Wu 2001). Punctuated equilibrium had thus taken on the baggage of contentious debates about the genetics of speciation, but it was not in a position to adjudicate that particular set of issues.

In hindsight, Steve himself acknowledged that the tethering of punctuated equilibrium to particular genetic correlates of speciation proved to be a mixed blessing. As he explains in *The Structure of Evolutionary Theory* (2002c), the tethering came through a series of ironic contingencies, most notably the desire to reunite mainstream evolutionary theory with paleontology in the early 1970s. Mainstream models of speciation at that time flowed largely from (or through) Ernst Mayr, who favored conceptions of speciation that were accompanied by unusual genetic events. Here, irony is stacked upon irony: Mayr's predilection for genetic revolutions at speciation may also have derived from his urge to make "speciation theory" its own thing, rather than a descriptive phenomenon that could be fully subsumed by population genetics (Mayr 1944, 1963, 1982; Mayr et al. 1957). In the end, however, it was Doug Futuyma's proposal that speciation acted not as the generator of rapid morphological change, but as its preserver (protecting it from the homogenizing effects of gene flow) (Futuyma 1987, 1988) that struck Steve as the most likely explanation for the correlation (2002c, 800).

In any event, I think it fair to say that Steve felt a certain impatience towards population genetics. That impatience was not directed at the versatility or elasticity of the machinery of population genetics. He was scrupulous in avoiding the chestnut that claimed that because population genetics models could explain everything, they explained nothing (an argument that I would occasionally hear articulated by zealous foot soldiers in the punctuated equilibrium army). He understood that population genetics theory was both powerful and central to evolutionary thought. What mattered, he insisted, was the likelihood and frequency with which the parameter values that produced patterns of punctuated equilibrium actually applied. What really frustrated Steve about population genetics was his perception that its practitioners were close-minded. He sensed in certain population geneticists an *a priori* belief that any evolutionary mechanism that could not be modeled with the clay of population genetics was at best unimportant and at worse, fatally flawed.

But there is a second, and perhaps less obvious point of friction hovering over the development of macroevolutionary theory, or at least Steve's version of it. Population genetics, given its domain of application, is essentially a continuationist discipline. The task of population genetics is to account for the pattern and distribution of extant genetic variation in natural populations, and to elucidate the mechanisms that govern the transmission of heritable information from generation to generation. Seen in this way, the null hypotheses of population genetics are always about continuity, both in space (for instance in the powerful homogenizing effects of gene flow) and in time (Hardy-Weinberg equilibrium is, after all, "Newton's first law" for genetics). Departures from these continuities required postulating additional forces that were either deterministic (e.g., selection) or stochastic (e.g., founder effects).

This continuationist bent—no sudden moves from generation to generation, please—made population geneticists reluctant to contemplate punctuated equilibrium as a distinct phenomenon. Steve felt, correctly, I think, that the continuationist tradition of population genetics had transmogrified into an insistence on gradualism at all scales—no sudden moves anywhere, please.

The traditional perspective from population genetics, extrapolated to the fossil record, thus predicted slow anagenetic change as the dominant motif. Alternatively, strong stabilizing selection in unusually static environments could result in little directional change in populations. The rapid appearance of new morphologies in the record could not be easily reconciled with the expectations of population genetics (but see Newman, Cohen et al. 1985; Lande 1986). In fact, population geneticists were not eager to lend support to either of the patterns that Eldredge and Gould had initially proposed. Stasis, population geneticists argued, was a rare occurrence, limited to habitats where selective forces remained unchanged for millions of years (Stenseth and Smith 1984; Vermeij and Dietl 2006). If selective forces remained unchanged, then organisms, passive executors of the selective will, would remain unchanged as well. But it was hard to imagine millions of years of unchanging stabilizing selection. The alternative, assuming stasis was in fact really all that common, was to treat it as an artifact of low resolution, and hence not in need of a separate explanation. Similarly, the rapid appearance of new morphologies was, according to population geneticists, simply "artifactual": slow directional change over 10,000 generations would be compressed into a single stratigraphic horizon, appearing sudden, but in fact not sudden at all.

The proponents of macroevolution were, of course, not oblivious to the limitations of the fossil record. Many of these proponents were paleontologists, trained to recognize the idiosyncrasies of the record, and dedicated to quantifying its limitations. I remember Steve's frustration on this score rather vividly. He wrote frequently about punctuated equilibrium, spending a considerable amount of time clarifying what was *not* being claimed, including a perfect fossil record that would preserve morphological excursions generation by generation. He would repeatedly reiterate that he meant "rapid" or "sudden" in geological terms, and was perfectly comfortable with the assertion that the morphological change associated with speciation would not have appeared sudden if we had been there watching it happen (e.g., 1982f, 1993j, 2002c, 768; 2007). Time and again, the friction with population genetics centered not so much on the continuationist aspects detailed earlier, but rather on the extrapolation of those central tendencies onto the timescale and texture of the fossil record. Steve, and others, understood that an independent macroevolution was unlikely to succeed if its mechanisms and predictions ran counter to those of population genetics (Eldredge et al. 2005). By the same token, however, the elegance and power of population genetics theory did not give it *de facto* primacy as the explanation of pattern in the history of life. Furthermore, if microevolutionary theory (and in particular population genetics) predicted slow steady change as populations tracked slowly changing environments, then why stasis? Conversely, if the lack of directional trends in the driving environmental (extrinsic) factors predicted stasis, then what were we to make of the rapid appearance of new forms coexisting with ancestral forms? Once again, we could imagine an alternation of slow and rapid environmental changes, and such alternations undoubtedly occur (Hunt 2007). But to argue that every case of punctuation reflected an unusual alternation of selective regimes carried a whiff of special pleading.

In the end, however, the real tension between population genetics and macroevolutionary proposals would not be about the tempo and mode of selective regime change. Instead, the more heated—and ultimately more interesting—discussions would center on the character of the *inputs* into the evolutionary process:

in short, on the nature of mutations. While Steve's thinking and writing on this issue certainly was in flux and did change over time, he remained consistent about two central issues. The first concerned the distinction between the "randomness" of mutations and the isotropy of mutations. Briefly, all evolutionary biologists from 1859 on understood that the Darwinian mechanism required variation to arise independently of the needs of the organism. In that sense, variation was said to be "random"; this was perhaps an unfortunate choice of vocabulary, but no biologist ever uses that term in its strict statistical meaning when speaking about variation. What is more contentious, however, is the notion that variation is both so copious and so without pattern that it could not meaningfully constrain evolutionary change. Darwinism appeared to require that whatever directional change we perceive, in experiments, from observations in living nature, or in the fossil record, could only come about from the hand of selection acting on unpatterned underlying variation.

Yet from his earliest writings, Steve felt it important to underscore the notion that variation could be patterned, and even show directionality, and still remain true to Darwinian logic, provided that such variation did not arise preferentially in response to organismal needs. His conviction that variation was not isotropic (equally probable and abundant in all directions) stemmed from his interest in the patterning—the "constraints"—imposed by organismal development and evolutionary history (see Smith, Burian et al. 1985). He thought the entirely externalist perspective that had come to dominate Anglo-American evolutionary thinking was disappointingly simplistic. Instead, he favored the structuralist and internalist perspectives that had characterized pre-evolutionary continental biology beginning in the early nineteenth century (1989c).

Conceptually, the exploration of the possible patterning and directionality imposed by the raw materials of evolutionary change seemed to many of us a fascinating undertaking. But that undertaking would soon become irretrievably intertwined, perhaps accidentally, with a second more contentious issue: the fate of mutations of large effect. In search of a mechanism that displayed punctuation at other, smaller, scales (and perhaps in a defense of scientists he felt had been unfairly dismissed), Steve sought to reintroduce Goldschmidtian mutations (Goldschmidt 1940) into the vocabulary of evolutionary thinking. The appeal for Steve of this class of mutations, with their large effects on phenotype, was obvious. They were not a prerequisite for punctuated equilibrium, but they did underscore the potential discontinuity between genotypic and phenotypic change. But the fate of these mutations of large effect in any population would equally obviously be guided not by their inherent interest, but by their effect on fitness. Steve's insistence on the role of extremely rare demographic or genetic phenomena (extreme bottlenecks, breakdown of genetic complexes, polyploidization and the like) that would lead—extremely rarely—to the fixation of deleterious macromutations was technically accurate. But it would also give opponents of macroevolution a chance to link punctuated equilibrium to saltation. To be sure, extremely rare events played out over billions of years become virtual certainties, but this argument seemed to miss population genetics' fundamental mistrust of dead ends. Later on, as the molecular underpinnings of development started coming into focus, the argument for mutations of discontinuous effect would become more interesting and more subtle (more on this in the next section).

In the early years of macroevolutionary theory, Steve's attempt to salvage the work of Goldschmidt would stir up the population genetics nest. For one thing, population genetics itself was beginning to accumulate impressive empirical evidence about change at the molecular level (Coyne, Eanes et al. 1979; Kreitman 1983; Lewontin 1985; Kreitman and Wayne 1994). Those data suggested overwhelmingly that change arose primarily in the form of single nucleotide changes, some fraction of which changed the amino acid sequence of proteins. Even when proteins changed, selection appeared not to notice in some proportion of cases, but was surprisingly sensitive and discerning in others. Little seemed to support the notion that mutations of large effect occurred, let alone that they determined the trajectory of evolutionary change.

Much, of course, has changed since, both in the pattern of the data and in its interpretation. We now understand that the techniques we used in the early days of population genetics (e.g., electrophoretic analysis of alleles at enzyme-encoding loci) was biased toward detecting small changes over large ones. The analysis of whole genomes, where all of the data are collected without bias, has revealed a far broader array of possible changes. Duplications,

deletions, expansions, and contractions of significant fractions of
the genome are now commonly described (see for instance Wolfe
and Shields 1997; Friedman and Hughes 2001a, b). More important, our more nuanced understanding of the genetic basis of
morphological change has made us somewhat more circumspect
about the linkage between genotypic and phenotypic change.

Nonetheless, Steve's inclusion of Goldschmidtian hopeful
monsters in early macroevolutionary discussions (e.g., 1980v,
1982h) seems to me a curious political miscalculation. The
political faux pas consisted in letting punctuated equilibrium be
saddled with (or even come close to) Goldschmidt's discredited
speculations about genetic mechanisms. At a time when macroevolution was already fighting on many fronts, opening up yet
another battlefield stretched the troops thin. To be fair, Eldredge
and Gould did not intend to invoke saltational mutations as the
mechanisms for punctuated equilibrium. They remained consistent from the outset in arguing that speciation, and not hopeful
monsters, led to rapid change in the fossil record. The discussion
of Goldschmidt was part of a broader exploration of mechanisms
that could generate rapid change or discontinuity at various levels
of biological organization. Perhaps the campaign to reevaluate
Goldschmidt was Steve at his most provocative (see also Allmon,
this volume). Although in later writings Steve would link his
infatuation for Goldschmidt to the clear articulation of developmental constraints in *The Material Basis of Evolution* (Goldschmidt
1940), I think for a while saltation seemed genuinely attractive to
him (though unnecessary to the operation of punctuated equilibrium). But once again, Steve the intellectual optimist assumed
that the papers in which he mentioned Goldschmidt and hopeful
monsters would be read with as much care as had gone into their
writing. For the most part, they were not. To population geneticists, appeals to hopeful monsters seemed a provocation, and
any theory making even cautious use of the concept was doomed
from the start.

Evo-devo and the return of the organism

In the end, macroevolution's focus on phenotypes, and Steve's
impatience with conservative fitness-driven arguments, set the

stage for the protracted cold war with classical population genetics (e.g., Smith 1983). In contrast, his relationship with developmental genetics, with its emphasis on mechanisms and phenotypic outcomes (rather than on the fate of mutations in populations) tended to be far more cordial.

The ways in which genetic instructions for the making of proteins gave rise to functioning, evolving, interacting organisms was (and is) perhaps the central problem of post-Darwinian biology. The resolution of the "genotype-phenotype problem" was critical to the macroevolutionary enterprise, and specifically to any theory that sought to introduce some breathing room or discontinuity between changes at the DNA level and their expression in phenotypes. At one level, we all knew the glib answer to "the genotype-phenotype problem": development. The process of development, after all, is where sense was made of genetic instructions, where the individual notes (genes) came together as a symphony (the organism). Yet beneath the certainty that "development" explained all lay considerable ignorance. To be sure, the rich traditions of embryology and comparative anatomy had produced an impressive body of description. From there, a set of underlying similarities and regularities had begun to emerge, as had the notion that the evolutionary history of organisms could be discerned from their development (e.g., Haeckel 1866; Cope 1887; 1977e). That recapitulationist perspective would prove didactically useful, but misleading in detail. By the 1980s, however, development had begun to yield to reductionist attack, and developmentally important genes and pathways were being elucidated (McGinnis, Garber et al. 1984; Shepherd, McGinnis et al. 1984; Harding, Wedeen et al. 1985; McGinnis, Kuziora et al. 1990).

The rise of molecular developmental biology was a source of great delight to Steve (1997m). As the resident neontologist in his lab group in the mid-1980s, I was often called upon to present at lab meeting on some sequence-containing paper that had caught Steve's attention. Many of those papers, as I recall, dealt with transcription factors or with alterations in regulatory regions that resulted in large changes in the expression or operation of particular genes. But nothing generated as much excitement as the trickle of papers (soon to be a flood) that dealt with homeobox, or "Hox," genes. Even in the 1980s, most biologists realized that Hox genes were telling us something important about development and evolution, but the discovery of this class of transcription factors had particular meaning for our understanding of body plans.

Homeotic mutations in *Drosophila*—mutations that transformed one body segment into another, or that resulted in the appearance of complete appendages in inappropriate locations (e.g., legs where antennae should be)—had been known since the early 1900s (Bridges and Brehme 1944). By the 1950s, E. B. Lewis (Lewis 1951), Richard Goldschmidt (yes, the same one) (Goldschmidt 1952) and others had identified developmentally important genes; mutations in those genes led to dramatic alterations in phenotype. Beginning in the 1970s, the genes responsible for these homeotic transformations were being identified and characterized. The picture that was emerging was shockingly unexpected: the genes in *Drosophila* that determined segment identity were transcription factors, arranged in a cluster that contained multiple Hox genes. Furthermore, the arrangement of genes in the cluster was collinear with their pattern of expression: genes at the front end of the cluster were expressed in the "front end" of the developing fly embryo; genes at the back end of the cluster were expressed at the "back end" of the developing fly. Finally, the temporal order of expression also proceeded from the anterior (3') end of the cluster to the posterior (5' end) (Kmita and Duboule 2003). The pattern of expression underscored the fact that elements of the cluster were regulating one another; the consequences of mutation in a Hox gene suggested that these genes sat at the apex of a regulatory cascade that would ultimately determine the fate and decoration of individual segments in the developing fly. By the mid-1980s, the Hox cluster was no longer a fascinating idiosyncrasy in a model organism. Instead, homologous genes (and clusters) were now being identified and isolated in vertebrates. The conserved sequence of the homeobox motif and the conserved organization of the cluster left little doubt that this homology was profound in both senses of the word: dating back deep in time and of major import (Garcia-Fernandez 2005; Lemons and McGinnis 2006).

The significance of this discovery for our understanding of the emergence of biological form was hard to overestimate: here was a generalized and flexible tool for the coordinated control of expression in developmentally important genes. This precise tool had been pressed into use in the determination of segmental identity in arthropods and in the specification of the anterior-posterior axis in vertebrates. After more than 500 million years of divergence, the tool remained recognizably similar in both lineages and was put to similar use. The implications of this discovery delighted Steve, who felt that perhaps we had been too quick to dismiss the homology between arthropod segments and vertebrate rhombomeres first suggested 175 years earlier by Geoffroy de St. Hilaire (Geoffroy Saint-Hilaire 1837). And the surprises kept on coming: the utility and versatility of the Hox cluster made it the exaptation of choice when "position" need to be specified. In vertebrates, for instance, a Hox cluster would be recruited to help organize the shoulder-to-fingertip axis of the developing vertebrate limb (see, e.g., Panganiban, Sebring et al. 1995; Panganiban, Irvine et al. 1997; Shubin, Tabin et al. 1997).

More important, the discovery of Hox clusters, their conservation, and their distribution across all animal phyla provided unexpected support for the notion of patterned or constrained morphological innovation. The picture emerging from the new discipline of evolutionary-developmental biology—"evo-devo"— seemed explicitly Chomskian. Beginning in the 1950s, linguistic models proposed by Noam Chomsky and others postulated the existence of a generative grammar that underlay the immense (but nonisotropic) diversity of human languages (Chomsky 1965). Here was a similar apparatus to account for diversity of biological forms: a single generative engine (the Hox cluster) capable of providing extensive, but constrained morphological variety (Carroll 1995; Warren and Carroll 1995). For someone (like Steve) who was fascinated by the possibility that developmental pathways did more than just crank out isotropic variation on which selection could act, the Hox cluster seemed like a gift.

The antiquity and versatility of the Hox cluster was also powerful ammunition in the debate about the explosion of body plans that occurred in the early Cambrian (Knoll and Carroll 1999). The traditional accretionist argument assumed that the emergence of a new body plan required the painstaking assembly of a series of genetic innovations that would eventually result in novel architecture (refs). This model was plausible, but inevitably led to difficulties when trying to account for the near-simultaneous origin of multiple, and radically different body plans. This conundrum required postulating either: (1) that the Cambrian explosion was not real, but simply the eventual culmination of a process of differentiation that had been set in motion far earlier; (2) that something dramatic had occurred at the beginning of the Cambrian, hastening the origin and assembly of multiple *Baüplane*-specific gene complexes, or (3) that the dramatic differentiation of *Baüplane* was easier to achieve than we had imagined, as it arose from tinkering (the ancestral Hox cluster). The discovery of the Hox cluster and its operation suggested that the emergence of new body plans did not depend on the slow accretion of new genetic information. Instead, the rampant diversification emerged from changes in the regulation of expression of one or more members of the Hox cluster, a cluster already present in a common bilaterian ancestor long before the Cambrian (Hughes 2000; 1989b). Changes in the timing and domain of expression of one or more members of the Hox cluster had important and dramatic consequences on form. In the comparatively under-occupied world of the Cambrian seas, a thousand forms could bloom (Knoll and Carroll 1999; Carroll, Grenier et al. 2005).

While I was his student in the mid-late 1980s, Steve would occasionally express understandable frustration with what he saw as the thoughtless stampede of the life sciences towards molecular and mechanistic biology, at the expense of organismal and comparative approaches. But his dismay had more to do with the sociology of the field and with the occasional patronizing tone of some of our molecular colleagues. In truth, he found the data emerging from molecular investigations fascinating, even if he frequently was amazed at how authors seemed to miss the most interesting implications of their own work. He understood early on that macroevolutionary theory at the very least needed to be compatible with the insights of molecular biology. More importantly, he understood that phenomenology without mechanism could only go so far. Evo-devo seemed to Steve (and of course to others as well) to promise a way out of a descriptive emphasis on outcomes and to offer a path towards an exploration of the underlying mechanisms responsible for both the conspicuous similarities and the startling differences in the form of organisms.

Even in its early phase, evo-devo was already offering up insights that were crucial to any theory of evolutionary change. Thus, for instance, the mutational analysis of developmentally important genes suggested that small changes in the sequence of these genes (or of their regulatory regions) could result in dramatic and unexpected phenotypic effects (Lewis 1994; Carroll, Grenier et al. 2005). Although, as we stated earlier, these data did not necessarily address the fitness consequences of such dramatic phenotypic shifts, the redundancy of developmental pathways began to suggest a way in which fitness effects could be masked or buffered. Secondly, control mechanisms like the Hox cluster gave immediate meaning to somewhat elusive notions of constraint. Given the rules of Hox operation, the identity of specific segments could be changed by altering the domain of expression of particular genes, but in an orderly and predictable (a.k.a. constrained) way. Given the temporal and colinear order of Hox gene expression within a cluster, shutting off one or more genes in a cluster causes anterior segments to move to the rear of the developing organism. Conversely, increasing the expression of one or more genes in the cluster will usually result in posterior segments moving towards the front. Considerable evidence suggested that many of the morphological innovations propitiated by alterations in the expression of the Hox cluster depended on reductions or restrictions in the domain of expression of Hox genes. This reduction and specialization of expression domains, coupled with the combinatorial potential of Hox clusters, seemed to supply an explanation for many of the major morphological transitions in the history of life.

I understand the pleasure that Steve took in finding our preconceptions tweaked by the discovery of how things actually worked. It might make sense to imagine that increasingly complex body plans required increasingly complex generating mechanisms (in the form of additional autapomorpic Hox genes), but it ain't necessarily so (Gilbert, Opitz et al. 1996; Popodi, Kissinger et al. 1996). He was even delighted when the notion that Hox clusters had remained together, intact and collinear—an article of faith in molecular developmental biology—began to fall apart as additional organisms were examined in detail. One of our last conversations touched on this point. So many surprises out there once your start looking, he suggested, and so many implications to the idea that the organization of Hox clusters themselves might be evolving in unexpected ways. We now know that Hox clusters are dynamic, as individual elements and clusters are lost or duplicated. On occasion, individual Hox genes escape the tyranny of colinearity and either preserve their ancestral function or are recruited into a novel developmental scheme (Hsia and McGinnis 2003; Seo, Edvardsen et al. 2004; Cameron, Rowen et al. 2006).

Steve would go to some pains to explain that the existence of the Hox cluster did not result in "constraint" in the vernacular sense (of something desirable that is prevented from happening). Rather, the developmental system, internal to the organism, guides or channels variation. Selection still has a huge role to play, eliminating the unfit (most *Drosophila* homeotic mutants are very, very unhappy) and sorting among the available variants based on their fitness. But this is no longer the organism-as-billiard-ball, free to move in any and all directions and guided solely by the strike of the cue. Instead, using Steve's favorite metaphor (borrowed from Darwin's cousin, Francis Galton) we are in the domain of organism-as-polyhedron. You can still push a polyhedron around a pool table, but not in any direction, and not without resistance (1992f, 2002c).

Steve's fascination with evo-devo was clearly linked to an intellectual agenda that hoped to replace the sterile caricature of organisms as collections of adaptations reflecting the power of natural selection with a more majestic and integrated conception of organisms as active players in the evolutionary game. Ironically, macroevolutionary theory had at the same time been attracting significant criticism for its efforts to "individuate" species. Any suggestion that biological entities above the level of the organism exhibited coherence over evolutionary time—and could thus be the target of sorting mechanisms—worried traditionalists. Macroevolution as a process appeared to challenge a cornerstone of modern evolutionary theory: the centrality of the organism as the only target of natural selection. Although that very orthodoxy was being challenged from below by notions of "the gene" as the coherent unit of study, efforts to imagine sorting or selection above the individual level were viewed with distinct suspicion (an argument best articulated by G. C. Williams's 1966 *Adaptation and Natural Selection*) (Williams 1966). Yet despite the ostensible primacy of the individual organism as the crucible of evolutionary forces, the conception of the organism seemed strangely two-dimensional. This externalist conception—where forces outside of the organism played the main shaping role—struck Steve as depauperate. Evo-devo held out the promise that we would understand the mechanisms of development in a comparative setting. In so doing, the ways in which an internal process (development) played a role in shaping variation and in guiding evolutionary trajectories would finally be revealed (Bolker and Raff 1996).

Steve expressed particular joy at the discoveries surrounding the genetic underpinnings of eye development. The revelation that an ectopically activated *Drosophila* homeotic gene (*eyeless*) could result in the formation of "eyes" at various locations in the adult *Drosophila* body (legs, tips of antennae, etc.) was exciting in itself (Halder, Callaerts et al. 1995). But the demonstration that the same effect could be achieved by the ectopic expression of the mouse homolog (*Pax-6*) raised the stakes (Gehring 1996). This was now treading on sacred ground: the eye had, since Darwin, been the quintessential example of evolutionary convergence (Salvini-Plawen and Mayr 1977). Convergence, in turn, was seen as compelling evidence for natural selection's power to continually find the globally optimal solution even when presented with vastly different starting points. Alternatively, if notions of constraint were to enter the narrative, they too were extrinsic to the organism. Eyes converged on similar architectures because the physics of light, optics and transduction left selection little room to maneuver. But if a homolog of *eyeless* could be found in mice, and that homolog could still trigger the developmental cascade that resulted in a *Drosophila* eye, the narrative required dramatic revision. Suddenly, the convergence embodied by arthropod eyes, cephalopod eyes, vertebrate eyes seemed the result of deep, shared developmental pathways (Gehring 1996; Tomarev, Callaerts et al. 1997). Convergence had become parallelism. To be sure, selection acted powerfully on both the pathways and their outcomes. But the options set before natural selection now reflected the operation of shared developmental constraints. The eyes in different phyla, similar in overall structure, fascinatingly different in detail, were no longer the refection of formless lumps of clay consistently and independently molded by selection into similar shapes. The organism, its history, and its developmental processes all mattered once again.

As argued elsewhere in this volume, almost all of Steve's thinking and writing was connected via a fairly small number of themes.

Nowhere is this more clearly illustrated than in evo-devo. In retrospect we can see that the very nature of development—cascades,
levels of regulation, small changes amplified (or dampened) by the
complexity of the system itself—all appealed to Steve's hierarchical
conception of the evolutionary process. As he saw it, molecular
biology, the most reductionist of approaches, was yielding up the
most compelling examples of properties and outcomes that could
not be predicted from the behavior of individual components.
Steve viewed evo-devo as an unexpected but welcome vindication
of his view of life: richly layered, contingent, and maddeningly, but
not hopelessly, complex.

References

Barton, N. H., and B. Charlesworth. 1984. Genetic revolutions founder
    effects and speciation. *Annual Review of Ecology and Systematics* 15:
    133–64.
Bolker, J. A., and R. A. Raff. 1996. Developmental genetics and tradi-
    tional homology. *Bioessays* 18(6): 489–94.
Bridges, C. B., and K. S. Brehme. 1944. The mutants of *Drosophila mela-
    nogaster*. Washington, DC: Carnegie Institution.
Cameron, R. A., L. Rowen, R. Nesbitt, S. Bloom, J. P. Rast, K. Berney,
    L. C. Arenas-Mena, P. Martinez, S. Lucas, P. M. Richardson, E. H.
    Davidson, K. J. Peterson, and L. Hood. 2006. Unusual gene order
    and organization of the sea urchin hox cluster. *Journal of Experimental
    Zoology*. Part B: Molecular and Developmental Evolution 306(1):
    45–58.
Carroll, S. B., J. K. Grenier, and S. D Weatherbee. 2005. *From DNA to
    diversity: molecular genetics and the evolution of animal design*. Malden,
    MA: Blackwell Publishers.
Cheetham, A. H. 1986. Tempo of evolution in a Neogene bryozoan rates
    of morphologic change within and across species boundaries. *Paleo-
    biology* 12(2): 190–202.
Chomsky, N. 1965. *Aspects of the theory of syntax*. Cambridge, MA: M.I.T.
    Press.
Cope, E. D. 1887. *The origin of the fittest: Essays on evolution*. New York:
    Appleton.
Coyne, J. A. 1994. Ernst Mayr and the origin of species. *Evolution* 48:
    19–30.
Coyne, J. A., W. F. Eanes, and R. C. Lewontin. 1979. The genetics of elec-
    trophoretic variation. *Genetics* 92(1): 353–60.
Eldredge, N., J. N. Thompson, P. M. Brakefield, S. Gavrilets, D. Jablonski,
    J. B. C. Jackson, R. E. Lenski, B. S. Lieberman, M. A. McPeek, and
    W. Miller III. 2005. The dynamics of evolutionary stasis. *Paleobiology*
    31: 133–45.

Fortey, R. A. 1985. Gradualism and punctuated equilibria as competing and complementary theories. *Special Papers in Paleontology* 33: 17–28.

Friedman, R., and A. L. Hughes. 2001a. Gene duplication and the structure of eukaryotic genomes. *Genome Research* 11(3): 373–81.

———. (2001b). Pattern and timing of gene duplication in animal genomes. *Genome Research* 11(11): 1842–47.

Futuyma, D. J. 1987. On the role of species in anagenesis. *The American Naturalist* 130: 465–73.

———. 1988. Sturm und Drang and the evolutionary synthesis. *Evolution* 42: 217–26.

Garcia-Fernandez, J. 2005. The genesis and evolution of homeobox gene clusters. *Nature Review of Genetics* 6(12): 881–92.

Gavrilets, S., and A. Hastings. 1996. Founder effect speciation: a theoretical reassessment. *The American Naturalist* 147(3): 466–91.

Gehring, W. J. 1996. The master control gene for morphogenesis and evolution of the eye. *Genes to Cells* 1(1): 11–15.

Geoffroy Saint-Hilaire, I. 1837. Histoire générale et particulière des anomalies de l'organisation chez l'homme et les animaux. ou, Traité de tératologie. Bruxelles, Société belge de librairie, etc. Hauman, Cattoir et ce.

Gilbert, S. F., J. M. Opitz, and R. A. Raff. 1996. Resynthesizing evolutionary and developmental biology. *Developmental Biology* 173(2): 357–72.

Goldschmidt, R. B. 1940. *The material basis of evolution.* New Haven, CT: Yale University Press.

Haeckel, E. 1866. *Generelle Morphologie der Organismen: allgemeine Grundzüge der organischen Formen-Wissenschaft, mechanisch begründet durch die von Charles Darwin reformirte Descendenz-Theorie.* Berlin: G. Reimer.

Halder, G., P. Callaerts, and W. J. Gehring. 1995. Induction of ectopic eyes by targeted expression of the eyeless gene in *Drosophila*. *Science* 267: 1788–92.

Harding, K., C. Wedeen, W. McGinnis, and M. Levine. 1985. Spatially regulated expression of homeotic genes in *Drosophila*. *Science* 229: 1236–42.

Hsia, C. C., and W. McGinnis. 2003. Evolution of transcription factor function. *Current Opinions in Genetics and Development* 13(2): 199–206.

Hughes, N. C. 2000. The rocky road to Mendel's play. *Evolution and Development* 2(2): 63–66.

Hunt, G. 2007. The relative importance of directional change, random walks, and stasis in the evolution of fossil lineages. *Proceedings of the National Academy of Sciences USA* 104(47): 18404–8.

Kmita, M., and D. Duboule. 2003. Organizing axes in time and space: 25 years of colinear tinkering. *Science* 301: 331–33.

Knoll, A. H., and S. B. Carroll. 1999. Early animal evolution: emerging views from comparative biology and geology. *Science* 284: 2129–37.

Kreitman, M. 1983. Nucleotide polymorphism at the alcohol dehydrogenase locus of *Drosophila melanogaster*. *Nature* 304: 412–17.

Kreitman, M., and M. L. Wayne. 1994. Organization of genetic variation at the molecular level: lessons from *Drosophila*. *Exs* (Supplement to *Experientia*) 69: 157–83.

Lande, R. 1980. Genetic variation and phenotypic evolution during allopatric speciation. *The American Naturalist* 116: 463–79.

——— 1985. Expected time for random genetic drift of a population between stable phenotypic states. *Proceedings of the National Academy of Sciences, USA* 82(22): 7641–45.

———. 1986. The dynamics of peak shifts and the pattern of morphological evolution. *Paleobiology* 12(4): 343–54.

Lemons, D., and W. McGinnis. 2006. Genomic evolution of Hox gene clusters. *Science* 313: 1918–22.

Lewin, R. 1986. Punctuated equilibrium is now old hat: The rapid changes seen in the fossil record can be accounted for by traditional explanations from population genetics, according to two recent mathematical models. *Science* 231: 672–73.

Lewis, E. B. 1951. Pseudoallelism and gene evolution. *Cold Spring Harbor Symposia in Quantitative Biology* 16: 159–74.

———. 1994. Homeosis: the first 100 years. *Trends in Genetics* 10(10): 341–43.

Lewontin, R. C. 1985. Population genetics. *Annual Review of Genetics* 19: 81–102.

Mayr, E. 1944. *Systematics and the origin of species*. New York: Columbia University Press.

———, ed. 1957. The species problem; a symposium presented at the Atlanta meeting of the American Association for the Advancement of Science, December 28–29, 1955. Washington, DC: AAAS.

———. 1963. *Animal species and evolution*. Cambridge: Belknap Press of Harvard University Press.

———. 1982. Speciation and macroevolution. *Evolution* 36(6): 1119–32.

McGinnis, N., M. A. Kuziora, and M. McGinnis. 1990. Human Hox-4.2 and *Drosophila* deformed encode similar regulatory specificities in *Drosophila* embryos and larvae. *Cell* 63(5): 969–76.

McGinnis, N., R. L. Garber, and W. McGinnis. 1984. A homologous protein-coding sequence in *Drosophila* homeotic genes and its conservation in other metazoans. *Cell* 37(2): 403–8.

Newman, C. M., J. E. Cohen, and C. Kipnis. 1985. Neo-darwinian evolution implies punctuated equilibria. *Nature* 315: 400–401.

Panganiban, G., S. M. Irvine, C. Lowe, H. Roehl, L. S. Corley, B. Sherbon, J. K. Grenier, J. F. Fallon, J. Kimble, M. Walker, G. A. Wray, B. J. Swalla, M. Q. Martindale, and S. B. Carroll. 1997. The origin and evolution of animal appendages. *Proceedings of the National Academy of Sciences, USA* 94: 5162–66.

Panganiban, G., A. Sebring, L. M. Nagy, and S. Carroll. 1995. The development of crustacean limbs and the evolution of arthropods. *Science* 270: 1363–66.

Popodi, E., J. C. Kissinger, M. E. Andrews, and R. Raff. 1996. Sea urchin Hox genes: insights into the ancestral Hox cluster. *Molecular Biology and Evolution* 13(8): 1078–86.

Salvini-Plawen, L. V., and E. Mayr. 1977. On the evolution of photoreceptors and eyes. *Evolutionary Biology* 10: 207–63.

Seo, H. C., R. B. Edvardsen, A. D. Maeland, M. Bjordal, M. F. Jensen, A. Hansen, M. Flaat, J. Weissenbach, H. Lehrach, P. Wincker, R. Reinhardt, and D. Chourrout. 2004. Hox cluster disintegration with persistent anteroposterior order of expression in *Oikopleura dioica*. *Nature* 431: 67–71.

Shepherd, J. C., W. McGinnis, A. E. Carrasco, E. M. De Robertis, and W. J. Gehring. 1984. Fly and frog homoeo domains show homologies with yeast mating type regulatory proteins. *Nature* 310: 70–71.

Shubin, N., C. Tabin, and S. Carroll. 1997. Fossils, genes and the evolution of animal limbs. *Nature* 388: 639–48.

Slatkin, M. 1996. In defense of founder-flush theories of speciation. *The American Naturalist* 147(4): 493–505.

Smith, J. M., R. Burian, S. Kauffman, P. Alberch, J. Campbell, B. Goodwin, R. Lande, D. M. Raup, and L. Wolpert. 1985. Developmental constraints and evolution: A perspective from the mountain lake conference on development and evolution. *Quarterly Review of Biology* 60(3): 265–87.

Somit, A., and S. A. Peterson, eds. 1992. *The dynamics of evolution: the punctuated equilibrium debate in the natural and social sciences*. Ithaca, NY: Cornell University Press.

Stanley, S. M. 1975. A theory of evolution above the species level. *Proceedings of the National Academy of Sciences, USA* 72(2): 646–50.

Stenseth, N. C., and J. M. Smith. 1984. Coevolution in ecosystems: red queen evolution or stasis? *Evolution*, 38(4): 870–80.

Tomarev, S. I., P. Callaerts, L. Kos, R. Zinovieva, G. Halder, W. Gehring, and J. Piatigorsky. 1997. Squid Pax-6 and eye development. *Proceedings of the National Academy of Sciences, USA* 94(6): 2421–26.

Turelli, M., N. H. Barton, & J. A. Coyne. 2001. Theory and speciation. *Trends in Ecology and Evolution* 16(7): 330–43.

Vermeij, G. J., and G. P. Dietl. 2006. Majority rule: adaptation and the long-term dynamics of species. *Paleobiology* 32: 173–78.

Warren, R., and S. B. Carroll. 1995. Homeotic genes and diversification of the insect body plan. *Current Opinions in Genetics and Development* 5(4): 459–65.

Williams, G. C. 1966). *Adaptation and natural selection: A critique of some current evolutionary thought*. Princeton, NJ: Princeton University Press.

Wolfe, K. H., and D. C. Shields. 1997. Molecular evidence for an ancient duplication of the entire yeast genome. *Nature* 387: 708–13.

Wu, C.-I. 2001. The genic view of the process of speciation. *Journal of Evolutionary Biology* 14(6): 851–65.

❧ 15 ❧

Bibliography: Stephen Jay Gould

Compiled by Warren D. Allmon

This bibliography includes all of Steve Gould's publications that are known to me other than abstracts. It is based on the list maintained for more than thirty years by his secretary, Agnes Pilot, with additions and corrections from other sources. It is important to note that most of the essays (especially those that originally appeared in *Natural History* magazine) that were later included in the eleven compilations published during Gould's lifetime usually appeared in these volumes revised to greater or lesser degrees and/or including new footnotes or postscripts, and thus often were effectively new publications; titles of many of the essays were also changed in the compilations. I have noted title changes and the volume in which essays were reprinted, but have not otherwise indicated editorial alterations. Also not included are most of the many other reprints or republications of Gould's articles in volumes written or edited by others, nor the many foreign printings and editions of his books. The resulting list includes 814 titles, of which at least 154 (19 percent) were peer-reviewed.

I am grateful to Emily Butler, Fred Collier, Kelly Cronin, Jessica Cundiff, Andrea Kreuzer, and Rosamond Purcell for valuable assistance in assembling this bibliography.

1965a. Interpretation of the coefficient in the allometric equation. *American Naturalist* 99: 5–18. (J. F. White & S. J. Gould)

1965b. Is uniformitarianism necessary? *American Journal of Science* 263: 223–28.

1965c. Reply to C. R. Longwell's criticism of "Is uniformitarianism necessary?" *American Journal of Science* 263: 919–21.

1966a. Notes on shell morphology and classification of the Siliquariidae (Gastropoda): The protoconch and slit of *Siliquaria squamata* Blainville. *American Museum of Natural History Novitates* 2263: 1–13.

1966b. Allometry in Pleistocene land snails from Bermuda: The influence of size upon shape. *Journal of Paleontology* 40: 1131–41.

1966c. Allometry and size in ontogeny and phylogeny. *Biological Reviews* 41: 587–640.

1966d. Asiatic Mesonychidae (Mammalia, Condylarthra). *Bulletin of the American Museum of Natural History* 132: 127–74. (F. S. Szalay & S. J. Gould)

1967a. Evolutionary patterns in pelycosaurian reptiles: A factor-analytic study. *Evolution* 21: 385–401.

1967b. Comments on "The adaptive significance of gastropod torsion." *Evolution* 21: 405–6. (R. L. Batten, H. B. Rollins & S. J. Gould)

1967c. Pleistocene history of Bermuda. *Bulletin of the Geological Society of America* 78: 993–1006. (L .S. Land, F. MacKenzie, & S. J. Gould)

1967d. Is uniformitarianism useful? *Journal of Geological Education* 15: 149–50.

1968a. *Trigonia* and the origin of species. *Journal of the History of Biology* 1: 41–56.

1968b. Ontogeny and the explanation of form: An allometric analysis. In D. B. Macurda, ed., *Paleobiological aspects of growth and development, a symposium.* Paleontological Society Memoir 2 (*Journal of Paleontology* 42(5), suppl.): 81–98.

1968c. Phenotypic reversion to ancestral form and habit in a marine snail. *Nature* 220: 804.

1968d. The molluscan fauna of an unusual Bermudian pond: A natural experiment in form and composition. *Museum of Comparative Zoology Breviora* 308: 1–13.

1969a. Review of M. T. Ghiselin, *Triumph of the Darwinian method. American Scientist* 58: 110.

1969b. Smithsonian's albatross (letter). *Science* 164: 497.

1969c. Character variation in two land snails from the Dutch Leeward Islands: Geography, environment, and evolution. *Systematic Zoology* 18: 185–200.

1969d. Levels of integration in mammalian dentitions: An analysis of correlations in *Nesophontes micrus* (Insectivora) and *Oryzomys couesi* (Rodentia). *Evolution* 23: 276–300. (S. J. Gould & R. A. Garwood)

1969e. Ecology and functional significance of uncoiling in *Vermicularia spirata*: An essay on gastropod form. *Bulletin of Marine Science* 19: 432–45.

1969f. The byssus of trigonian clams: Phylogenetic vestige or functional organ? *Journal of Paleontology* 43: 1125–29.

1969g. An evolutionary microcosm: Pleistocene and Recent history of the land snail *P.* (*Poecilozonites*) in Bermuda. *Bulletin of the Museum of Comparative Zoology* 138: 407–532.

1969h. Land snail communities and Pleistocene climates in Bermuda: A multivariate analysis of microgastropod diversity. *Proceedings of the North American Paleontological Convention*, pt. E, 486–521.

1970a. History vs. prophecy. *American Journal of Science* 268: 187–89.

1970b. Coincidence of climatic and faunal fluctuations in Pleistocene Bermuda. *Science* 168: 572–73.

1970c. Evolutionary paleontology and the science of form. *Earth Science Reviews* 6: 77–119.

1970d. Private thoughts of Lyell on progression and evolution. *Science* 169: 663–64.

1970e. Dollo on Dollo's law: Irreversibility and the status of evolutionary laws. *Journal of the History of Biology* 3(2): 189–212.

1970f. Seeking confluence. Review of G. de Beer, *Streams of culture*. *Science* 168: 717.

1970g. New biological determinism. *Harvard Independent* 2 (9): 8.

1971a. Tübingen meeting on form (report). *Journal of Paleontology* 45: 1042–43.

1971b. D'Arcy Thompson and the science of form. *New Literary History* 2(2): 229–58.

1971c. Environmental control of form in land snails: A case of unusual precision. *Nautilus* 84(3): 86–93.

1971d. The paleontology and evolution of *Cerion*, II: Age and fauna of Indian shell middens on Curaçao and Aruba. *Museum of Comparative Zoology Breviora* 372: 1–26.

1971e. Geometric similarity in allometric growth: A contribution to the problem of scaling in the evolution of size. *American Naturalist* 105: 113–36.

1971f. Muscular mechanics and the ontogeny of swimming in scallops. *Palaeontology* 14(1): 61–94.

1971g. Darwin's "Retreat." Review of P. J. Vorzimmer, *Charles Darwin the years of controversy. The Origin of Species and its critics, 1859–1882*. *Science* 172: 677–78.

1971h. Paleontology. In *McGraw-Hill Encyclopedia of Science and Technology*. Chicago: McGraw-Hill Professional, 588–591.

1971i. Precise but fortuitous convergence in Pleistocene land snails from Bermuda. *Journal of Paleontology* 45: 409–18.

1972a. Jurassic. In *McGraw-Hill Yearbook for Science and Technology*. Chicago: McGraw-Hill Professional, 265–66.

1972b. Zealous advocates. Review of A. Koestler, *The Case of the Midwife Toad*. *Science* 176: 623–25.

1972c. Alpheus Hyatt. In *Dictionary of Scientific Biography*, vol. 6, 613–14. New York: Charles Scribner's Sons.

1972d. Allometric fallacies and the evolution of *Gryphaea*: A new interpretation based on White's criterion of geometric similarity. *Evolutionary Biology* 6: 91–118.

1972e. Punctuated equilibria: An alternative to phyletic gradualism. In T. J. M. Schopf, ed., *Models in paleobiology*, 82–115. San Francisco: Freeman, Cooper & Co. (N. Eldredge & S. J. Gould)

1972f. Geographic variation. *Annual Review of Ecology and Systematics* 3: 457–98. (S. J. Gould & R. F. Johnson)

1973a. Review of A.V. Carozzi (ed. & trans.), *Lucien Cayeux, past and present causes in geology*. *Journal of Geology* 81(1): 125.

1973b. Review of R. A. Reyment, *Quantitative paleoecology*. *Earth Science Reviews* 9(1): 75–76.

1973c. Review of J. Needham, *Order and life*. *Leonardo* 6: 267.

1973d. The misnamed, mistreated, and misunderstood Irish Elk. *Natural History* 82(3): 10–19. (Reprinted in *Ever since Darwin* [1977f]. New York: W. W. Norton, 79–90.)

1973e. T. H. Huxley. Review of C. Bibby, *Scientist extraordinary. The life and scientific work of Thomas Henry Huxley, 1825–1895*. *Science* 181: 47–48.

1973f. Positive allometry of antlers in the "Irish Elk." *Megaloceros giganteus*. *Nature* 244: 375–76.

1973g. Factor analysis of caseid pelycosaurs. *Journal of Paleontology* 47(5): 886–91. (S. J. Gould & J. Littlejohn)

1973h. Review of M. J. S. Rudwick, *The meaning of fossils: episodes in the history of palaeontology*. *Palaeobotany and Palynology* 16: 210–12.

1973i. Systematic pluralism and the uses of history. *Systematic Zoology* 22(3): 322–24.

1973j. Stochastic models of phylogeny and the evolution of diversity. *Journal of Geology* 81(5): 525–42 (D. M. Raup, S. J.Gould, T. J. M. Schopf, & D. S. Simberloff).

1973k. The shape of things to come. *Systematic Zoology* 22(4): 401–4.

1974a. The first decade of numerical taxonomy. Review of P. H. A. Sneath and R. R. Sokal, *The principles and practice of numerical classification*. *Science* 183: 739–40.

1974b. Growth and variation in *Eurypterus remipes* DeKay. *Bulletin of the Geological Institute, University of Uppsala, N.S.* 4(6): 81–114 (H. E. Andrews, J. C. Brower, S. J. Gould, & R. A. Reyment).

1974c. The pallial ridge of *Neotrigonia*: Functional siphons without mantle fusion. *Veliger* 17(1): 1–7. (S. J. Gould & C. C. Jones)

1974d. The origin and function of "bizarre" structures: Antler size and skull size in the "Irish Elk," *Megaloceros giganteus*. *Evolution* 28: 191–220.

1974e. Joseph Augustine Cushman. In *Dictionary of American Biography* 9(2): 205–7.

1974f. On biological and social determinism. *History of Science* 11: 212–20.

1974g. Size and scaling in human evolution. *Science* 186: 892–901. (D. Pilbeam & S. J. Gould)

1974h. Reply to Morphological transformation, the fossil record, and the mechanism of evolution: A debate. Pt. 1. The statement and critique, by M. K. Hecht. *Evolutionary Biology* 7: 295–308. (N. Eldredge & S. J. Gould)

1974i. Stochastic simulation and evolution of morphology—towards a nomothetic paleontology. *Systematic Zoology* 23(3): 305–22. (D. M. Raup & S. J. Gould)

1974j. Evolutionary theory and the rise of American paleontology. *Syracuse University Geological Contribution* 3: 1–16.

1974k. Genetics and morphometrics of *Cerion* at Pongo Carpet: A new systematic approach to this enigmatic land snail. *Systematic Zoology* 23(4): 518–35 (S. J. Gould, D. S. Woodruff, & J. P. Martin).

1974l. Editorial introduction to G.O.G. Greiner, "Environmental factors controlling the distribution of Recent benthonic Foraminifera." *Museum of Comparative Zoology Breviora* 420: 1–2. (S. J. Gould & A. D. Hecht)

1974m. A man who returned man to nature. Review of Howard E. Gruber, *Darwin on man. New York Times Book Review*, July 14.

1974n. Evolution of the brain and intelligence (letter). *Science* 185: 400.

1974o. Issues raised by biology. Review of D. Hull, *Philosophy of biological science. Science* 186: 45–46.

1974p. Size and shape. *Natural History* 83(1): 20–26. (Reprinted in *Ever since Darwin* [1977f]. New York: W. W. Norton, 171–78.)

1974q. Sizing up human intelligence. *Natural History* 83(2): 10–14. (Reprinted in *Ever since Darwin* [1977f]. New York: W. W. Norton, 179–85.)

1974r. The race problem. *Natural History* 83(3): 8–14. (Reprinted [as "Why we should not name human races—a biological view"] in *Ever since Darwin* [1977f]. New York: W. W. Norton, 231–36.)

1974s. The nonscience of human nature. *Natural History* 83(4): 21–24. (Reprinted in *Ever since Darwin* [1977f]. New York: W. W. Norton, 237–42.)

1974t. Racist arguments and I.Q. *Natural History* 83(5): 24–29. (Reprinted in *Ever since Darwin* [1977f]. New York: W. W. Norton, 243–47.)

1974u. Darwin's dilemma. *Natural History* 83(6): 16–22. (Reprinted [as "Darwin's dilemma: the odyssey of evolution"] in *Ever since Darwin* [1977f]. New York: W. W. Norton, 34–38.)

1974v. On heroes and fools in science. *Natural History* 83(7): 30–32. (Reprinted in *Ever since Darwin* [1977f]. New York: W. W. Norton, 201–6.)

1974w. The great dying. *Natural History* 83(8): 22–27. (Reprinted in *Ever since Darwin* [1977f]. New York: W. W. Norton, 134–38.)

1974x. An unsung single celled hero. *Natural History* 83(9): 33–42. (Reprinted in *Ever since Darwin* [1977f]. New York: W. W. Norton, 119–25.)

1974y. Darwin's delay. *Natural History* 83(10): 68–70. (Reprinted in *Ever since Darwin* [1977f]. New York: W. W. Norton, 21–27.)

1974z. The central role of retardation and neoteny in the evolution of man. Paper prepared in advance for participants in Burg Wartenstein Symposium No. 61, Phylogeny of the primates: an interdisciplinary approach. Wenner–Gren Foundation for Anthropological Research.

1975a. Allometry in primates, with emphasis on scaling and the evolution of the brain. In F. Szalay, ed. *Approaches to primate paleobiology. Contributions to Primatology* (Basel: S. Karger). Vol. 5, 244–92.

1975b. Disruption of ideal geometry in the growth of receptaculitids: A natural experiment in theoretical morphology. *Paleobiology* 1(1): 1–20. (S. J. Gould & M. Katz)

1975c. Genomic versus morphologic rates of evolution: Influence of morphologic complexity. *Paleobiology* 1(1): 63–70. (T. J. M. Schopf, D. M. Raup, S. J. Gould & D.S. Simberloff)

1975d. Review of H. J. Jerison, *Evolution of the brain and intelligence. Paleobiology* 1(1): 125–29.

1975e. On the scaling of tooth size in mammals. *American Zoologist* 15: 351–62.

1975f. The evolution of British and American Middle and Upper Jurassic *Gryphaea*: A biometric study. *Proceedings of the Royal Society of London*, B 189: 511–42. (A. Hallam & S. J. Gould)

1975g. Darwin's "big book." Review of R. C. Stauffer, ed., *Charles Darwin's Natural Selection. Being the Second Part of His Big Species Book, written from 1856 to 1858. Science* 188: 824–26.

1975h. *Lift up thy voice with strength: The first 100 years of the Cecilia Society.* Boston: Red Sun Press.

1975i. Size and shape. *Harvard Magazine*, Oct., 43–50.

1975j. The other scientific method. Review of P. Boffey, *The brain bank of America. An inquiry into the politics of science. New York Times*, May 4.

1975k. By the evidence. Review of L. S. B. Leakey, *Memoirs, 1932–1951. New York Times Book Review*, Nov. 17.

1975l. Allometry and early hominids. Reply. *Science* 189: 64.

1975m. Evolution of the brain and intelligence (comment). *Current Anthropology* 16: 409–10.

1975n. We should eliminate prizes to protect scientists from themselves. *Boston Globe*, Nov. 6.

1975o. Introduction. In J. Burns, *Biograffiti*, xiii–xv. New York: W. W. Norton.

1975p. Stochastic simulation of cladistic and phenetic patterns in phylogeny. In G. F. Estabrook, ed., *Proceedings 8th International Conference, Numerical Taxonomy*, 72–75. San Francisco: W. H. Freeman.

1975q. Review of T. F. Glick, *The comparative reception of Darwinism*. *American Scientist* 63: 482.

1975r. A gardener touched with genius. Review of P. Dreyer, *The life of Luther Burbank*. *New York Times Book Review*, Nov. 16, 6.

1975s. Evolution and the brain. *Natural History* 84(1): 24–26. (Reprinted [as "History of the vertebrate brain"] in *Ever since Darwin* [1977f]. New York: W. W. Norton, 186–91.)

1975t. Catastrophe and steady state Earth. *Natural History* 84(2): 14–18. (Reprinted [as "Uniformity and catastrophe"] in *Ever since Darwin* [1977f]. New York: W. W. Norton, 147–52.)

1975u. Velikovsky in collision. *Natural History* 84(3): 20–26. (Reprinted in *Ever since Darwin* [1977f]. New York: W. W. Norton, 153–59.)

1975v. Reverend Burnet's dirty little planet. *Natural History* 84(4): 26–28. (Reprinted in *Ever since Darwin* [1977f]. New York: W. W. Norton, 141–46.)

1975w. The child as man's real father. *Natural History* 84(5): 18–22. (Reprinted in *Ever since Darwin* [1977f]. New York: W. W. Norton, 63–69.)

1975x. Racism and recapitulation. *Natural History* 84(6): 18–25. (Reprinted in *Ever since Darwin* [1977f]. New York: W. W. Norton, 214–21.)

1975y. Man and other animals. *Natural History* 84(7): 24–30. (Reprinted [as "A matter of degree"] in *Ever since Darwin* [1977f]. New York: W. W. Norton, 49–55.)

1975z. Diversity through time. *Natural History* 84(8): 24–32.

1975aa. Posture maketh the man. *Natural History* 84(9): 38–44. (Reprinted in Ever since Darwin [1977f]. New York: W. W. Norton, 207–13.)

1975bb. A threat to Darwinism. *Natural History* 84(10): 4–9.

1976a. Weight and shape. COSPAR Symposium, Life Sciences and Space Research XIV. Berlin Akademie-Verlag, 57–68.

1976b. The genomic metronome as a null hypothesis. *Paleobiology* 2(2): 177–79.

1976c. Palaeontology plus ecology as palaeobiology. In R. M. May, ed., *Theoretical ecology. Principles and applications*, 218–36. Philadelphia: W. B. Saunders Co. (See also 1981z.)

1976d. Grades and clades revisited. In R. B. Masterton, W. Hodos, and H. Jerison, eds., *Evolution, brain and behavior*, 115–22. Hillsdale, NJ: Lawrence Erlbaum Associates.

1976e. In defense of the analog: A commentary to N. Hotton. In R. B. Masterton et al., *Evolution, brain and behavior*, 175–79. Hillsdale, NJ: Lawrence Erlbaum Associates.

1976f. The stately mansions of the Radiolaria. *Horizon* 18(2): 79–81.

1976g. The procession begins. In *Our continent. A natural history of North America*, 52–73.Washington: National Geographic Society.

1976h. Reply to P. Dreyer. *New York Times Book Review*, Jan. 11, 28.

1976i. Review of A. J. Boucot, *Evolution and extinction rate controls. Sedimentary Geology* 15: 213.

1976j. Darwin and the Captain. *Natural History* 85(1): 32–34. (Reprinted [as "Darwin's sea change, or five years at the Captain's table"] in *Ever since Darwin* [1977f]. New York: W. W. Norton, 28–33.)

1976k. Human babies as embryos. *Natural History* 85(2): 22–26. (Reprinted in *Ever since Darwin* [1977f]. New York: W. W. Norton, 70–75.)

1976l. Criminal man revived. *Natural History* 85(3): 16–18. (Reprinted [as "The criminal as nature's mistake, or the ape in some of us"] in *Ever since Darwin* [1977f]. New York: W. W. Norton, 222–28.)

1976m. Ladders, bushes, and human evolution. *Natural History* 85(4): 24–31. (Reprinted [as "Bushes and ladders in human evolution"] in *Ever since Darwin* [1977f]. New York: W. W. Norton, 56–62.)

1976n. Biological potential vs. biological determinism. *Natural History* 85(5): 12–22. (Reprinted in *Ever since Darwin* [1977f]. New York: W. W. Norton, 251–59.)

1976o. The five kingdoms. *Natural History* 85(6): 30–37. (Reprinted [as "The pentagon of life"] in *Ever since Darwin* [1977f]. New York: W. W. Norton, 113–18.)

1976p. The interpretation of diagrams. *Natural History* 85(7): 18–28. (Reprinted [as "Is the Cambrian explosion a sigmoid fraud?"] in *Ever since Darwin* [1977f]. New York: W. W. Norton, 126–33.)

1976q. Darwin's untimely burial. *Natural History* 85(8): 24–30. (Reprinted in *Ever since Darwin* [1977f]. New York: W. W. Norton, 39–45.)

1976r. So cleverly kind an animal. *Natural History* 85(9): 32–36. (Reprinted in *Ever since Darwin* [1977f]. New York: W. W. Norton, 260–67.)

1976s. The advantages of eating Mom. *Natural History* 85(10): 24–31. (Reprinted [as "Organic wisdom, or why should a fly eat its mother from inside"] in *Ever since Darwin* [1977f]. New York: W. W. Norton, 91–96.)

1977a. The shape of evolution: A comparison of real and random clades. *Paleobiology* 3(1): 23–40 (S. J. Gould, D. M. Raup, J. J. Sepkoski Jr., T. J. M. Schopf, & D. S. Simberloff).

1977b. Eternal metaphors of palaeontology. In A. Hallam, ed., *Patterns of evolution as illustrated by the fossil record*, 1–26. Amsterdam: Elsevier.

1977c. Punctuated equilibria: The tempo and mode of evolution reconsidered. *Paleobiology* 3(2): 115–51. (S. J. Gould & N. Eldredge)

1977d. Evolutionary models and biostratigraphic strategies. In E. G. Kauffman & J. E. Hazel, eds., *Concepts and methods of biostratigraphy*,

25–40. Stroudsburg, PA: Dowden, Hutchinson & Ross. (N. Eldredge & S. J. Gould)

1977e. *Ontogeny and phylogeny*. Cambridge, MA: Harvard University Press.

1977f. *Ever since Darwin*. New York: W. W. Norton.

1977g. Biological interaction between fossil species: Character displacement in Bermudian land snails. *Paleobiology* 3: 259–69. (D. E. Schindel & S. J. Gould)

1977h. Natural history of *Cerion*. VII. Geographic variation of Cerion (Mollusca: Pulmonata) from the eastern end of its range (Hispaniola to the Virgin Islands): Coherent patterns and taxonomic simplification. *Museum of Comparative Zoology Breviora* 445: 1–24. (S. J. Gould & C. Paull)

1977i. Review of A. Nisbett, *Konrad Lorenz*. *New York Times Book Review*, Feb. 27, 2, 22.

1977j. Progressionism. Review of P. J. Bowler, *Fossils and progress*. *Science* 196: 517–18.

1977k. Review of P. B. and J. S. Medawar, *The life science*. *New York Times Book Review*, May 22, 11, 20.

1977l. Review of P. H. Barrett, *The collected papers of Charles Darwin*. *The Sciences* 17(5): 28.

1977m. Review of A. G. Roper, *Ancient eugenics*. *Isis* 68: 626–27.

1977n. The problem of perfection. *Natural History* 86(1): 32–35. (Reprinted [as "The problem of perfection, or how can a clam mount a fish on its rear end?"] in *Ever since Darwin* [1977f]. New York: W. W. Norton, 103–10.)

1977o. The continental drift affair. *Natural History* 86(2): 12–18. (Reprinted [as "The validation of continental drift"] in *Ever since Darwin* [1977f]. New York: W. W. Norton, 160–67.)

1977p. Twin-engined spaceship Earth. *Natural History* 86(3): 72–76. (Reprinted [as "Planetary sizes and surfaces"] in *Ever since Darwin* [1977f]. New York: W. W. Norton, 192–98.)

1977q. The 120–year bamboo clock. *Natural History* 86(4): 8–16. (Reprinted [as "Of bamboos, cicadas, and the economy of Adam Smith"] in *Ever since Darwin* [1977f]. New York: W. W. Norton, 97–102.)

1977r. Evolution's erratic pace. *Natural History* 86(5): 12–16. (Reprinted [as "The episodic nature of evolutionary change"] in *The panda's thumb* [1980h]. New York: W. W. Norton, 179–85.)

1977s. The return of hopeful monsters. *Natural History* 86(6): 22–30. (Reprinted [as "The return of the hopeful monster"] in *The panda's thumb* [1980h]. New York: W. W. Norton, 186–93.)

1977t. Our allotted lifetimes. *Natural History* 86(7): 34–41. (Reprinted in *The panda's thumb* [1980h]. New York: W. W. Norton, 299–305.)

1977u. Sticking up for marsupials. *Natural History* 86(8): 22–30. (Reprinted in *The panda's thumb* [1980h]. New York: W. W. Norton, 289–95.)

1977v. The telltale wishbone. *Natural History* 86(9): 26–36. (Reprinted in *The panda's thumb* [1980h]. New York: W. W. Norton, 267–77.)

1977w. Caring groups and selfish genes. *Natural History* 86(10): 20–24. (Reprinted in *The panda's thumb* [1980h]. New York: W. W. Norton, 85–92.)

1977x. Review of Y. Coppens, F. C. Howell, G. L. Isaac, and R. E. F. Leakey, *Earliest man and environments in the Lake Rudolf Basin: Stratigraphy, paleoecology, and evolution. Quarterly Review of Biology* 52(4): 464–65.

1978a. Generality and uniqueness in the history of life: An exploration with random models. *BioScience* 28(4): 277–81.

1978b. Koestler's solution. Review of A. Koestler, *Janus: A summing up. New York Review of Books* 25(6): 35–37.

1978c. Morton's ranking of races by cranial capacity. *Science* 200: 503–9.

1978d. The finagle factor. *Human Nature* 1 (July): 80–87.

1978e. Natural history of *Cerion*. VIII: Little Bahama Bank—a revision based on genetics, morphometrics, and geographic distribution. *Bulletin of the Museum of Comparative Zoology* 148(8): 371–415. (S. J. Gould & D. S. Woodruff)

1978f. Heroes in nature. Review of J. Kastner, *A species of eternity. New York Review of Books* 25(14): 31–32.

1978g. Review of E. O. Wilson, *On human nature. Human Nature*, Oct. 1978, 20–28.

1978h. Sociobiology: The art of storytelling. *New Scientist* 80(1129): 530–33.

1978i. Nature's odd couples. *Natural History* 87(1): 38–41. (Reprinted in *The panda's thumb* [1980h]. New York: W. W. Norton, 278–88.)

1978j. An early start. *Natural History* 87(2): 10–24. (Reprinted in *The panda's thumb* [1980h]. New York: W. W. Norton, 217–26.)

1978k. Crazy old Randolph Kirkpatrick. *Natural History* 87(3): 20–26. (Reprinted in *The panda's thumb* [1980h]. New York: W. W. Norton, 227–35.)

1978l. Bathybius meets Eozoon. *Natural History* 87(4): 16–22. (Reprinted in *The panda's thumb* [1980h]. New York: W. W. Norton, 236–44.)

1978m. Were dinosaurs dumb? *Natural History* 87(5): 9–16. (Reprinted in *The panda's thumb* [1980h]. New York: W. W. Norton, 259–66.)

1978n. Flaws in a Victorian veil. *Natural History* 87(6): 16–26. (Reprinted in *The panda's thumb* [1980h]. New York: W. W. Norton, 169–76.)

1978o. The great scablands debate. *Natural History* 87(7): 12–18. (Reprinted in *The panda's thumb* [1980h]. New York: W. W. Norton, 194–203.)

1978p. Women's brains. *Natural History* 87(8): 44–50. (Reprinted in *The panda's thumb* [1980h]. New York: W. W. Norton, 152–59.)

1978q. The panda's peculiar thumb. *Natural History* 87(9): 20–30. (Reprinted [as "The panda's thumb"] in *The panda's thumb* [1980h]. New York: W. W. Norton, 19–26.)

1978r. Senseless signs of history. *Natural History* 87(10): 22–28. (Reprinted in *The panda's thumb* [1980h]. New York: W. W. Norton, 27–34.)

1978s. Review of T. J. Pedley, ed., *Scale effects in animal locomotion. Quarterly Review of Biology* 53(4): 473–74.

1979a. Exultation and explanation. Review of G. E. Hutchinson, *The kindly fruits of the Earth: Recollections of an embryo ecologist* and *An introduction to population ecology. New York Review of Books* 26(8): 3–6. (Reprinted in *An urchin in the storm* [1987f]. New York: W. W. Norton, 180–88.)

1979b. Review of L. Thomas, *The medusa and the snail. More notes of a biology watcher. New York Times Book Review*, May 6, 1, 32–33.

1979c. On the importance of heterochrony for evolutionary biology. *Systematic Zoology* 28(2): 224–26.

1979d. Darwin vindicated! Review of L. Eiseley, *Darwin and the mysterious Mr. X: New light on the evolutionists. New York Review of Books* 26(13): 36–38. (Reprinted [as "Misserving memory"] in *An urchin in the storm* [1987f]. New York: W. W. Norton, 51–61.)

1979e. The morphology of a "hybrid zone" in *Cerion*: Variation, clines, and an ontogenetic relationship between two "species" in Cuba. *Evolution* 33(2): 714–27. (L. Galler & S. J. Gould)

1979f. Agassiz's marginalia in Lyell's *Principles*, or the perils of uniformity and the ambiguity of heroes. In W. Coleman & C. Limoges, eds., *Studies in history of biology*, 119–38. Baltimore: The Johns Hopkins University Press.

1979g. Agassiz's later, private thoughts on evolution: His marginalia in Haeckel's "Natürliche Schöpfungsgeschichte" (1868). In C. J. Schneer, ed., *Two hundred years of geology in America*, 277–82. Hanover, NH: University Press of New England.

1979h. Dreamer. Review of F. Dyson, *Disturbing the universe. New York Review of Books* 26(15): 3–4. (Reprinted [as "Pleasant dreams"] in *An urchin in the storm* [1987f]. New York: W. W. Norton, 199–207.)

1979i. Size and shape in ontogeny and phylogeny. *Paleobiology* 5(3): 296–317. (P. Alberch, S. J. Gould, G. Oster & D. Wake)

1979j. An allometric interpretation of species–area curves: The meaning of the coefficient. The *American Naturalist* 114(3): 335–43.

1979k. The spandrels of San Marco and the Panglossian paradigm: A critique of the adaptationist programme. *Proceedings of the Royal Society of London*, B 205: 581–98. (S. J. Gould & R. C. Lewontin)

1979l. *The history of paleontology*. First announcement of 34 books in 43 volumes. S. J. Gould, advisory ed.; T. J. M. Schopf, consulting ed. New York: Arno Press.

1979m. A Darwinian paradox. *Natural History* 88(1): 32–44. (Reprinted [as "Double trouble"] in *The panda's thumb* [1980h]. New York: W. W. Norton, 35–44.)

1979n. Wide hats and narrow minds. *Natural History* 88(2): 34–40. (Reprinted in *The panda's thumb* [1980h]. New York: W. W. Norton, 145–51.)

1979n. Piltdown revisited. *Natural History* 88(3): 86–97. (Reprinted in *The panda's thumb* [1980h]. New York: W. W. Norton, 108–24.)

1979o. Time's vastness. *Natural History* 88(4): 18–27. (Reprinted in *The panda's thumb* [1980h]. New York: W. W. Norton, 315–23.)

1979p. Mickey Mouse meets Konrad Lorenz. *Natural History* 88(5): 30–36. (Reprinted [as "A biological homage to Mickey Mouse"] in *The panda's thumb* [1980h]. New York: W. W. Norton, 95–107.)

1979q. Our greatest evolutionary step. *Natural History* 88(6): 40–44. (Reprinted in *The panda's thumb* [1980h]. New York: W. W. Norton, 125–33.)

1979r. A quahog is a quahog. *Natural History* 88(7): 18–26. (Reprinted in *The panda's thumb* [1980h]. New York: W. W. Norton, 204–13.)

1979s. Shades of Lamarck. *Natural History* 88(8): 22–28. (Reprinted in *The panda's thumb* [1980h]. New York: W. W. Norton, 76–84.)

1979t. Perceptive bees, birds, and bacteria. *Natural History* 88(9): 25–30. (Reprinted [as "Natural attraction: bacteria, the birds and the bees"] in *The panda's thumb* [1980h]. New York: W. W. Norton, 306–14.)

1979u. Darwin's middle road. *Natural History* 88(10): 27–3l. (Reprinted in *The panda's thumb* [1980h]. New York: W. W. Norton, 59–68.)

1979v. The politics of sociobiology. *New York Review of Books* 26(9), May 31: 32–35. (B. Bruce, E. Egelman, F. Salzman, H. Inouye, J. Beckwith, J. Alper, L. Leibowitz, R. C. Lewontin, R. Lange, R. Crompton, R. Hubbard, S. J. Gould, & V. Dusek)

1979w. Review of E. Blechschmidt and R. F. Gasser, eds., *Biokinetics and biodynamics of human differentiation: Principles and applications. Quarterly Review of Biology* 54(4): 455–56.

1979x. Making these bones live. *New York Times*, Dec. 9.

1979y. The siabon: interesting but probably not fruitful. *New York Times*, Aug. 5.

1980a. Jensen's last stand. Review of A. R. Jensen, *Bias in mental testing. New York Review of Books* 27(7): 38–40. (Reprinted in *An urchin in the storm* [1987f]. New York: W. W. Norton, 124–44.)

1980b. The promise of paleobiology as a nomothetic, evolutionary discipline. *Paleobiology* 6(1): 96–118.

1980c. Is a new and general theory of evolution emerging? *Paleobiology* 6(1): 119–30.

1980d. The evolutionary biology of constraint. *Daedalus*, 109(2): 39–52.

1980e. Evolution theology, and the Victorian scientist. Review of N. C. Gillespie, *Charles Darwin and the problem of creation. Nature* 285: 343–44.

1980f. Tilly Edinger. In B. Sicherman & C. H. Green, eds., *Notable American women*, 218–19. Cambridge, MA: Harvard University Press.

1980g. Introduction. In B. Kurtén, *The dance of the tiger*, xii–xix. New York: Pantheon.

1980h. *The panda's thumb.* New York: W. W. Norton.

1980i. In an evolutionary crucible. Review of T. D. Moore, *Galapagos: Islands lost in time. New York Review of Books* 27(18): 23–24.

1980j. Clams and brachiopods—ships that pass in the night. *Paleobiology* 6: 383–96. (S. J. Gould & C. B. Calloway)

1980k. Geographic differentiation and speciation in *Cerion*—a preliminary discussion of patterns and processes. *Biological Journal of the Linnaean Society* 14: 389–416. (D. S. Woodruff & S. J. Gould)

1980l. G. G. Simpson, paleontology, and the modern synthesis. In E. Mayr & W. B. Provine, eds., *The evolutionary synthesis*, 153–72. Cambridge, MA: Harvard University Press.

1980m. Wallace's fatal flaw. *Natural History* 89(1): 26–40. (Reprinted [as "Natural selection and the human brain Darwin vs. Wallace"] in *The panda's thumb* [1980h]. New York: W. W. Norton, 47–58.)

1980n. In the midst of life. *Natural History* 89(2): 34–40. (Reprinted in *The panda's thumb* [1980h]. New York: W. W. Norton, 134–42.)

1980o. Death before birth. *Natural History* 89(3): 32–37. (Reprinted in *The panda's thumb* [1980h]. New York: W. W. Norton, 69–75.)

1980p. Dr. Down's syndrome. *Natural History* 89(4): 142–48. (Reprinted in *The panda's thumb* [1980h]. New York: W. W. Norton, 160–68.)

1980q. The first forebear. *Natural History* 89(5): 20–28. (Reprinted [as "Might we fit inside a sponge's cell"] in *The panda's thumb* [1980h]. New York: W. W. Norton, 245–56.)

1980r. The belt of an asteroid. *Natural History* 89(6): 26–33. (Reprinted in *Hen's teeth and horse's toes* [1983d]. New York: W. W. Norton, 320–31.)

1980s. Hen's teeth and horse's toes. *Natural History* 89(7): 24–28. (Reprinted in *Hen's teeth and horse's toes* [1983d]. New York: W. W. Norton, 177–86.)

1980t. The Piltdown conspiracy. *Natural History* 89(8): 8–28. (Reprinted in *Hen's teeth and horse's toes* [1983d]. New York: W. W. Norton, 201–26.)

1980u. Vision with a vengeance. *Natural History* 89(9): 16–20. (Reprinted [as "Our natural place"] in *Hen's teeth and horse's toes* [1983d]. New York: W. W. Norton, 241–50.)

1980v. Hopeful monsters. *Natural History* 89(10): 6–15. (Reprinted [as "Helpful monsters"] in *Hen's teeth and horse's toes* [1983d]. New York: W. W. Norton, 187–98.)

1980w. Chance riches. *Natural History* 89(11): 36–44. (Reprinted in *Hen's teeth and horse's toes* [1983d]. New York: W. W. Norton, 332–42.)

1980x. Science and Jewish immigration. *Natural History* 89(12): 14–19. (Reprinted in *Hen's teeth and horse's toes* [1983d]. New York: W. W. Norton, 291–302.)

1980y. Phyletic size decrease in Hershey bars. In C. J. Rubin, D. Rollert, J. Farago, R. Stark, & J. Etra. *Junk food*, 178–79. New York: Dial Press. (Reprinted in *Hen's teeth and horse's toes* [1983d]. New York: W. W. Norton, 313–19.)

1980z. What is intelligence? Reply to H. J. Eysenck. *New York Review of Books* 27(20): 66–67

1980aa. Jensen and bias: An exchange. Reply to H. J. Eysenck and Nathan P. Glazer. *New York Review of Books* 27(16): 52–53

1980bb. The middle way. Review of P. B. Medawar, *Advice to a young scientist. New York Review of Books* 26(21, 22): 47

1980cc. Introduction. In S. J. Gould, ed., *The evolution of* Gryphaea. New York: Arno Press.

1981a. The ghost of Protagoras. Review of J. T. Bonner, *The evolution of culture in animals* and P. J. Wilson, *Man, the promising primate. New York Review of Books* 27(22, 23): 42–44. (Reprinted in *An urchin in the storm* [1987f]. New York: W. W. Norton, 62–72.)

1981b. *Cerion*, snail of many shells. *Bahamas Naturalist* 5: 2–3.

1981c. *A view of life.* New York: Benjamin/Cummings. (S.E. Luria, S. J. Gould & S. Singer).

1981d. Darwin novelized. Review of I. Stone, *The origin. A biographical novel of Charles Darwin. Science* 211: 270–71.

1981e. Deep time and ceaseless motion. Review of J. McPhee, *Basin and range,* and W. Hobbs, *The Earth generated and anatomized. New York Review of Books* 28(8): 25–28. (Reprinted in *An urchin in the storm* [1987f]. New York: W. W. Norton, 93–103.)

1981f. Evolution as fact and theory. *Discover* 2(5): 34–37. (Reprinted in *Hen's teeth and horse's toes* [1983d]. New York: W. W. Norton, 253–62.)

1981g. *The rise of Neo-Lamarckism in America.* International Colloquium on Lamarck. Paris: Librairie Philosophique J. Vrin.

1981h. Born again creationism. *Science for the People* 13(5): 10–11, 37.

1981i. But not Wright enough: Reply to Orzack. *Paleobiology* 7(1): 131–34.

1981j. Of dinosaurs and asteroids. *Encyclopedia Britannica 1982 Yearbook of Science and the Future,* 122–33.

1981k. Misunderstood monsters. *Psychology Today* 15(12): 97–103.

1981l. *The mismeasure of man.* New York: W. W. Norton. (See also 1996j.)

1981m. The politics of census. *Natural History* 90(1): 20–24. (Reprinted in *Hen's teeth and horse's toes* [1983d]. New York: W. W. Norton, 303–9.)

1981n. Hyena myths and realities. *Natural History* 90(2): 16–24. (Reprinted in *Hen's teeth and horse's toes* [1983d]. New York: W. W. Norton, 147–57.)

1981o. Kingdoms without wheels. *Natural History* 90(3): 42–48. (Reprinted in *Hen's teeth and horse's toes* [1983d]. New York: W. W. Norton, 158–65.)

1981p. A most chilling statement. *Natural History* 90(4): 14–21. (Reprinted [as "A hearing for Vavilov"] in *Hen's teeth and horse's toes* [1983d]. New York: W. W. Norton, 134–44.)

1981q. The titular bishop of Titiopolis. *Natural History* 90(5): 20–24. (Reprinted in *Hen's teeth and horse's toes* [1983d]. New York: W. W. Norton, 69–78.)

1981r. Piltdown in letters. *Natural History* 90(6): 12–30.

1981s. What, if anything, is a zebra? *Natural History* 90(7): 6–12. (Reprinted in *Hen's teeth and horse's toes* [1983d]. New York: W. W. Norton, 355–65.)

1981t. What color is a zebra? *Natural History* 90(8): 16–22. (Reprinted in *Hen's teeth and horse's toes* [1983d]. New York: W. W. Norton, 366–75.)

1981u. Quaggas, coiled oysters, and flimsy facts. *Natural History* 90(9): 16–26. (Reprinted in *Hen's teeth and horse's toes* [1983d]. New York: W. W. Norton, 376–85.)

1981v. A visit to Dayton. *Natural History* 90(10): 8–22. (Reprinted in *Hen's teeth and horse's toes* [1983d]. New York: W. W. Norton, 263–79.)

1981w. The ultimate parasite. *Natural History* 90(11): 7–14. (Reprinted [as "What happens to bodies if genes act for themselves?"] in *Hen's teeth and horse's toes* [1983d]. New York: W. W. Norton, 166–76.)

1981x. Agassiz in the Galapagos. *Natural History* 90(12): 7–14. (Reprinted in *Hen's teeth and horse's toes* [1983d]. New York: W. W. Norton, 107–19.)

1981y. Foreword. In L. Margulis & K. V. Schwartz, *Five kingdoms*, ix–x. New York: W. H. Freeman and Company. (See also 1998z.)

1981z. Palaeontology plus ecology as palaeobiology. In R. M. May, ed., *Theoretical ecology. Principles and applications*, 295–317. 2nd ed. Oxford: Blackwell Scientific Publication. (Revision of 1976c.)

1982a. Change in developmental timing as a mechanism of macroevolution. In J. T. Bonner, ed., *Evolution and development*, 333–46. Dahlem Konferenzen. Berlin: Springer Verlag.

1982b. In praise of Charles Darwin. *Discover* 3(2): 20–25. (Reprinted as Preface in B. Farrington, *What Darwin really said*. New York, Schocken Books, ix–xxi.)

1982c. The quack detector. Review of M. Gardner, *Science: good, bad and bogus*. *New York Review of Books* 29(1): 31–32. (Reprinted in *An urchin in the storm* [1987f]. New York: W. W. Norton, 240–46.)

1982d. Will man become obsolete? Review of R. Jastrow, *The enchanted loom: Mind in the universe*. *New York Review of Books* 29(6): 26–27. (Reprinted [as "The perils of hope"] in *An urchin in the storm* [1987f]. New York: W. W. Norton, 208–15.)

1982e. Exaptation—a missing term in the science of form. *Paleobiology* 8(1): 4–15. (S. J. Gould & E. S. Vrba)

1982f. The meaning of punctuated equilibrium and its role in validating a hierarchical approach to macroevolution. In R. Milkman, ed., *Perspectives on evolution*, 83–104. Sunderland, MA: Sinauer Associates. (Abridgement by Jeremy Cherfas published [as "Punctuated equilibrium. A different way of seeing"] in *New Scientist* 15: 137–41, 1982, and in J. Cherfas, ed., *Darwin up to date*. London: New Scientist Publications, 1982, 20–30.)

1982g. Darwinism and the expansion of evolutionary theory. *Science* 216: 380–87.

1982h. The uses of heresy. Introduction. In R. Goldschmidt, *The material basis of evolution*, xiii–xlii. New Haven: Yale University Press.

1982i. Introduction. In T. Dobzhansky, *Genetics and the origin of species*, xvii–xli. New York: Columbia University Press.

1982j. On paleontology and prediction. *Discover* 3(7): 56–57.

1982k. Creationism: genesis vs. geology. *Atlantic Monthly*, Sept., 10–17.

1982l. The guano ring. *Natural History* 91(1): 12–19. (Reprinted in *Hen's teeth and horse's toes* [1983d]. New York: W. W. Norton, 46–55.)

1982m. Nonmoral nature. *Natural History* 91(2): 19–26. (Reprinted in *Hen's teeth and horse's toes* [1983d]. New York: W. W. Norton, 32–45.)

1982n. Moon, Mann, and Otto. *Natural History* 91(3): 4–10. (Reprinted in *Hen's teeth and horse's toes* [1983d]. New York: W. W. Norton, 280–90.)

1982o. The importance of trifles. *Natural History* 91(4): 16–23. (Reprinted [as "Worm for a century, and all seasons"] in *Hen's teeth and horse's toes* [1983d]. New York: W. W. Norton, 120–33.)

1982p. Hutton's purposeful view. *Natural History* 91(5): 6–12. (Reprinted [as "Hutton's purpose"] in *Hen's teeth and horse's toes* [1983d]. New York: W. W. Norton, 79–93.)

1982q. The stinkstones of Oeningen. *Natural History* 91(6): 6–13. (Reprinted in *Hen's teeth and horse's toes* [1983d]. New York: W. W. Norton, 94–106.)

1982r. The oddball human male. *Natural History* 91(7): 14–22. (Reprinted [as "Big fish, little fish"] in *Hen's teeth and horse's toes* [1983d]. New York: W. W. Norton, 21–31.)

1982s. Free to be extinct. *Natural History* 91(8): 12–16. (Reprinted [as "O grave, where is thy victory?"] in *Hen's teeth and horse's toes* [1983d]. New York: W. W. Norton, 343–52.)

1982t. Solitary origins of some societies. *Natural History* 91(9): 6–12. (Reprinted [as "Quick lives and quirky changes"] in *Hen's teeth and horse's toes* [1983d]. New York: W. W. Norton, 56–65.)

1982u. The Hottentot Venus. *Natural History* 91(10): 20–27. (Reprinted in *The flamingo's smile* [1985y]. New York: W. W. Norton, 291–305.)

1982v. Living with connections. *Natural History* 91(11): 18–22. (Reprinted in *The flamingo's smile* [1985y]. New York: W. W. Norton, 64–77.)

1982w. Of wasps and WASP's. *Natural History* 91(12): 8–15. (Reprinted in *The flamingo's smile* [1985y]. New York: W. W. Norton, 155–66.)

1983a. Losing the edge: The extinction of the .400 hitter. *Vanity Fair*, March, 120, 264–78. (Reprinted [as "Losing the edge"] in *The flamingo's smile* [1985y]. New York: W. W. Norton, 215–29.)

1983b. SETI and the wisdom of Casey Stengel. *Discover* 4(3): 62–65. (Reprinted in *The flamingo's smile* [1985y]. New York: W. W. Norton, 403–13.)

1983c. Utopia, limited. Review of F. Capra, *The turning point. New York Review of Books* 30(3): 23–25. (Reprinted in *An urchin in the storm* [1987f]. New York: W. W. Norton, 216–28.)

1983d. *Hen's teeth and horse's toes*. New York: W. W. Norton.

1983e. Genes on the brain. Review of C. J. Lumsden & E. O. Wilson, *Promethean fire*. *New York Review of Books* 30(11): 5–10. (Reprinted [in *An urchin in the storm* [1987f]. New York: W. W. Norton, 107–23.)

1983f. Dix-huit points au sujet des equilibres ponctués. *Colloque* 330 C.N.R.S., 39–41.

1983g. Calling Dr. Thomas. Review of L. Thomas, *The youngest science: Notes of a medicine-watcher*. *New York Review of Books* 30(12): 12–14. (Reprinted in *An urchin in the storm* [1987f]. New York: W. W. Norton, 189–96.)

1983h. Left holding the bat. *Vanity Fair*, Aug. 27–31.

1983i. Unorthodoxies in the first formulation of natural selection. *Evolution* 37(4): 856–58.

1983j. Irrelevance, submission, and partnership: The changing role of palaeontology in Darwin's three centennials, and a modest proposal for macroevolution. In D. S. Bendall, ed., *Evolution from molecules to men*, 347–66. Cambridge: Cambridge University Press.

1983k. The hardening of the modern synthesis. In M. Greene, ed., *Dimensions of Darwinism. Themes and counterthemes in twentieth-century evolutionary theory*, 71–93. Cambridge Cambridge University Press.

1983l. Thwarted genius. Review of K. R. Manning, *Black Apollo of science: The life of Ernest Everett Just*. *New York Review of Books* 30(18): 3, 6–8. (Reprinted in *An urchin in the storm* [1987f]. New York: W. W. Norton, 169–79.)

1983m. Of crime, cause, and correlation. *Discover* 4(12): 34, 38, 40, 43.

1983n. On original ideas. *Natural History* 92(1): 26–33. (Reprinted [as "Hannah West's left shoulder and the origin of natural selection"] in *The flamingo's smile* [1985y]. New York: W. W. Norton, 335–46.)

1983o. For want of a metaphor. *Natural History* 92(2): 14–19. (Reprinted in *The flamingo's smile* [1985y]. New York: W. W. Norton, 139–51.)

1983p. Unconnected truths. *Natural History* 92(3): 22–28. (Reprinted [as "Freezing Noah"] in *The flamingo's smile* [1985y]. New York: W. W. Norton, 114–25.)

1983q. Opus 100. *Natural History* 92(4): 10–21. (Reprinted in *The flamingo's smile* [1985y]. New York: W. W. Norton, 167–84.)

1983r. Mind and supermind. *Natural History* 92(5): 34–38. (Reprinted in *The flamingo's smile* [1985y]. New York: W. W. Norton, 392–402.)

1983s. A life and death tail. *Natural History* 92(6): 12–16.

1983t. Nature's great era of experiments. *Natural History* 92(7): 12–21. (Reprinted [as "Reducing riddles"] in *The flamingo's smile* [1985y]. New York: W. W. Norton, 245–60.)

1983u. Sex and size. *Natural History* 92(8): 24–27. (Reprinted in *The flamingo's smile* [1985y]. New York: W. W. Norton, 56–63.)

1983v. Darwin at sea. *Natural History* 92(9): 14–20. (Reprinted [as "Darwin at sea—and the virtues of port"] in *The flamingo's smile* [1985y]. New York: W. W. Norton, 347–59.)

1983w. False premise, good science. *Natural History* 92(10): 20–26. (Reprinted in *The flamingo's smile* [1985y]. New York: W. W. Norton, 126–38.)

1983x. Bound by the great chain. *Natural History* 92(11): 20–24. (Reprinted in *The flamingo's smile* [1985y]. New York: W. W. Norton, 281–90.)

1983y. Chimp on the chain. *Natural History* 92(12): 18–27. (Reprinted [as "To show an ape"] in *The flamingo's smile* [1985y]. New York: W. W. Norton, 263–80.)

1983z. Darwin's gradualism. *Systematic Zoology* 32(4): 444–45. (N. Eldredge & S. J. Gould)

1983aa. Long-term biological consequences of nuclear war. *Science* 222: 1283–92. (P. R. Ehrlich, J. Harte, M. A. Harwell, P. H. Raven, C. Sagan, G. M. Woodwell, J. Berry, E. S. Ayensu, A. H. Ehrlich, T. Eisner, S. J. Gould, H. D. Grover, R. Herrera, R. M. May, E. Mayr, C. P. McKay, H. A. Mooney, N. Myers, D. Pimentel, & J. M. Teal)

1984a. Balzan Prize to Ernst Mayr. *Science* 223: 255–57.

1984b. Geology of New Providence Island, Bahamas. *Bulletin of the Geological Society of America* 95(2): 209–20. (P. Garrett & S. J. Gould)

1984c. Sex, drugs, disasters, and the extinction of dinosaurs. *Discover* 5(3): 67–72. (Reprinted in *The flamingo's smile* [1985y]. New York: W. W. Norton, 417–26.)

1984d. Triumph of a naturalist. Review of E. Fox Keller, *A feeling for the organism: The life and work of Barbara McClintock. New York Review of Books* 31(5): 3–6. (Reprinted in *An urchin in the storm* [1987f]. New York: W. W. Norton, 157–68.)

1984e. Splendor in the city. *GEO* 6(3): 29–33.

1984f. Who has donned Lysenko's mantle?—contretemps with B. Davis. *Public Interest* 75: 148–51.

1984g. One man's approach to a basic course in geological sciences. *Journal of Geological Education* 32: 120–22.

1984h. Toward the vindication of punctuational change. In W. A. Berggren & J. A. Van Couvering, eds., *Catastrophes and Earth History*, 9–34. Princeton, NJ: Princeton University Press.

1984i. Covariance sets and ordered geographic variation in *Cerion* from Aruba, Bonaire and Curaçao: A way of studying nonadaptation. *Systematic Zoology* 33(2): 217–37.

1984j. Morphological channeling by structural constraint: Convergence in styles of dwarfing and gigantism in *Cerion*, with a description of two new fossil species and a report on the discovery of the largest *Cerion*. *Paleobiology* 10(2): 172–94.

1984k. The life and work of T. J. M. Schopf (1939–1984). *Paleobiology* 10(2): 280–85.

1984l. A biological comment on Erikson's notion of pseudospeciation. *Yale Review* 73(4) (Summer): 487–90.

1984m. Between you and your genes. Review of R. C. Lewontin, S. Rose, and L. J. Kamin, *Not in our genes: Biology, ideology, and human nature. New York Review of Books* 31(13): 30–32. (Reprinted [as "Nurturing nature"] in *An urchin in the storm* [1987f]. New York: W. W. Norton, 145–54.)

1984n. Evolutionäre flexibilität und menschliches Bewusstsein. In *Evolution und Freiheit.* Civitas Resultate, Band 5. Stuttgart: Hirzel Verlag, 24–35.

1984o. Cambiamenti punctuativi e realta delle species. In M. P. Palmarini, ed., *Livelli di realta,* 256–81. Feltrinelli, Italy.

1984p. Smooth curve of evolutionary rate: A psychological and mathematical artifact (Reply to Gingerich). *Science* 226: 994–995.

1984q. Just in the middle. *Natural History* 93(1): 24–33. (Reprinted in *The flamingo's smile* [1985y]. New York: W. W. Norton, 377–91.)

1984r. The Ediacaran experiment. *Natural History* 93(2): 14–23. (Reprinted [as "Death and transfiguration"] in *The flamingo's smile* [1985y]. New York: W. W. Norton, 230–44.)

1984s. A short way to corn. *Natural History* 93(3): 12–20. (Reprinted in *The flamingo's smile* [1985y]. New York: W. W. Norton, 360–73.)

1984t. Continuity. *Natural History* 93(4): 4–10. (Reprinted in *The flamingo's smile* [1985y]. New York: W. W. Norton, 427–37.)

1984u. Singapore's patrimony (and matrimony). *Natural History* 93(5): 22–29. (Reprinted in *The flamingo's smile* [1985y]. New York: W. W. Norton, 319–32.)

1984v. Adam's navel. *Natural History* 93(6): 6–14. (Reprinted in *The flamingo's smile* [1985y]. New York: W. W. Norton, 99–113.)

1984w. Carrie Buck's daughter. *Natural History* 93(7): 14–18. (Reprinted in *The flamingo's smile* [1985y]. New York: W. W. Norton, 306–18.)

1984x. The cosmic dance of Siva. *Natural History* 93(8): 14–19. (Reprinted in *The flamingo's smile* [1985y]. New York: W. W. Norton, 438–50.)

1984y. Only his wings remained. *Natural History* 93(9): 10–18. (Reprinted in *The flamingo's smile* [1985y]. New York: W. W. Norton, 40–55.)

1984z. The rule of five. *Natural History* 93(10): 14–23. (Reprinted in *The flamingo's smile* [1985y]. New York: W. W. Norton, 199–211.)

1984aa. Human equality as a contingent fact of history. *Natural History* 93(11): 26–33. (Reprinted in *The flamingo's smile* [1985y]. New York: W. W. Norton, 185–98.)

1984bb. A most ingenious paradox. *Natural History* 93(12): 20–29. (Reprinted in *The flamingo's smile* [1985y]. New York: W. W. Norton, 78–95.)

1984cc. Bernhard Kummel. Memorial minute adopted by the Faculty of Arts and Sciences, Harvard University, Dec. 13, 1983. *Harvard University Gazette,* 79(25) (S. J. Gould, C. Frondel, C. S. Hurlbut, R. Siever & J. B. Thompson)

1984dd. Evolutionary theory and human origins. In K. J. Isselbacher, ed., *Medicine, science and society. Symposia celebrating the Harvard medical school bicentennial*, 19–33. New York: John Wiley & Sons.

1984ee. Relationship of individual and group change. Ontogeny and phylogeny in biology. *Human Development* 27: 233–39.

1984ff. Challenges to neo–Darwinism and their meaning for a revised view of human consciousness. The Tanner Lectures on Human Values, Clare Hall, Cambridge University, Apr. 30 and May 1, 1984, 55–73.

1985a. On the origin of specious critics. Review of J. Rifkin, *Algeny. Discover* 6(1): 34–42. (Reprinted [as "Integrity and Mr. Rifkin"] in *An urchin in the storm* [1987f]. New York: W. W. Norton, 229–39.)

1985b. Recording marvels: The life and work of George Gaylord Simpson. *Evolution* 39(1): 229–32.

1985c. An improper taxonomy of death. Review of M. H. Nitecki, ed., *Extinctions. Nature* 313: 505–6.

1985d. Molluscan paleobiology—as we creep and scrape towards the millennium. In D. J. Bottjer, C. E. Hickman, & P. D. Ward, eds., *Mollusks: Paleontological society short courses*, 258–67. Knoxville: University of Tennessee, Department of Geological Sciences, Studies in Geology 13.

1985e. Subjectivity in science. *Encyclopedia Britannica Year Book of Science 1985*, 236–48.

1985f. The paradox of the first tier: An agenda for paleobiology. *Paleobiology* 11(1): 2–12.

1985g. The consequences of being different: Sinistral coiling in *Cerion. Evolution* 39(6): 1364–79. (S. J. Gould, N. D. Young, & B. Kasson)

1985h. Mysteries of the panda. Review of G. B. Schaller, Hu Jinchu, Pan Wenshi, & Zhu Jing, *The giant pandas of Wolong. New York Review of Books* 32(13): 12, 14. (Reprinted [as "How does a panda fit?"] in *An urchin in the storm* [1987f]. New York: W. W. Norton, 19–25.)

1985i. The median isn't the message. *Discover* 6(6): 40–42. (Reprinted in *Bully for Brontosaurus* [1991a]. New York: W. W. Norton, 473–78.)

1985j. All the news that's fit to print and some opinions that aren't. *Discover* 6(11): 86–91.

1985k. The most compelling pelvis since Elvis. *Discover* 6(12): 54–58.

1985l. Phenetics and phylogeny. Review of C. Oxnard, *The order of man: A biomathematical anatomy of the primates. Paleobiology* 11(4): 451–54.

1985m. The sinister and the trivial. *Natural History* 94(1): 16–26.

1985n. Nasty little facts. *Natural History* 94(2): 14–25. (Reprinted [as "A foot soldier for evolution"] in *Eight little piggies* [1993l]. New York: W. W. Norton, 439–55.)

1985o. The flamingo's smile. *Natural History* 94(3): 6–19. (Reprinted in *The flamingo's smile* [1985y]. New York: W. W. Norton, 23–39.)

1985p. A clock of evolution. *Natural History* 94(4): 12–25.

1985q. Red wings in the sunset. *Natural History* 94(5): 12–24. (Reprinted in *Bully for Brontosaurus* [1991a]. New York: W. W. Norton, 209–28.)

1985r. Fleeming Jenkin revisited. *Natural History* 94(6): 14–20. (Reprinted in *Bully for Brontosaurus* [1991a]. New York: W. W. Norton, 340–53.)

1985s. Here goes nothing. *Natural History* 94(7): 12–19. (Reprinted in *Bully for Brontosaurus* [1991a]. New York: W. W. Norton, 294–306.)

1985t. To be a platypus. *Natural History* 94(8): 10–15. (Reprinted in *Bully for Brontosaurus* [1991a]. New York: W. W. Norton, 269–80.)

1985u. Bligh's bounty. *Natural History* 94(9): 2–10. (Reprinted in *Bully for Brontosaurus* [1991a], New York: W. W. Norton, 281–93.)

1985v. Not necessarily a wing. *Natural History* 94(10): 12–25. (Reprinted in *Bully for Brontosaurus* [1991a]. New York: W. W. Norton, 139–51.)

1985w. Geoffroy and the homeobox. *Natural History* 94(11): 12–23.

1985x. Treasures in a taxonomic wastebasket. *Natural History* 94(12): 22–33.

1985y. *The flamingo's smile.* New York: W. W. Norton.

1985z. Comment. *Modern Geology* 9: 314–15.

1986a. Evolution and the triumph of homology, or why history matters. *American Scientist* 74(1): 60–69.

1986b. Fuzzy Wuzzy was a bear, Andy Panda, too. *Discover* 7(2): 40–48.

1986c. A triumph of historical excavation. Review of M. J. S. Rudwick, *The great Devonian controversy: The shaping of scientific knowledge among gentlemanly specialists. New York Review of Books* 33(3): 9–15. (Reprinted [as "The power of narrative"] in *An urchin in the storm* [1987f]. New York: W. W. Norton, 75–92.)

1986d. Evolution and systematics of *Cerion* (Mollusca: Pulmonata) on New Providence Island: A radical revision. *Bulletin of the American Museum of Natural History* 182(4): 389–490. (S. J. Gould & D. S. Woodruff)

1986e. We first stood on our own two feet in Africa, not Asia. *Discover* 7(5): 52–56.

1986f. Reflections from an interior world. Review of P. Medawar, *Memoir of a thinking radish. An autobiography. Nature* 320: 647–48.

1986g. The architectural panorama. *Architecture*, May, 168–71.

1986h. The hierarchical expansion of sorting and selection: Sorting and selection cannot be equated. *Paleobiology* 12(2): 217–28. (E. S. Vrba & S. J. Gould)

1986i. Punctuated equilibrium at the third stage. *Systematic Zoology* 35(1): 143–48. (S. J. Gould & N. Eldredge)

1986j. Cardboard Darwinism. Review of P. Kitcher, *Vaulting ambition,* A. Fausto-Sterling, *Myths of gender,* and B. Kevles, *Females of the species: Sex and survival in the animal kingdom. New York Review of Books* 33(14): 47–54. (Reprinted in *An urchin in the storm* [1987f]. New York: W. W. Norton, 26–50.)

1986k. Entropic homogeneity isn't why no one hits .400 any more. *Discover* 7(8): 60–66.

1986l. Mickey Mantle. *Sport*, Dec., 74–81.

1986m. *Illuminations. A bestiary.* New York: W. W. Norton & Co. (R. W. Purcell & S .J. Gould)

1986n. The nature of the fossil record: A biological perspective. In D. M. Raup & D. Jablonski, eds., *Patterns and processes in the history of life*, 7–22. Dahlem Konferenzen 1986. Berlin: Springer Verlag,. (D. Jablonski, S. J. Gould, & D. M. Raup)

1986o. Directions in the history of life. Group report. In D. M. Raup & D. Jablonski, eds., *Patterns and processes in the history of life*, 47–67. Dahlem Konferenzen 1986. Berlin: Springer Verlag. (D. B. Wake, E. F. Connor, A. J. de Ricqlès, J. Dzik, D.C. Fisher, S. J.Gould, M. LaBarbera, D. A. Meeter, V. Mosbrugger, W.-E. Reif, R. M. Rieger, A. Seilacher, & G. P. Wagner)

1986p. A short way to big ends. *Natural History* 95(1): 18–28.

1986q. Play it again, life. *Natural History* 95(2): 18–26.

1986r. On rereading Edmund Halley. *Natural History* 95(3): 14–21. (Reprinted in *Eight little piggies* [1993]. New York: W. W. Norton, 168–80.)

1986s. Soapy Sam's logic. *Natural History* 95(4): 16–24.

1986t. Knight takes bishop? *Natural History* 95(5): 18–33. (Reprinted in *Bully for Brontosaurus* [1991a]. New York: W. W. Norton, 385–401.)

1986u. Fortitude from heaven. *Natural History* 95(6): 12–18. (Reprinted in *Bully for Brontosaurus* [1991a]. New York: W. W. Norton, 489–98.)

1986v. The egg-a-day barrier. *Natural History* 95(7): 16–24. (Reprinted [as "A dog's life in Galton's polyhedron"] in *Eight little piggies* [1993]. New York: W. W. Norton, 382–95.).

1986w. Linnaean limits. *Natural History* 95(8): 16–23.

1986x. The Archaeopteryx flap. *Natural History* 95(9): 16–25.

1986y. Archetype and adaptation. *Natural History* 95(10): 16–27.

1986z. Of kiwi eggs and the Liberty Bell. *Natural History* 95(11): 20–29. (Reprinted in *Bully for Brontosaurus* [1991a]. New York: W. W. Norton, 109–23.)

1986aa. Glow, big glowworm. *Natural History* 95(12): 10–165. (Reprinted in *Bully for Brontosaurus* [1991a]. New York: W. W. Norton, 255–68.)

1986bb. Review of W. A. Calder III, *Size, function, and life history. Quarterly Review of Biology* 61(1): 77.

1986cc. Whimsey takes the field. *Boston Globe*, Oct. 17.

1986dd. Still in my dinosaur phase. *New York Times Book Review*, Oct. 12, 36.

1986ee. Foreword. In K. Mather, *The permissive universe*, ix–xi. Albuquerque: University of New Mexico Press.

1987a. Darwinism defined: The difference between fact and theory. *Discover* 8(1): 64–70.

1987b. Asymmetry of lineages and the direction of evolutionary time. *Science* 236: 1437–41. (S. J. Gould, N. L. Gilinsky, & R. Z. German)

1987c. *Time's arrow, time's cycle*. Cambridge, MA: Harvard University Press.

1987d. Systematics and levels of covariation in *Cerion* from the Turks and Caicos. *Bulletin of the Museum of Comparative Zoology* 151(6): 321–63. (S. J. Gould & D. S. Woodruff)

1987e. Fifty years of interspecific hybridization: Genetics and morphometrics of a controlled experiment on the land snail *Cerion* in the Florida Keys. *Evolution* 41(5): 1022–45. (D. S. Woodruff & S. J. Gould)

1987f. *An urchin in the storm*. New York: W. W. Norton.

1987g. The terrifying normalcy of AIDS. *New York Times Magazine*, Apr. 19, 32–33.

1987h. Animals and us. Review of J. Serpell, *In the company of animals: A study of human–animal relationship*; V. Hearne, *Adam's task: Calling animals by name*; L. Mighetto, *Among the animals: The wildlife writings of John Muir*; and J. Goodall, *The chimpanzees of Gombe: Patterns of behavior*. *New York Review of Books* 34(11): 20–25.

1987i. The verdict on creationism. *New York Times Magazine*, July 19, 32–33.

1987j. James Hampton's throne and the dual nature of time. *Smithsonian Studies in American Art* 1(1): 47–57.

1987k. The panda's thumb of technology. *Natural History* 96(1): 14–23. (Reprinted in *Bully for Brontosaurus* [1991a]. New York: W. W. Norton, 59–75.)

1987l. Freudian slip. *Natural History* 96(2): 14–21. (Reprinted [as "Male nipples and clitoral ripples"] in *Bully for Brontosaurus* [1991a]. New York: W. W. Norton, 124–38.)

1987m. Hatracks and theories. *Natural History* 96(3): 12–23.

1987n. Life's little joke. *Natural History* 96(4): 16–25. (Reprinted in *Bully for Brontosaurus* [1991a]. New York: W. W. Norton, 168–81.)

1987o. Empire of the apes. *Natural History* 96(5): 20–25. (Reprinted in *Eight little piggies* [1993l]. New York: W. W. Norton, 284–95.)

1987p. Bushes all the way down. *Natural History* 96(6): 12–19.

1987q. Petrus Camper's angle. *Natural History* 96(7): 12–18. (Reprinted in *Bully for Brontosaurus* [1991a]. New York: W. W. Norton, 229–40.)

1987r. An universal freckle. *Natural History* 96(8): 14–20.

1987s. The godfather of disaster. *Natural History* 96(9): 20–29. (Reprinted in *Bully for Brontosaurus* [1991a]. New York: W. W. Norton, 367–81.)

1987t. Justice Scalia's misunderstanding. *Natural History* 96(10): 14–21. (Reprinted in *Bully for Brontosaurus* [1991a]. New York: W. W. Norton, 448–60.)

1987u. William Jennings Bryan's last campaign. *Natural History* 96(11): 16–26. (Reprinted in *Bully for Brontosaurus* [1991a]. New York: W. W. Norton, 416–31.)

1987v. Freud's phylogenetic fantasy. *Natural History* 96(12): 10–197. (Reprinted [as "Freud's evolutionary fantasy"] in *I have landed* [2001m]. New York: Harmony Books, 147–58.)

1987w. Letter. *Natural History* 96(4): 6.

1987x. The lesson of the dinosaurs: Evolution didn't inevitably lead to us. *Discover* 8(3): 51.

1987y. Fossil Endpapers. A parade of shrimp. *Harvard Magazine* 89(1) (Sept.–Oct.): 67. (S. J. Gould & R. Purcell)

1987z. Fossil Endpapers. Animal, vegetable, or mineral? *Harvard Magazine* 89(2) (Nov.–Dec.): 74–75. (S. J. Gould & R. Purcell)

1988a. Pussycats and the owl. Review of H. Ritvo, *The animal astate: The English and other creatures in the Victorian age*, and B. Heinrich, *One man's owl. New York Review of Books* 35(3): 7–10.

1988b. Prolonged stability in local populations of *Cerion agassizi* (Pleistocene–Recent) on Great Bahama Bank. *Paleobiology* 14(1): 1–18.

1988c. Trends as changes in variance: A new slant on progress and directionality in evolution (presidential address). *Journal of Paleontology* 62(3): 319–29.

1988d. The ontogeny of Sewall Wright and the phylogeny of evolution. Review of W. B. Provine, *Sewall Wright and evolutionary biology. Isis* 79: 273–96.

1988e. Mighty Manchester. Review of F. J. Dyson, *Infinite in all directions. New York Review of Books* 35(16): 32, 34–35.

1988f. Strike up the choir! *New York Times Magazine*, Nov. 6, 100–103. (Reprinted [as "Madame Jeanette"] in *Bully for Brontosaurus* [1991a]. New York: W. W. Norton, 201–8.)

1988g. On replacing the idea of progress with an operational notion of directionality. In M. H. Nitecki, ed., *Evolutionary Progress*, 319–38. Chicago: University of Chicago Press.

1988h. The streak of streaks. Review of M. Seidel, *Streak: Joe DiMaggio and the summer of '41. New York Review of Books* 35(13): 8–12. (Reprinted in *Bully for Brontosaurus* [1991a]. New York: W. W. Norton, 463–72.)

1988i. The uses of heterochrony. In M. L. McKinney, ed., *Heterochrony in evolution*, 1–13. New York: Plenum Press.

1988j. The case of the creeping fox terrier clone. *Natural History* 97(1): 16–24. (Reprinted in *Bully for Brontosaurus* [1991a]. New York: W.W. Norton, 155–67.)

1988k. The heart of terminology. *Natural History* 97(2): 24–31.

1988l. Honorable men and women. *Natural History* 97(3): 16–20.

1988m. Pretty pebbles. *Natural History* 97(4): 14–25. (Reprinted [as "Shades of expectation—and actuality"] in *Eight little piggies* [1993l]. New York: W. W. Norton, 409–26.)

1988n. A tale of three pictures. *Natural History* 97(5): 14–21. (Reprinted in *Eight little piggies* [1993l]. New York: W. W. Norton, 427–38.)

1988o. A novel notion of Neanderthal. *Natural History* 97(6): 16–21.

1988p. Kropotkin was no crackpot. *Natural History* 97(7): 12–21. (Reprinted in *Bully for Brontosaurus* [1991a]. New York: W. W. Norton, 325–39.)

1988q. In a jumbled drawer. *Natural History* 97(8): 12–19. (Reprinted in *Bully for Brontosaurus* [1991a]. New York: W. W. Norton, 309–24.)

1988r. Genesis and geology. *Natural History* 97(9): 12–20. (Reprinted in *Bully for Brontosaurus* [1991a]. New York: W. W. Norton, 402–15.)

1988s. A web of tales. *Natural History* 97(10): 16–23.

1988t. The ant and the plant. *Natural History* 97(11): 18–24. (Reprinted in *Bully for Brontosaurus* [1991a]. New York: W. W. Norton, 479–88.)

1988u. Ten thousand acts of kindness. *Natural History* 97(12): 12–17. (Reprinted in *Eight little piggies* [1993l]. New York: W. W. Norton, 275–83.)

1988v. Foreword. In G. Larson, *The Far Side gallery 3*. Riverside, New Jersey: Andrews McMeel Publishing, v–vii.

1988w. Fossil Endpapers. Part, Counterpart. *Harvard Magazine* 90(3) (Jan.–Feb.): 51. (S. J. Gould & R. Purcell)

1988x. Fossil Endpapers. Superposition. *Harvard Magazine* 90(4) (Mar.–Apr.): 47. (S. J. Gould & R. Purcell)

1988y. Fossil Endpapers. Less Stately Mansions. *Harvard Magazine* 90(5) (May–June): 79. (S. J. Gould & R. Purcell)

1988z. Fossil Endpapers. Symmetries. *Harvard Magazine* 90(6) (July–Aug.): 47. (S. J. Gould & R. Purcell)

1988aa. Fossil Endpapers. The Old Red Sandstone. *Harvard Magazine* 90(1) (Sept.–Oct.): 70–71. (S. J. Gould & R. Purcell)

1989a. A way with words. *Nature* 338: 385–86.

1989b. Winning and losing: It's all in the game. *Rotunda*, Spring, 25–3l.

1989c. A developmental constraint in *Cerion*, with comments on the definition and interpretation of constraint in evolution. *Evolution* 43(3): 516–39.

1989d. *Wonderful life: The Burgess Shale and the nature of history*. New York: W. W. Norton.

1989e. Punctuated equilibrium in fact and theory. *Journal of Social and Biological Structures* 12: 117–36.

1989f. Dark outcasts, cast in bronze. *New England Monthly*, Oct., 26–28, 100–102.

1989g. An asteroid to die for. *Discover* 10(10): 60–65.

1989h. Church, Humboldt, and Darwin The tension and harmony of art and science. In F. Kelly, *Frederic Edwin Church*, 94–107. Washington, DC: National Gallery of Art and the Smithsonian Institution Press. (Reprinted [as "Art meets science in The Heart of the Andes: Church paints, Humboldt dies, Darwin writes, and nature blinks in the fateful year of 1859"] in *I have landed* [2001m]. New York: Harmony Books, 90–109.)

1989i. An essay on a pig roast. *Natural History* 98(1): 14–25. (Reprinted in *Bully for Brontosaurus* [1991a]. New York: W. W. Norton, 432–47.)

1989j. Grimm's greatest tale. *Natural History* 98(2): 20–28. (Reprinted in *Bully for Brontosaurus* [1991a]. New York: W. W. Norton, 32–41.)

1989k. The wheel of fortune and the wedge of progress. *Natural History* 98(3): 14–21. (Reprinted in *Eight little piggies* [1993l]. New York: W. W. Norton, 300–312.)

1989l. Tires to sandals. *Natural History* 98(4): 8–15. (Reprinted in *Eight little piggies* [1993l]. New York: W. W. Norton, 313–24.)

1989m. George Canning's left buttock and the *Origin of Species*. *Natural History* 98(5): 18–23. (Reprinted in *Bully for Brontosaurus* [1991a]. New York: W. W. Norton, 21–31.)

1989n. The passion of Antoine Lavoisier. *Natural History* 98(6): 16–25. (Reprinted in *Bully for Brontosaurus* [1991a]. New York: W. W. Norton, 354–66.)

1989o. The chain of reason vs. the chain of thumbs. *Natural History* 98(7): 12–21. (Reprinted in *Bully for Brontosaurus* [1991a]. New York: W. W. Norton, 182–97.)

1989p. The dinosaur rip-off. *Natural History* 98(8): 14–18. (Reprinted in *Bully for Brontosaurus* [1991a]. New York: W. W. Norton, 94–106.)

1989q. Through a lens, darkly. *Natural History* 98(9): 16–24.

1989r. Full of hot air. *Natural History* 98(10): 28–38. (Reprinted in *Eight little piggies* [1993l]. New York: W. W. Norton, 109–20.)

1989s. The creation myths of Cooperstown. *Natural History* 98(11): 14–24. (Reprinted in *Bully for Brontosaurus* [1991a]. New York: W. W. Norton, 42–58.)

1989t. The horn of Triton. *Natural History* 98(12): 18–27. (Reprinted in *Bully for Brontosaurus* [1991a]. New York: W. W. Norton, 499–512.)

1989u. The tragedy of AIDS. *New York Review of Books* 35(21–22): 59–60.

1989v. A touch of glass. *Harvard Magazine* 91(3) (Jan.–Feb.): 44–45. (S. J. Gould & R. Purcell)

1989v. Pickle of the litter. *Harvard Magazine* 91(5) (May–June): 60–61. (S. J. Gould & R. Purcell)

1989w. Consider the lilies. *Harvard Magazine* 91(1) (Sept.–Oct.): 46–47. (S. J. Gould & R. Purcell)

1990a. Taxonomy as politics. *Dissent*, Winter, 73–78.

1990b. Down on the farm. Review of D. O. Henry, *From foraging to agriculture: The Levant at the end of the Ice Age. New York Review of Books* 36(21): 26–27.

1990c. History as a cause of area effects: an illustration from *Cerion* on Great Inagua, Bahamas. *Biological Journal of the Linnaean Society* 40: 67–98. (S. J. Gould & D. S. Woodruff)

1990d. Lack of significant associations between allozyme heterozygosity and phenotypic traits in the land snail *Cerion*. *Evolution* 44(1): 210–13. (C. L. Booth, D. S. Woodruff, & S. J. Gould)

1990e. Speciation and sorting as the source of evolutionary trends; or "things are seldom what they seem." In K. McNamara, ed., *Evolutionary trends*, 3–27. London: Belhaven Press.

1990f. Polished pebbles, pretty shells: an appreciation of OTSOG. In J. Clark, C. Modgil, & S. Modgil, eds., *Robert K. Merton, Consensus and Controversy*, 35–47. London: Falmer Press.

1990g. The virtues of nakedness. Review of H. Seymour, *Baseball: The people's game*, G. F. Will, *Men at work: The craft of baseball*; and L. Heiman et al., *When the cheering stops: Former major leaguers talk about their game and their lives. New York Review of Books* 37(15): 3–7.

1990h. Staying the course with honor. In C. Fadiman, ed., *Living philosophies, The reflections of some eminent men and women of our time*, 138–44. New York: Doubleday.

1990i. The individual in Darwin's world. The Second Edinburgh Medal Address. Edinburgh: Edinburgh University Press.

1990j. Dusty Rhodes. In D. Peary, ed., *Cult baseball players*, 222–25. New York: Simon & Schuster.

1990k. Counters and cable cars. *Natural History* 99(1): 18–23. (Reprinted in *Eight little piggies* [1993l]. New York: W. W. Norton, 238–46.)

1990l. Bully for *Brontosaurus. Natural History* 99(2): 16–24. (Reprinted in *Bully for Brontosaurus* [1991a]. New York: W. W. Norton, 79–93.)

1990m. An earful of jaw. *Natural History* 99(3): 12–23. (Reprinted in *Eight little piggies* [1993l]. New York: W. W. Norton, 95–108.)

1990n. Men of the thirty-third division. *Natural History* 99(4): 12–24. (Reprinted in *Eight little piggies* [1993l]. New York: W. W. Norton, 124–37.)

1990o. Bent out of shape. *Natural History* 99(5): 12–25. (Reprinted in *Eight little piggies* [1993l]. New York: W. W. Norton, 79–94.)

1990p. Everlasting legends. *Natural History* 99(6): 12–17. (Reprinted [as "Literary bias on the slippery slope"] in *Bully for Brontosaurus* [1991a]. New York: W. W. Norton, 241–52.)

1990q. In touch with Walcott. *Natural History* 99(7): 10–19. (Reprinted in *Eight little piggies* [1993l]. New York: W. W. Norton, 220–37.)

1990r. Muller Bros. Moving & Storage. *Natural History* 99(8): 12–16. (Reprinted in *Eight little piggies* [1993l]. New York: W. W. Norton, 198–205.)

1990s. The golden rule—a proper scale for our environmental crisis. *Natural History* 99(9): 24–30. (Reprinted in *Eight little piggies* [1993l]. New York: W. W. Norton, 41–51.)

1990t. Enigmas of the small shellies. *Natural History* 99(10): 6–17. (Reprinted [as "Defending the heretical and the superfluous"] in *Eight little piggies* [1993l]. New York: W. W. Norton, 326–41.)

1990u. Darwin and Paley meet the invisible hand. *Natural History* 99(11): 8–16. (Reprinted in *Eight little piggies* [1993l]. New York: W. W. Norton, 138–52.)

1990v. Shoemaker and morning star. *Natural History* 99(12): 14–20. (Reprinted in *Eight little piggies* [1993l]. New York: W. W. Norton, 206–17.)

1990w. Foreword. In R. M. Ross & W. D. Allmon, eds., *Causes of evolution. A paleontological perspective,* vii–xi. Chicago: University of Chicago Press.

1990x. Foreword. In R. Milner, *The encyclopedia of evolution. Humanity's search for its origins,* v–vi. New York: Facts on File.

1990y. Confessions of a N.Y. Yankees fan. *Boston Globe,* Apr. 9.

1991a. *Bully for Brontosaurus.* New York: W. W. Norton.

1991b. Achieving the impossible dream: Ted Williams and .406. In D. Johnson, ed., *Ted Williams. A portrait in words and pictures,* 65–70. New York: Walker & Company.

1991c. The birth of the two-sex world. Review of T. Laqueur, *Making sex: body and gender from the Greeks to Freud. New York Review of Books* 38(11): 11–13.

1991d. Exaptation: A crucial tool for an evolutionary psychology. *Journal of Social Issues* 47(3): 43–65.

1991e. Armchair revolutionary. *Art & Antiques,* Sept. 1991, 56–61, 94.

1991f. Creativity in evolution and human innovation. In Creativity and culture. The Inaugural Mansfield American–Pacific Lectures, 1989–1990, 11–25; Follow-up dialogue between Stephen Jay Gould and Hiroshi Inose, 49–57.

1991g. Ode to Cecilia. *Boston Globe Magazine,* Oct. 20, 16–26, 31.

1991h. The H and Q of baseball. Review of M. Miller, *A whole different ball game: The sport and business of baseball;* D. Johnson, ed., *Ted Williams: A Portrait in Words and Pictures;* M. Mantle, *My favorite summer 1956;* R. Robinson, *The home run heard 'round the world: The dramatic story of the 1951 Giants–Dodgers pennant race. New York Review of Books* 38(17): 47–52.

1991i. The disparity of the Burgess Shale arthropod fauna and the limits of cladistic analysis: Why we must strive to quantify morphospace. *Paleobiology* 17(4): 411–23.

1991j. Eight (or fewer) little piggies. *Natural History* 100(1): 22–29. (Reprinted [as *"Eight little piggies"*] in *Eight little piggies* [1993l]. New York: W. W. Norton, 63–78.)

1991k. More light on leaves. *Natural History* 100(2): 16–23. (Reprinted in *Eight little piggies* [1993l]. New York: W. W. Norton, 153–65.)

1991l. The great seal principle. *Natural History* 100(3): 4–12. (Reprinted in *Eight little piggies* [1993l]. New York: W. W. Norton, 371–81.)

1991m. What the immaculate pigeon teaches the burdened mind. *Natural History* 100(4): 12–21. (Reprinted in *Eight little piggies* [1993l]. New York: W. W. Norton, 355–70.)

1991n. Abolish the recent. *Natural History* 100(5): 16–21.

1991o. On the loss of a limpet. *Natural History* 100(6): 22–27. (Reprinted in *Eight little piggies* [1993l]. New York: W. W. Norton, 52–60.)

1991p. Of mice and mosquitoes. *Natural History* 100(7): 12–20.

1991q. Opus 200. *Natural History* 100 (8): 12–18.

1991r. Unenchanted evening. *Natural History* 100(9): 4–13. (Reprinted in *Eight little piggies* [1993l]. New York: W. W. Norton, 23–40.)

1991s. The moral state of Tahiti—and of Darwin. *Natural History* 100(10): 12–19. (Reprinted in *Eight little piggies* [1993l]. New York: W. W. Norton, 262–74.)

1991t. Fall in the House of Ussher. *Natural History* 100(11): 12–21. (Reprinted in *Eight little piggies* [1993l]. New York: W. W. Norton, 181–93.)

1991u. The smoking gun of eugenics. *Natural History* 100(12): 8–17. (Reprinted in *Dinosaur in a haystack* [1995k]. New York: Harmony Books, 296–308.)

1991v. Introduction. In D. M. Raup, *Extinction. Bad genes or bad luck?*, xiii–xvii. New York: W. W. Norton.

1991w. The last chance continent. Review of P. Matthiesen, *African silences. New York Times Book Review*, Aug. 18, 3–4.

1992a. The paradox of genius. Review of A. Desmond and J. Moore, *Darwin. Nature* 355: 215–16.

1992b. Form and scale in nature and culture; modern landscape as necessary integration. In *Between home and heaven. Contemporary American landscape photography*, 74–83. Washington, DC: National Museum of American Art, Smithsonian Institution with University of New Mexico Press.

1992c. Ontogeny and phylogeny—revisited and reunited. *BioEssays* 14(4): 275–79.

1992d. Impeaching a self-appointed judge. Review of P. E. Johnson, *Darwin on trial. Scientific American* 226 (July): 118–21.

1992e. Dreams that money can buy. Review of P. Lefcourt, *The Dreyfus affair: A love story*; W. P. Kinsella, *Box Socials*; L. Salisbury, *The Cleveland Indian: The legend of King Saturday*; A. Zimbalist, *Baseball and billions: A probing look inside the big business of our national pastime*; N. J. Sullivan, *The diamond revolution: The prospects for baseball after the collapse of its ruling class*; P. C. Bjarkman, *The Brooklyn Dodgers*; Lawrence S. Ritter, *Lost ballparks: A celebration of baseball's legendary fields*; P. Dixon, *The Negro Baseball Leagues: A photographic history*; D. Weil, D. Halberstam & P. Richmond, *Baseball: The perfect game. New York Review of Books* 39(18): 41–45.

1992f. On "genomenclature": A comprehensive (and respectful) taxonomy for pseudogenes and other "junk DNA." *Proceedings of the National Academy of Sciences USA* 89: 10706–10. (J. Brosius & S. J. Gould)

1992g. Heterochrony. In E. Fox Keller & E. A. Lloyd, eds., *Keywords in evolutionary biology*, 158–65. Cambridge, MA: Harvard University Press.

1992h. The confusion over evolution. Review of H. Cronin, *The ant and the peacock: Altruism and sexual selection from Darwin to today*; N. Eldredge, *The miner's canary*; P. D. Ward, *On Methuselah's trail: Living fossils and the great extinctions. New York Review of Books* 39(19): 47–54.

1992i. What is a species? *Discover* 13(12): 40–44.

1992j. Constraint and the square snail: life at the limits of a covariance set. The normal teratology of *Cerion disforme*. *Biological Journal of the Linnaean Society* 47: 341–47.

1992k. Cambrian and Recent morphological disparity. *Science* 258: 1816. (M. J. Foote & S. J. Gould)

1992l. *Finders, keepers*. New York: W. W. Norton. (R. W. Purcell & S. J. Gould)

1992m. The reversal of *Hallucigenia*. *Natural History* 101(1): 12–20. (Reprinted in *Eight little piggies* [1993l]. New York: W. W. Norton, 342–52.)

1992n. Mozart and modularity. *Natural History* 101(2): 8–16. (Reprinted in *Eight little piggies* [1993l]. New York: W. W. Norton, 249–61.)

1992o. Dinosaur in a haystack. *Natural History* 101(3): 2–13. (Reprinted in *Dinosaur in a haystack* [1995k]. New York: Harmony Books, 147–58.)

1992p. Does the stoneless plum instruct the thinking reed? *Natural History* 101(4): 16–25. (Reprinted in *Dinosaur in a haystack* [1995k]. New York: Harmony Books, 285–95.)

1992q. The most unkindest cut of all. *Natural History* 101(5): 2–11. (Reprinted in *Dinosaur in a haystack* [1995k]. New York: Harmony Books, 309–19.)

1992r. We are all monkeys' uncles. *Natural History* 101(6): 14–21. (Reprinted [as "If kings can be hermits, then we are all monkeys' uncles"] in *Dinosaur in a haystack* [1995k]. New York: Harmony Books, 388–400.)

1992s. A humongous fungus among us. *Natural History* 101(7): 10–18. (Reprinted in *Dinosaur in a haystack* [1995k]. New York: Harmony Books, 335–43.)

1992t. Sweetness and light. *Natural History* 101(8): 10–17. (Reprinted in *Dinosaur in a haystack* [1995k]. New York: Harmony Books, 76–87.)

1992u. Magnolias from Moscow. *Natural History* 101(9): 10–18. (Reprinted in *Dinosaur in a haystack* [1995k]. New York: Harmony Books, 401–11.)

1992v. Life in a punctuation. *Natural History* 101(10): 10–21.

1992w. Red in tooth and claw. *Natural History* 101(11): 14–23. (Reprinted [as "The tooth and claw centennial"] in *Dinosaur in a haystack* [1995k]. New York: Harmony Books, 63–75.)

1992x. Columbus cracks an egg. *Natural History* 101(12): 4–11.

1992y. Foreword. In D. W. Thompson (J. Bonner, ed.), *On growth and form*, ix–xiii. Cambridge: Cambridge University Press.

1993a. Species selection on variability. *Proceedings of the National Academy of Sciences USA* 90: 595–99. (E. A. Lloyd & S. J. Gould)

1993b. Prophet for the Earth. Review of E. O. Wilson, *The diversity of life*. *Nature* 361: 311–12.

1993c. The inexorable logic of the punctuational paradigm: Hugo de Vries on species selection. In D. R. Lees and D. Edwards, eds., *Evolu-

tionary Patterns and Processes, 3–18. London: The Linnaean Society of London.

1993d. The pathway and pattern in the evolution of life. Muratec Human Talk '93, Kyoto, Japan, 1–41.

1993e. Dinomania. Review of *Jurassic Park*, written by Michael Crichton and directed by Steven Spielberg. *New York Review of Books* 40(14): 51–56. (Reprinted in *Dinosaur in a haystack* [1995k]. New York: Harmony Books, 221–37.)

1993f. Flowers of glass. *Discover* 14(11): 92–95.

1993g. Baseball: joys and lamentations. Review of W. Sheed, *My life as a fan*; B. Edwards, *Fridays with Red: A radio friendship*; R. Kahn, *The era, 1947–1957: When the Yankees, the Giants, and the Dodgers ruled the world*; I. Berkow, *The gospel according to Casey: Casey Stengel's inimitable, instructional, historical baseball book*; T. Peyer, ed., *O holy cow! The selected verse of Phil Rizzuto. New York Review of Books* 40(18): 60–65.

1993h. Preface. In S. J. Gould, ed., *The book of life*, 9–12. New York: W. W. Norton. (See also 2001c.)

1993i. Fulfilling the spandrels of world and mind. In J. Selzer, ed., *Understanding scientific prose*, 310–36. Madison: University of Wisconsin Press.

1993j. Punctuated equilibrium comes of age. *Nature* 366: 223–27. (S. J. Gould & N. Eldredge)

1993k. How to analyze Burgess Shale disparity—a reply to Ridley. *Paleobiology* 19(4): 522–23.

1993l. *Eight little piggies*. New York: W. W. Norton.

1993m. A special fondness for beetles. *Natural History* 102(1): 4–12. (Reprinted in *Dinosaur in a haystack* [1995k]. New York: Harmony Books, 377–87.)

1993n. Cordelia's dilemma. *Natural History* 102(2): 10–18. (Reprinted in *Dinosaur in a haystack* [1995k]. New York: Harmony Books, 123–32.)

1993o. Modified grandeur. *Natural History* 102(3): 14–20.

1993p. The first unmasking of nature. *Natural History* 102(4): 14–21. (Reprinted in *Dinosaur in a haystack* [1995k]. New York: Harmony Books, 415–26.)

1993q. Four antelopes of the apocalypse. *Natural History* 102(5): 16–23. (Reprinted in *Dinosaur in a haystack* [1995k]. New York: Harmony Books, 272–81.)

1993r. The invisible woman. *Natural History* 102(6): 14–23. (Reprinted in *Dinosaur in a haystack* [1995k]. New York: Harmony Books, 187–201.)

1993s. Poe's greatest hit. *Natural History* 102(7): 10–19. (Reprinted in *Dinosaur in a haystack* [1995k]. New York: Harmony Books, 173–86.)

1993t. The gift of new questions. *Natural History* 102(8): 3–14.

1993u. Fungal forgery. *Natural History* 102(9): 12–21.

1993v. The Razumovsky duet. *Natural History* 102(10): 10–19. (Reprinted in *Dinosaur in a haystack* [1995k]. New York: Harmony Books, 260–71.)

1993w. The sexual politics of classification. *Natural History* 102(11): 20–29. (Reprinted [as "Ordering nature by budding and full-breasted sexuality"] in *Dinosaur in a haystack* [1995k]. New York: Harmony Books, 427–41.)

1993x. Four metaphors in three generations. *Natural History* 102(12): 12–22. (Reprinted in *Dinosaur in a haystack* [1995k]. New York: Harmony Books, 442–58.)

1993y. Foreword: In O. H. Schindewolf, *Basic questions in paleontology. Geologic time, organic Evolution and biological systematics*, ix–xiv. Chicago: University of Chicago Press.

1993z."Confusion over evolution": an exchange. Reply to Daniel C. Dennett and John Maynard Smith. *New York Review of Books* 40(1–2): 44.

1993aa. Foreword. In J. C. Smith, compiler, *Georges Cuvier: An annotated bibliography of his published works*, vii–xi. Washington, DC: Smithsonian Institution Press.

1993bb. Molecular constructivity. Nature 365: 102. (J. Brosius & S. J. Gould)

1994a. Une rencontre fructueuse entre la science et la fiction. In J.-P. Andrevon, S. Cadelo, S. J. Gould, *L'Homme aux Dinosaures*, 109–19. Paris: Éditions du Seuil.

1994b. Ernst Mayr and the centrality of species. *Evolution* 48(1): 31–35.

1994c. Tempo and mode in the macroevolutionary reconstruction of Darwinism. *Proceedings of the National Academy of Sciences USA* 91: 6764–71.

1994d. A plea and a hope for Martian paleontology. In V. Neal, ed., *Where next, Columbus? The future of space exploration*, 107–27. New York: Oxford University Press.

1994e. The promotion and prevention of recoiling in a maximally snail-like vermetid gastropod: a case study for the centenary of Dollo's Law. *Paleobiology* 20(3): 368–90. (S. J. Gould & B. A. Robinson)

1994f. *Petaloconchus sculpturatus alaminatus*, a new Pliocene subspecies of vermetid gastropods lacking its defining generic character, with comments on vermetid systematics in general. *Journal of Paleontology* 68(5): 1025–36.

1994g. Introduction: The coherence of history. In S. Bengtson, ed., *Early life on Earth*, 1–8. New York: Columbia University Press.

1994h. So near and yet so far. Review of E. Trinkaus and P. Shipman, *The Neandertals: changing the image of mankind*, and C. Stringer and C.Gamble, *In search of the Neanderthals: solving the puzzle of human origins. New York Review of Books* 41(17): 24–28.

1994i. The evolution of life. *Scientific American* 271(4): 63–69.

1994j. Curveball. *New Yorker*, no. 70 (Nov. 28): 139–49.

1994k. The geometer of race. *Discover* 15(11): 65–69. (Reprinted in *I have landed* [2001m]. New York: Harmony Books, 356–66).

1994l. Cabinet museums revisited. *Natural History* 103(1): 12–20. (Reprinted in *Dinosaur in a haystack* [1995k]. New York: Harmony Books, 238–47.)

1994m. In the mind of the beholder. *Natural History* 103(2): 14–23. (Reprinted in *Dinosaur in a haystack* [1995k]. New York: Harmony Books, 93–107.)

1994n. The persistently flat Earth. *Natural History* 103(3): 12–19. (Reprinted [as "The late birth of a flat Earth"] in *Dinosaur in a haystack* [1995k]. New York: Harmony Books, 38–50.)

1994o. Dousing diminutive Dennis's debate. *Natural History* 103(4): 4–12. (Reprinted in *Dinosaur in a haystack* [1995k]. New York: Harmony Books, 11–23.)

1994p. Hooking Leviathan by its past. *Natural History* 103(5): 8–15. (Reprinted in *Dinosaur in a haystack* [1995k]. New York: Harmony Books, 359–76.)

1994q. The power of this view of life. *Natural History* 103(6): 6–8.

1994r. The monster's human nature. *Natural History* 103(7): 14–21. (Reprinted in *Dinosaur in a haystack* [1995k]. New York: Harmony Books, 53–62.)

1994s. Happy thoughts on a sunny day in New York City. *Natural History* 103(8): 10–17. (Reprinted in *Dinosaur in a haystack* [1995k]. New York: Harmony Books, 3–10.)

1994t. Lucy on the Earth in stasis. *Natural History* 103(9): 12–20. (Reprinted in *Dinosaur in a haystack* [1995k]. New York: Harmony Books, 133–44.)

1994u. Jove's thunderbolts. *Natural History* 103(10): 6–12. (Reprinted in *Dinosaur in a haystack* [1995k]. New York: Harmony Books, 159–69.)

1994v. The celestial mechanic and the Earthly naturalist. *Natural History* 103(11): 4–15. (Reprinted in *Dinosaur in a haystack* [1995k]. New York: Harmony Books, 24–37).

1994w. Common pathways of illumination. *Natural History* 103(12): 10–20.

1994x. Faces are special. *The Sciences* 34(1) (Jan./Feb.): 20. (S. J. Gould & R. W. Purcell)

1994y. Pride of place: In Defense of Order. *The Sciences* 34(2) (Mar./Apr.): 24–25. (S. J. Gould & R. W. Purcell)

1994z. Scales of destruction. *The Sciences* 34(3) (May/June): 23–24. (S. J. Gould and R. W. Purcell)

1994aa. Revealing legs. *The Sciences* 34(4) (July/Aug.): 23. (S. J. Gould & R. W. Purcell)

1994bb. Exotic interiors. *The Sciences* 34(5) (Sept./Oct.): 16. (S. J. Gould & R. W. Purcell)

1994cc. Worldly skulls. *The Sciences* 34(6) (Nov./Dec.): 13. (S. J. Gould & R. W. Purcell)

1994dd. Foreword. In C. Cohen, *The fate of the mammoth. Fossils, myth, and history*, xiii–xvii. Chicago: University of Chicago Press.

1995a. Hors du temps dans les jungles pensives. In M. Arent Safir, ed., *Mélancolies du savoir: essais sur l'oeuvre de Michel Rio*, 37–67. Paris: Seuil.

1995b. The Darwinian body. Festschrift for A. Seilacher. *Neues Jahrbuch für Geologie und Paläontologie Abhandlungen* 195(1–3): 267–78.

1995c. Ladders and cones: constraining evolution by canonical icons. In
R. B. Silvers, ed., *Hidden histories of science*, 37–67. New York: *New York
Review of Books*.

1995d. A task for Paleobiology at the threshold of majority. *Paleobiology*
21(1): 1–14.

1995e. Boundaries and categories. In Alexis Rockman, *Second nature*,
26–43. Normal, IL: University Galleries of Illinois State University.

1995f. A wolf at the door. Environmentalism becomes an imperative. In
L. Glennon, ed., *Our times. The illustrated history of the 20th century*,
510–13. Atlanta: Turner Publishing, Inc..

1995g. Jurassic Park. In M. C. Carnes, ed., *Past imperfect. History according
to the movies*, 31–35. New York: Henry Holt & Co.

1995h. 'What is life?' as a problem in history. In M. P. Murphy and L. A.
J. O'Neill, eds., *What is life? The next fifty years*, 25–39. Cambridge:
Cambridge University Press.

1995i. Keynote Address given at conference "In the Company of Animals."
Social Research 62(3): 609–37.

1995j. Three facets of evolution. In J. Brockman and K. Matson, eds.,
How things are: A science toolkit for the mind, 81–86. New York: Wm.
Morrow & Co..

1995k. *Dinosaur in a haystack*. New York: Harmony Books.

1995l. Good sports & bad. Review of J. Bouton (L. Schecter, ed.) *Ball
four*; A. Stump, *My life in Baseball: The true record by Ty Cobb*; A. Stump,
Cobb: A biography; *Cobb*, a film written and directed by R. Shelton;
R. Robinson, *Matty: An American hero, Christy Mathewson and the New
York Giants*; E. Linn, *Hitter: The life and turmoils of Ted* Williams; J. Kruk,
"I ain't an athlete, lady": My well-rounded life and times; M. Ribowsky,
Don't look back: Satchel Paige in the shadows of baseball; N. Dawidoff,
The catcher was a spy: The mysterious life of Moe Berg; N. Trujillo, *The
meaning of Nolan Ryan. New York Review of Books* 42(4): 20–24.

1995m. Of tongue worms, velvet worms, and water bears. *Natural
History* 104(1): 6–15. (Reprinted in *Dinosaur in a haystack* [1995k].
New York: Harmony Books, 108–20.)

1995n. Ghosts of bell curves past. *Natural History* 104(2): 12–19.

1995o. Evolution by walking. *Natural History* 104(3): 10–15. (Reprinted
in *Dinosaur in a haystack* [1995k]. New York: Harmony Books,
248–59.)

1995p. Left snails and right minds. *Natural History* 104(4): 10–18.
(Reprinted in *Dinosaur in a haystack* [1995k]. New York: Harmony
Books, 202–17.)

1995q. Speaking of snails and scales. *Natural History* 104(5): 14–23.
(Reprinted in *Dinosaur in a haystack* [1995k]. New York: Harmony
Books, 344–55.)

1995r. Age–old fallacies of thinking and stinking. *Natural History* 104(6):
6–13. (Reprinted in *I have landed* [2001m]. New York: Harmony Books,
347–55.)

1995s. Spin doctoring Darwin. *Natural History* 104(7): 6–9, 70–7l. (Reprinted [as "Can we complete Darwin's revolution?"] in *Dinosaur in a haystack* [1995k]. New York: Harmony Books, 325–34.)

1995t. Boyle's law and Darwin's details. *Natural History* 104(8): 8–11, 68–71. (Reprinted in *Leonardo's mountain of clams and the diet of worms* [1998x]. New York: Harmony Books, 285–300.)

1995u. Reversing established orders. *Natural History* 104(9): 12–16. (Reprinted in *Leonardo's mountain of clams and the diet of worms* [1998x]. New York: Harmony Books, 393–404.)

1995v. The *Great Western* and the fighting *Temeraire*. *Natural History* 104(10): 16–19, 62–65. (Reprinted in *Leonardo's mountain of clams and the diet of worms* [1998x]. New York: Harmony Books, 45–56.)

1995w. A seahorse for all races. *Natural History* 104(11): 10–15, 72–75. (Reprinted in *Leonardo's mountain of clams and the diet of worms* [1998x]. New York: Harmony Books, 119–40.)

1995x. The anatomy lesson. *Natural History* 104(12): 10–15, 62–63. (Reprinted [as "The clam stripped bare by her naturalists, even"] in *Leonardo's mountain of clams and the diet of worms* [1998x]. New York: Harmony Books, 77–98.)

1995y. Inside dimensions. *The Sciences* 35(1) (Jan./Feb.): 20. (S. J. Gould & R. W. Purcell)

1995z. Two sides to every issue. *The Sciences* 35(3) (May/June): 20–21. (S. J. Gould & R. W. Purcell)

1995aa. Double entendre. *The Sciences* 35(4) (July/Aug.): 16. (S. J. Gould & R. W. Purcell)

1995bb. Placid in plaster. *The Sciences* 35(5) (Sept./Oct.): 32. (S. J. Gould and R. W. Purcell)

1995cc. Report to the librarian. *The Sciences* 35(6) (Nov./Dec.): 18. (S. J. Gould & R. W. Purcell)

1995dd. The case for the kid. Reply to Michael C. D. Macdonald. *New York Review of Books* 42 (15): 57–58.

1995ee. Great sport. Reply to Howard Webb. *New York Review of Books* 42(7): 72.

1995ff. Of it, not above it. *Nature* 377: 681–82.

1995gg. Asking big questions on science and meaning. Review of G. Johnson, *Fire in the mind: Science faith, and the search for order*. *New York Times*, Oct. 16.

1995hh. No more "wretched refuse." *New York Times*, June 7.

1996a. The truth of fiction: An exegesis of G. G. Simpson's dinosaur fantasy. Afterword to G. G. Simpson's *The dechronization of Sam Magruder*, 105–26. New York: St. Martin's Press.

1996b. A life's epistolary drama. Foreword. In F. Burkhardt, *Charles Darwin's letters. A selection 1825–1859*, ix–xx. Cambridge: Cambridge University Press.

1996c. The shape of life. *Art Journal*, Spring, 44–46.

1996d. *Full house: The spread of excellence from Plato to Darwin.* New York: Harmony Books.

1996e. Creating the creators. *Discover* 17(10): 43–54.

1996f. On telling, altering, and enriching stories: An unpublished Darwin letter from a key incident in his later life. In D. Jablonski, D. H. Erwin, and J. H. Lipps, eds., *Evolutionary paleobiology*, 437–60. Chicago: University of Chicago Press.

1996g. Paleontology and chronology of two evolutionary transitions by hybridization in the Bahamian land snail *Cerion. Science* 274: 1894–97. (G. A. Goodfriend & S. J. Gould)

1996h. Why Darwin? Review of J. Browne, *Voyaging: The life of Charles Darwin. New York Review of Books* 43(6): 10–14. (Reprinted [as "A sly dullard named Darwin recognizing the multiple facets of genius"] in *The lying stones of Marrakech* [2000k]. New York: Harmony Books, 169–82.)

1996i. Foreword. In Walter Gratzer, ed., *A bedside Nature: genius and eccentricity in science 1869–1953*, vii–x. London: MacMillan Magazines Ltd..

1996j. *The mismeasure of man.* Revised Edition. New York: W. W. Norton. (See also 1981l.)

1996k. Triumph of the root-heads. *Natural History* 105(1): 10–17. (Reprinted in *Leonardo's mountain of clams and the diet of worms* [1998x]. New York: Harmony Books, 355–74.)

1996l. What separated Darwin's transcendent greatness from James Dwight Dana's merely ordinary greatness? On a toothed bird's place in nature. *Natural History* 105(2): 23–28, 88–93. (Reprinted in *Leonardo's mountain of clams and the diet of worms* [1998x]. New York: Harmony Books, 99–118.)

1996m. Microcosmos. *Natural History* 105(3): 21–23, 66–68.

1996n. Can we truly know sloth and rapacity? *Natural History* 105(4): 18–23, 54–56. (Reprinted in *Leonardo's mountain of clams and the diet of worms* [1998x]. New York: Harmony Books, 375–92.)

1996o. The tallest tale. *Natural History* 105(5): 18–23, 54–57. (Reprinted in *Leonardo's mountain of clams and the diet of worms* [1998x]. New York: Harmony Books, 301–18.)

1996p. Mr. Sophia's pony. *Natural History* 105(6): 20–24, 66–69. (Reprinted in *Leonardo's mountain of clams and the diet of worms* [1998x]. New York: Harmony Books, 141–58.)

1996q. Up against a wall. *Natural History* 105(7): 16–22, 70–73. (Reprinted in *Leonardo's mountain of clams and the diet of worms* [1998x]. New York: Harmony Books, 161–78.)

1996r. A lesson from the old masters. *Natural History* 105(8): 16–22, 58–59. (Reprinted in *Leonardo's mountain of clams and the diet of worms* [1998x]. New York: Harmony Books, 179–96.)

1996s. The diet of worms and the defenestration of prague. *Natural History* 105(9): 18–24, 64–67. (Reprinted in *Leonardo's mountain of clams and the diet of worms* [1998x]. New York: Harmony Books, 251–65.)

1996t. A *Cerion* for Christopher. *Natural History* 105(10): 22–29, 78–79. (Reprinted in *Leonardo's mountain of clams and the diet of worms* [1998x]. New York: Harmony Books, 215–30.)

1996u. The dodo in the caucus race. *Natural History* 105(11): 22–33. (Reprinted in *Leonardo's mountain of clams and the diet of worms* [1998x]. New York: Harmony Books, 231–50.)

1996v. War of the worldviews. *Natural History* 105(12): 22–33. (Reprinted in *Leonardo's mountain of clams and the diet of worms* [1998x]. New York: Harmony Books, 339–54.)

1996w. Brothers under the hair. *The Sciences* 36(1) (Jan./Feb.): 32. (S. J. Gould and R. W. Purcell)

1996x. The yellow leaf road. *The Sciences* 36(3) (May/June): 14. (S. J. Gould & R. W. Purcell)

1996y. Preposterous. *The Sciences* 36(4) (July/Aug.): 19. (S. J. Gould & R. W. Purcell)

1996z. Big birds. *The Sciences* 36(5) (Nov./Dec.): 33. (S. J. Gould & R. W. Purcell)

1996aa. Foreword. In P. Medawar, *The strange case of the spotted mice: and other classic essays on science*, v–xi. Oxford: Oxford University Press.

1996bb. A voice with heart. *New York Times,* June 26.

1997a. Cope's rule as psychological artefact. *Nature* 385: 199–200.

1997b. Bright star among billions. *Science* 275: 599. (Reprinted in *The lying stones of Marrakech* [2000k]. New York: Harmony Books, 237–40.)

1997c. Redrafting the tree of life. *Proceedings of the American Philosophical Society* 141: 30–54.

1997d. Self–help for a hedgehog stuck on a molehill. Review of R. Dawkins, *Climbing Mount Improbable. Evolution* 51: 1020–23.

1997e. The taxonomy and geographic variation of *Cerion* on San Salvador (Bahama Islands). In J. L. Carew, ed., *Proceedings of the 8th Symposium on the Geology of the Bahamas and other Carbonate Regions*, 73–91. Bahamian Field Station, Ltd., San Salvador, Bahamas.

1997 f. The exaptive excellence of spandrels as a term and prototype. *Proceedings of the National Academy of Sciences USA* 94: 10750–55.

1997g. An evolutionary perspective on strengths, fallacies, and confusions in the concept of native plants. In J. Wolschke-Bulmahn, ed., *Nature and ideology*, 11–19. Washington, DC: Dumbarton Oaks Research Library and Collection. (Reprinted [as "An evolutionary perspective on the concept of native plants"] in *I have landed* [2000m]. New York: Harmony Books, 335–46.)

1997h. *Questioning the millennium. A rationalist's guide to a precisely arbitrary countdown.* New York: Harmony Books. (See also 1999o.)

1997i. Darwinian fundamentalism. *New York Review of Books* 44(10): 34–37.

1997j. Evolution: The pleasures of pluralism. *New York Review of Books* 44(11): 47–52.

1997k. Darwinian fundamentalism: An exchange. Reply to Robert Wright and Daniel C. Dennett. *New York Review of Books* 44(13): 64–65.

1997l. Evolutionary psychology: An exchange. *New York Review of Books* 44(15): 55–58.

1997m. As the worm turns. *Natural History* 106(2): 24–27, 68–73. (Reprinted [as "Brotherhood by inversion (or, as the worm turns)"] in *Leonardo's mountain of clams and the diet of worms* [1998x]. New York: Harmony Books, 319–35.)

1997n. Non-overlapping magisteria. *Natural History* 106(3): 16–22, 60–62. (Reprinted in *Leonardo's mountain of clams and the diet of worms* [1998x]. New York: Harmony Books, 269–84.)

1997o. Unusual unity. *Natural History* 106(4): 20–23, 69–71. (Reprinted in *Leonardo's mountain of clams and the diet of worms* [1998x]. New York: Harmony Books, 197–212.)

1997p. Theory of the living Earth. *Natural History* 106(5): 18–21, 58–64. (Reprinted [as "The upwardly mobile fossils of Leonardo's living Earth"] in *Leonardo's mountain of clams and the diet of worms* [1998x]. New York: Harmony Books, 17–44.)

1997q. Dolly's fashion and Louis's passion. *Natural History* 106(6): 18–23, 76. (Reprinted in *The lying stones of Marrakech* [2000k]. New York: Harmony Books, 287–98.)

1997r. Seeing eye to eye. *Natural History* 106(7–8): 14–18, 60–62. (Reprinted [as "Seeing eye to eye, through a glass clearly"] in *Leonardo's mountain of clams and the diet of worms* [1998x]. New York: Harmony Books, 57–73.)

1997s. Drink deep, or taste not the Pierian spring. *Natural History* 106(8): 24–25. (Reprinted in *The lying stones of Marrakech* [2000k]. New York: Harmony Books, 221–26.)

1997t. A tale of two worksites. *Natural History* 106(9): 18–22, 29, 62–68. (Reprinted in *The lying stones of Marrakech* [2000k]. New York: Harmony Books, 251–68.)

1997u. Room of one's own. *Natural History* 106(10): 22, 64–70. (Reprinted in *The lying stones of Marrakech* [2000k], New York: Harmony Books, 347–56).

1997v. The paradox of the visibly irrelevant. *Natural History* 106(11): 12–18, 60–66. (Reprinted in *The lying stones of Marrakech* [2000k]. New York: Harmony Books, 333–46.)

1997w. Pulling teeth. *The Sciences* 37(3) (May/June): 30–32. (S. J. Gould & R. W. Purcell)

1997x. Individuality. *The Sciences* 37(5) (Sept./Oct.): 14–16 (S. J. Gould & R. W. Purcell)

1997y. Foreword: The frontier remains endless. In R. M. Hazen, *Why aren't black holes black? The unanswered questions at the frontiers of science*, ix–xiv. New York: Anchor Books.

1997aa. Today is the day. *New York Times*, Oct. 23.

1997bb. On punctuated equilibria. *Science* 276: 338–339. (N. Eldredge & S. J. Gould)

1998a. The great asymmetry. *Science* 279: 812–13.

1998b. Gulliver's further travels: the necessity and difficulty of a hierarchical theory of selection. *Philosophical Transactions of the Royal Society of London,* B 353: 307–14.

1998c. On transmuting Boyle's Law to Darwin's revolution. In A. C. Fabian, ed., *Evolution society, science and the universe,* 4–27. Cambridge: Cambridge University Press.

1998d. Let's leave Darwin out of it. *New York Times,* May 29. (Reprinted [as "A Darwin for all reasons"] in *I have landed* [2001m]. New York: Harmony Books, 219–20.)

1998e. The amazing dummy. In S. Ware, ed., *Forgotten heroes,* 129–42. New York: Free Press.

1998f. Foreword. Honoring a great career in progress. In P. A. Johnston & J. W. Haggart, eds., *Bivalves: An eon of evolution,* vii–ix. Calgary, Alberta: University of Calgary Press.

1998g. L'évolution de Bibendum. *Parlez-nous de lui,* 98–103.

1998h. The man who invented natural history. Review of J. Roger, *Buffon. New York Review of Books* 45(16): 83–86, 90. (Reprinted [as "Inventing natural history in style"] in *The lying stones of Marrakech* [2000k]. New York: Harmony Books, 75–90.)

1998i. The athlete of the century. *American Heritage* 49(6): 14–17.

1998j. In gratuitous battle. *Civilization,* Oct./Nov., 86–88.

1998k. An awful, terrible dinosaurian irony. *Natural History* 107(2): 24–26, 61–68. (Reprinted in *The lying stones of Marrakech* [2000k]. New York: Harmony Books, 183–200.)

1998l. The internal brand of the scarlet W. *Natural History* 107(3): 22–25, 70–78. (Reprinted in *The lying stones of Marrakech* [2000k]. New York: Harmony Books, 269–86.)

1998m. The lying stones of Würzburg and Marrakech. *Natural History* 107(4): 16–21, 82–90. (Reprinted in *The lying stones of Marrakech* [2000k]. New York: Harmony Books, 9–26.)

1998n. The sharp-eyed lynx outfoxed by nature. Pt. 1. *Natural History* 107(5): 16–21, 70–72. (Reprinted [with 1998o] in *The lying stones of Marrakech* [2000k]. New York: Harmony Books, 27–52.)

1998o. The sharp-eyed lynx outfoxed by nature. Pt. 2. *Natural History* 107(6): 23–27, 69–73. (Reprinted [with 1998n] in *The lying stones of Marrakech* [2000k]. New York: Harmony Books, 27–52.)

1998p. On embryos and ancestors. *Natural History* 107(7/8): 20–22, 58–65. (Reprinted in *The lying stones of Marrakech* [2000k]. New York: Harmony Books, 317–32.)

1998q. Second-guessing the future. *Natural History* 107(9): 20–29, 64–66. (Reprinted in *The lying stones of Marrakech* [2000k]. New York: Harmony Books, 201–16.)

1998r. Above all, do no harm. *Natural History* 107(10): 16–24, 78–82. (Reprinted in *The lying stones of Marrakech* [2000k]. New York: Harmony Books, 299–314.)

1998s. Writing in the margins. *Natural History* 107(11): 16–20. (Reprinted [with 1998t] [as "The proof of Lavoisier's plates"] in *The lying stones of Marrakech* [2000k]. New York: Harmony Books, 91–114.)

1998t. Capturing the center. *Natural History* 107(12): 14–25. (Reprinted [with 1998s] [as "The proof of Lavoisier's plates"] in *The lying stones of Marrakech* [2000k]. New York: Harmony Books, 91–114.)

1998u. Divide and conquer. *The Sciences* 38(1) (Jan./Feb.): 30–32. (S. J. Gould & R. W. Purcell)

1998v. Stretching to fit. *The Sciences* 38(4) (July/Aug.): 12–15. (S. J. Gould & R. W. Purcell)

1998w. The allure of equal halves. *The Sciences* 38(6) (Nov./Dec.): 32–34. (S. J. Gould & R. W. Purcell)

1998x. *Leonardo's mountain of clams and the diet of worms.* New York: Harmony Books.

1998y. Foreword. In C. Pinto-Correia, *The ovary of Eve: Egg and sperm and preformation*, xiii–xviii. Chicago: University of Chicago Press.

1998z. Foreword. In L. Margulis & K. V. Schwartz, *Five kingdoms: An illustrated guide to the phyla of life on Earth*. 3rd ed., xi–xiv. New York: W. H. Freeman. (See also 1981y.)

1998aa. How the new sultan of swat measures up. *New York Times*, Sept. 10.

1998bb. At last, I love a parade. *New York Times*, Oct. 24.

1999a. Tragic optimism for a millennial dawning, commentary. *1999 Britannica Book of the Year*, 6–9. (Encyclopaedia Britannica Online. 13 June, 2008 www.britannica.com/eb/article-9123674>.

1999b. Introduction: The scales of contingency and punctuation in history. In J. Bintliff, ed., *Structure and contingency*, ix–xxii. London: Leicester University Press.

1999c. No science without fancy, no art without facts: The lepidoptery of Vladimir Nabokov. In G. Horowitz, ed., *Véra's butterflies by Sarah Funke*, 84–114. New York: Glenn Horowitz Bookseller, Inc.. (Reprinted in *I have landed* [2001m]. New York: Harmony Books, 29–53.)

1999d. Direct measurement of age in fossil *Gryphaea*: The solution to a classic problem in heterochrony. *Paleobiology* 25(2): 158–87. (D. S. Jones & S. J. Gould)

1999e. Darwin's more stately mansion. *Science* 284: 2087. (Reprinted in *I have landed* [2001m]. New York: Harmony Books, 216–18.)

1999f. The human difference. *New York Times*, July 2.

1999g. Dorothy, it's really Oz. Viewpoint. *Time*, Aug. 23, 59. (Reprinted [as "Darwin and the munchkins of Kansas"] in *I have landed* [2001m]. New York: Harmony Books, 213–15.)

1999h. Message from a mouse. Viewpoint. *Time*, Sept. 13, 62. (Reprinted [as "The without and within of smart mice"] in *I have landed* [2001m]. New York: Harmony Books, 233–34.)

1999i. Fiat money and booby birds. *Forbes* 164(9): 90–91.

1999j. Individuality and adaptation across levels of selection: how shall we name and generalize the unit of Darwinism? *Proceedings of the National Academy of Sciences USA* 96(21): 11904–909. (S. J. Gould & E. A. Lloyd)

1999k. Of two minds and one nature. *Science* 286: 1093–94. (R. R. Shearer & S. J. Gould)

1999l. The green box stripped bare: Marcel Duchamp's 1934 "Facsimiles" yield surprises. *Tout-fait News* 1(1): 1–8 (R. R. Shearer & S. J. Gould)

1999m. Hidden in plain sight: Duchamp's 3 standard stoppages, more truly a "stoppage" (an invisible mending) than we ever realized. *Tout-fait News* 1(1): 1–8 (R. R. Shearer & S. J. Gould)

1999n. *Rocks of ages: science and religion in the fullness of life.* New York: Ballantine.

1999o. New preface for 2nd ed. of *Questioning the Millennium,* 13–49. New York: Harmony Books. (See also 1997h.)

1999p. Il mito teleologico. *Argonauti nella noosfera.* Vol. 2, no. 118, Dec., 86–91.

1999q. A division of worms. Pt. 1. *Natural History* 108(2): 18–22, 76–81. (Reprinted [with 1999r] [as "A tree grows in Paris: Lamarck's division of worms and revision of nature"] in *I have landed* [2001m]. New York: Harmony Books, 115–43.)

1999r. Branching through a wormhole. Pt. 2. *Natural History* 108(3): 24–27, 84–89. (Reprinted [with 1999q] [as "A tree grows in Paris: Lamarck's division of worms and revision of nature"] in *I have landed* [2001m]. New York: Harmony Books, 115–43.)

1999s. Lyell's pillars of wisdom. Pt. 1. *Natural History* 108(4): 28–34, 87–89. (Reprinted [with 1999t] [as "Lyell's pillars of wisdom"] in *The lying stones of Marrakech* [2000k]. New York: Harmony Books, 147–68.)

1999t. Pozzuoli's pillars revisited. Pt. 2. *Natural History* 108(5): 24, 81–83, 88–91. (Reprinted [with 1999s] [as "Lyell's pillars of wisdom"] in *The lying stones of Marrakech* [2000k]. New York: Harmony Books, 147–68.)

1999u. Bacon, brought home. *Natural History* 108(6): 28–33, 72–78. (Reprinted [as "How the vulva stone became a brachiopod"] in *The lying stones of Marrakech* [2000k]. New York: Harmony Books, 53–71.)

1999v. The great physiologist of Heidelberg. *Natural History* 108(7/8): 26–29, 62–70. (Reprinted in *I have landed* [2001m]. New York: Harmony Books, 367–83.)

1999w. A Darwinian gentleman at Marx's funeral. *Natural History* 108(9): 32–33, 56–66. (Reprinted in *I have landed* [2001m]. New York: Harmony Books, 113–29.)

1999x. When fossils were young. *Natural History* 108(10): 24–26, 70–75. (Reprinted in *I have landed* [2001m]. New York: Harmony Books, 175–91.)

1999y. The pre-Adamite in a nutshell. *Natural History* 108(11): 24–27, 72–77. (Reprinted in *I have landed* [2001m]. New York: Harmony Books, 130–46.)

1999z. Boats and deckchairs. *Natural History* 108(12): 32–44. (S. J. Gould & R. R. Shearer)

1999aa. Introduction to the revised edition. In J. Goodall, *In the shadow of man*, v–viii. Gloucester, MA: Peter Smith Publisher.

1999aa. Layering. *The Sciences*, Mar./Apr., 39(2), 30–32. (S. J. Gould & R. W. Purcell)

1999bb. Infiltration. *The Sciences*, July/Aug., 39(4), 32–34. (S. J. Gould & R. W. Purcell)

1999cc. Transform or perish. *The Sciences* 39(6) (Nov./Dec.): 30–32. (S. J. Gould & R. W. Purcell)

1999dd. Foreword. In S. G. Kohlstedt, M. M. Sokal & B. V. Lewenstein, eds., *The establishment of science in America: 150 years of the American Association for the Advancement of Science*, vii–x. New Brunswick: Rutgers University Press.

1999ee. Take another look. *Science* 286: 899.

2000a. Deconstructing the "science wars" by reconstructing an old mold. *Science* 287: 253–61.

2000b. The true embodiment of everything that's excellent. *American Scholar* 69: 35–49. (Reprinted in *I have landed* [2001m]. New York: Harmony Books, 71–89.)

2000c. The substantial ghost: towards a general exegesis of Duchamp's artful wordplays. *Tout-fait News* 1(2): 1–21.

2000d. Scales of change. In *Forces of change: A new view of nature*, 95–111. Washington, DC: Smithsonian Press & The National Geographic Society.

2000e. Will we figure out how life began? *Time* 155(14): 93.

2000f. Of coiled oysters and big brains: how to rescue the terminology of heterochrony, now gone astray. *Evolution and Development* 2: 241–48.

2000g. Only human. *Forbes ASAP*, Oct. 2, 261–64.

2000h. Beyond competition. *Paleobiology* 26: 1–6.

2000i. Indecision 2000: Heads or tails? *Boston Globe*, Nov. 30.

2000j. Drawing the maxim from the minim: the unrecognized source of Niceron's influence upon Duchamp. *Tout-fait News* 1(3): 1–12.

2000k. *The lying stones of Marrakech*. New York: Harmony Books

2000l. *Crossing over: Where art and science meet*. New York: Three Rivers Press. (S. J. Gould & R. W. Purcell)

2000m. What does the dreaded "E" word mean, anyway? *Natural History* 109(1): 28–44. (Reprinted in *I have landed* [2000m]. New York: Harmony Books, 241–56.)

2000m. Abscheulich! (atrocious!). *Natural History* 109(2): 42–49. (Reprinted in *I have landed* [2001m]. New York: Harmony Books, 305–20.)

2000n. The first day of the rest of our life. *Natural History* 109(3): 32–38, 82–86. (Reprinted in *I have landed* [2001m]. New York: Harmony Books, 257–70.)

2000o. Jim Bowie's letter and Bill Buckner's legs. *Natural History* 109(4): 26–40. (Reprinted in *I have landed* [2001m]. New York: Harmony Books, 54–70.)

2000p. The Jew and the Jewstone. *Natural History* 109(5): 26–39. (Reprinted in *I have landed* [2001m]. New York: Harmony Books, 161–74.)

2000q. The narthex of San Marco and the pangenetic paradigm. *Natural History* 109(6): 24–37. (Reprinted in *I have landed* [2001m]. New York: Harmony Books, 271–84.)

2000r. Linnaeus's luck? *Natural History* 109(7): 18–25, 66–76. (Reprinted in *I have landed* [2001m]. New York: Harmony Books, 287–304.)

2000s. Syphilis and the shepherd of Atlantis. *Natural History* 109(8): 38–42, 74–82. (Reprinted in *I have landed* [2001m]. New York: Harmony Books, 192–207.)

2000t. Tails of a feathered tail. *Natural History* 109(9): 32–42. (Reprinted in *I have landed* [2001m]. New York: Harmony Books, 321–32.)

2000u. *I have landed. Natural History* 109(10): 46–59. (Reprinted in *I have landed* [2001m]. New York: Harmony Books, 13–25.)

2000v. Endnote on a new beginning. *Natural History* 109(10): 118.

2000w. Typecasting a bit part. *The Sciences* 40(2) (Mar./Apr.): 34–35. (S. J. Gould & R. W. Purcell)

2000x. Dance, dance, wherever you may be. *The Sciences* 40(4) (July/Aug.): 35. (S. J. Gould & R. W. Purcell)

2000y. Introduction. In E. Asinof, *Eight men out: The Black Sox and the 1919 World Series*, xv–xviii. New York: W. H. Freeman/Owl Books.

2000z. The natural status of science as literature. Introduction to the Modern Library Science Series. In F. Bacon, *The advancement of learning* (The Oxford Francis Bacon), xi–xxxv. Oxford: Oxford University Press.

2000aa. Foreword. In M. Bekoff, ed., *The smile of a dolphin: Remarkable accounts of animal emotions*, 13–21. New York: Discovery Books.

2000bb. Foreword. In L. Chang, ed., *Scientists at work: Profiles of today's ground-breaking scientists from Science Times*, xi–xiii. New York: McGraw-Hill.

2000cc. Freud at the ballpark. *New York Times*, Oct. 19.

2000cc. The brain of brawn. *New York Times*, June 25.

2000dd. A taxonomist's taxonomist. *Whole Earth*, Fall, 53–58.

2001a. Size matters and function counts. In D. Falk & K. R. Gibson, eds., *Evolutionary anatomy of the primate cerebral cortex*, xiii–xvii. Cambridge: Cambridge University Press.

2001b. Contingency. In D. E. G. Briggs & P .R. Crowther, eds., *Palaeobiology 2*, 195–98. London: Blackwell Science.

2001c. A flawed work in progress. A new introduction. In S. J. Gould, ed., *The book of life*, 1–21. Rev. ed. New York: W. W. Norton. (See also 1993h.)

2001d. Humbled by the genome's mysteries. *New York Times*, Feb. 19. (Reprinted [as "When less is truly more"] in *I have landed* [2001m]. New York: Harmony Books, 225–26.)

2001e. What only the embryo knows. *New York Times*, Aug. 27.

2001f. The good people of Halifax. *Globe and Mail* (Toronto), Sept. 20. (Reprinted in *I have landed* [2001m]. New York: Harmony Books, 389–92.)

2001g. Apple brown betty. *New York Times*, Sept. 26. (Reprinted in *I have landed* [2001m]. New York: Harmony Books, 393–95.)

2001h. The evolutionary definition of selective agency, validation of the theory of hierarchical selection, and fallacy of the selfish gene. In R. S. Singh, C. B. Krimbas, D. B. Paul, & J. Beatty, eds., *Thinking about Evolution*, 208–34. Vol. 2. Cambridge: Cambridge University Press.

2001i. The interrelationship of speciation and punctuated equilibrium. In J. B. C. Jackson, S. Lidgard, & F. K. McKinney, eds., *Evolutionary patterns*, 196–217. Chicago: University of Chicago Press.

2001j. Introduction. In C. Zimmer, *Evolution: The triumph of an idea*, ix–xiv. New York: HarperCollins.

2001k. The man who set the clock back. Review of S. Winchester, *The map that changed the world: William Smith and the birth of modern geology. New York Review of Books* 48(15): 51–56.

2001l. The inner eye of outer reality. *The Sciences* 41(1) (Jan./Feb.): 33. (S. J. Gould & R. W. Purcell)

2001m. *I have landed.* New York: Harmony Books.

2001n. Foreword: *Life through our ages.* In C. R. Knight, *Life through the ages*, vii–x. Commemorative ed. Bloomington: Indiana University Press.

2001o. Foreword. In B. J. Alters & S. M. Alters, *Defending evolution: A guide to the evolution/creation controversy*, 1–4. Sudbury, MA: Jones and Bartlett Publishers.

2001p. Sept. 11, '01. *Boston Globe*, Sept. 30. (Reprinted in *I have landed* [2001m]. New York: Harmony Books, 399–401.)

2001q. A time of gifts. *New York Times*, Sept. 26.

2002a. Both neonate and elder: the first fossil of 1557. Conceptual index fossils: No. 1. *Paleobiology* 28(1): 1–8.

2002b. Drawing a gloriously false inference. Conceptual index fossils: No. 2. *Paleobiology* 28(2): 179–83.

2002c. *The structure of evolutionary theory.* Cambridge, MA: Harvard University Press.

2002d. Foreword. The positive power of skepticism. In M. Shermer, *Why people believe weird things: Pseudoscience, superstition, and other confusions of our time*, ix–xii. New York: W. H. Freeman/Owl Books.

2002e. Ground zero. *Antioch Review* 60(1): 150–54.

2002f. Foreword. In M. A. Mares, *A desert calling. Life in a forbidding landscape*, ix–xv. Cambridge, MA: Harvard University Press.

2002g. Preface. In A. Berry, ed., *Infinite tropics: An Alfred Russel Wallace anthology*, xi–xiv. New York: Verso.

2002h. Introduction: To open a millennium. In S. J. Gould & R. Atwan, eds., *Best American essays 2002*, xiii–xvii. Boston: Houghton Mifflin.

2002i. Macroevolution. In M. Pagel, ed., *Encyclopedia of evolution*, E23–E28. Vol. 1. New York: Oxford University Press.

2002j. Punctuated equilibrium. In M. Pagel, ed., *Encyclopedia of evolution*, 963–68. Vol. 2. New York: Oxford University Press.

2003a. *Triumph and tragedy in Mudville: A lifelong passion for baseball.* New York: W. W. Norton.

2003b. *The hedgehog, the fox, and the magister's pox: Mending the gap between science and the humanities.* New York: Harmony Books.

2004. Father Athanasius on the isthmus of a middle state: Understanding Kirchner's paleontology. In P. Findlen, ed., *Athanasius Kirchner. The last man who knew everything*, 207–37. New York: Routledge.

2006a. *The richness of life: The essential Stephen Jay Gould.* P. McGarr and S. Rose, eds., London: Jonathan Cape.

2006b. *The richness of life: The essential Stephen Jay Gould.* S. Rose, ed. New York: W. W. Norton & Co.

2007. *Punctuated equilibrium.* Cambridge, MA: Harvard University Press.

Notes

Chapter 1

1. Steve seemed impatient with the work of others in which he could not perceive a coherent theme or thread of larger message. I recall that as a student, I heard him say that he had just received Nitecki's book *Extinctions* to review for *Nature*, but he couldn't figure out what united the chapters and so was (uncharacteristically) delaying writing the review. (He eventually decided that his theme was that there was no theme—titling his review [1985c] "An improper taxonomy of death"—which he felt was an important substantive conclusion consistent with his emerging ideas of the dissociation of mass and background extinction.) Similarly, he wrote in the introduction to a volume based on an international symposium on Precambrian life that he had he had struggled with "what I perceived as a lack of cohesion among the varied topics of this conference..." (1994g, 1). He eventually found it (inspired by wandering in Alfred Nobel's library): "Our point of union is the greatest and, in many ways, the most obvious subject of all: Nobel's personal favorite of *history itself*...Without this integration," he said, tellingly, "the papers of this conference make no sense as a unified inquiry. With this perspective, we convert a disparate set of studies (whatever their individual excellences) into a worthy, integral fabric" (emphasis in original) (1994g, 2–3).

2. Stearns (2002, 2345) suggests that the reverse might be true: "I now frequently encounter biology undergraduates, nonacademics, and even some nonevolutionary biologists who have read Steve and carry quite alarming misapprehensions about natural selection, adaptation, and microevolutionary mechanisms, but insist that they have understood evolution at the hand of a master and need go no

further." The question of who Steve's fans as opposed to his critics were and are (for example, what kind of educational and professional background they have) would make an interesting research topic in the sociology of science.

3. Steve's legendary productivity must be considered in any interpretation of his work and his legacy. Although he did leave work undone, he surely published a larger proportion of what he thought about than most capable people, at least in the fields with which I am familiar. This suggests that a comparison of his worldview to that of other scientists may be skewed simply by Steve's staggering volume of output. That is, many people have lots of interesting ideas, but Steve actually had the discipline and energy to write most of them out coherently. Perhaps many scientists have equally expansive, coherent, and multifaceted worldviews, but don't put them on paper. (See Shermer [2002] for a partial analysis of Steve's published writings by topic.)

4. Steve said that he thought one reason that so few people (by his estimation) had actually read Eldredge and Gould (1972) is because it was published in a relatively obscure and hard-to-find edited volume of a small press.

5. "The proponents of the synthetic theory maintain that all evolution is due to the accumulation of small genetic changes, guided by natural selection, and that transpecific evolution is nothing but an extrapolation and magnification of the events that take place within populations and species" (Mayr 1963, 586).

6. In a letter to Sewall Wright, dated January 26, 1981, Steve wrote: "My defense of Goldschmidt in *Natural History* [1977s] was only for his general concept of non-extrapolation across evolutionary levels, and not for hopeful monsters in his terms. I rejected his systemic mutations as based on the incorrect idea that the entire genome is physically interconnected so that point mutations, though mappable to a spot, can literally affect the entire genome. I did support the macro-effect upon phenotypes caused by small genetic changes that operate to change rates of development early in ontogeny.... But I supported them only for the discontinuous origin of key features, not for the sudden origin of entire taxa—for I believe that a successful new Bauplan must also be based on a large number of subsequent, adaptive accommodations to such a new key feature.... Ironically, a mistake was made in printing the title which, to my embarrassment, came out as 'hopeful monsters,' when it was supposed to be a sardonic commentary on Goldschmidt's phrase—'helpful monsters,' i.e., hopeless as newly evolved forms but helpful to us in our efforts to understand genetics and evolution."

7. Steve argued frequently that despite all of Darwin's fame, we have not yet fully come to grips with his "outrage" and accepted the personal and philosophical implications of natural selection (e.g., 1977f, 1995s, 1996d).

8. "I can access everything I've read," he once told an interviewer (Lessem 1986, 94). In another interview, not long before his death,

he was asked: "You quote so extensively from so many different works of literature. Do you honestly recall those passages?" Steve responded: "I grew up on baseball statistics. Everybody's got funny little skills. I know where to find things. It may not be in my head, but anything I've ever read, I could find. [Pause.] Well, that's ridiculous, of course. That's not literally true, but I think I pretty much can access things that I've come across. I don't have photographic memory, but I do have a good sense of where I've found something and where I can get it" (Monastersky 2002, A17).

9. Alexander Pope, *Essay on Man* (1734), Epistle 1, stanza viii:

> From Vast chain of being! which from God began;
> Natures ethereal, human, angel, man,
> Beast, bird, fish, insect, who no eye can see,
> No glass can reach; from infinite to thee;
> From thee to nothing.—On superior powers
> Were we to press, inferior might on ours;
> Or in the full creation leave a void,
> Where, one step broken, the great scale's destroyed:
> From Nature's chain whatever link you like,
> Tenth, or ten thousandth, breaks the chain alike.

10. Ruse (2001, 10) writes tellingly that Steve always struck him "as being closer to God than many conventional believers."

11. Although the Dinamation Corporation, which pioneered such exhibits, is no more (see "The Decline of the Dinamation Dinos: How One Man's Robots Became Passe," *Wall Street Journal*, May 21, 2001), the principle remains the same, perhaps even more so, with the advent of more sophisticated robotics and computerized special effects.

12. My memory of this paper also includes the somewhat surreal experience of watching Steve lecture on it—virtually reading the text of the paper—to his large lecture class of more than 300 mostly nonscience majors, and then of me and my fellow teaching assistants struggling to explain hierarchical evolutionary theory to some *very* puzzled Harvard undergraduates.

Chapter 4

1. Steve mentioned to one of us (WDA) more than once that if this was the best that paleoecology could do then it was clearly not going to be a very important part of evolutionary biology.

2. WDA, with Steve's encouragement, spent five months in the Brazilian Amazon in 1985 studying forest floor herpetofauna (see Allmon 1991), and Steve once called him "our resident ecologist," obviously not necessarily a compliment.

3. This passage is reminiscent of a famous description of Darwin's reaction to tropical rainforest, being amazing in its natural state but also enhanced by human modification, in *The Voyage of the Beagle*: "The land is one great wild, untidy, luxuriant hothouse, made by Nature for herself, but taken possession of by man, who has studded it with gay houses and formal gardens" (Darwin 1860, 494).

4. In the mid-1980s, when one of us (WDA) introduced as a section topic in Steve's large lecture course, Science B-16 (see Ross, this volume), the connection between modern extinction due to tropical deforestation and past extinctions in geological time (e.g., Lewin 1983), and it was soon adopted by other teaching assistants in the course, Steve exclaimed (with some mixture of real and mock dismay) during a TA meeting that the topic was "sweeping through the course like a transposon!"

Chapter 6

1. This statement is in no way meant to be pejorative; I respected Steve immensely as an advisor, teacher, and thinker and I came to know and like him as a person. He was proud of his accomplishments to the point of arrogance, something he freely admitted, but he was also the hardest working person I've ever met. Nevertheless, the length and digressive nature of *The Structure of Evolutionary Theory* will deter many reading it through.
2. According to Steve, Jack Sepkoski started his compendia as part of a lab group discussion when he was Steve's student in the mid '70s.
3. Steve enjoyed telling this story a lot and maybe it was true, but it also elides Walter Alvarez's and Earle Kauffman's (1984) primary roles as developers of the general mass extinction theory.
4. Forbidding in its imaginary physical aspect, but, to my mind, forbidding to differing viewpoints, too.

Chapter 7

1. This chapter was originally published in 2002 in vol. 54 of the *Monthly Review* and is reprinted here with the permission of the authors.

Chapter 8

1. Originally published in *Biology and Philosophy* 19 (2004): 1–15.
2. His activities in this regard were not so much a matter of writing articles—although "Darwin's Untimely Burial" (1976q) was an early attempt to squash some popular mainstays of the nascent creationist movement—as in giving lectures and interviews, especially in popular magazines, newspapers, radio, and television, and, of course, in testifying in the Little Rock trial.
3. Since I am again praising this book, it is worth responding briefly to the recent complaints by Neven Sesardic to the effect that Steve's claims about craniometry have been refuted, and that philosophers have been credulous in following Steve and overlooking the refutation (Sesardic 2000, 2003). The truth of the matter is that Steve's interpretations of Samuel Morton's cranial data have been *questioned* by John S. Michael, who, as an undergraduate student at Macalester College, remeasured the skulls as part of an honors project (Michael 1988).

It is not entirely evident that one should prefer the measurements of an undergraduate to those of a professional paleontologist whose own specialist work included some very meticulous measurements of fossil snails. But Sesardic leaps from the relatively modest differences between Steve's measurements and Michael's to a much less nuanced conclusion than that which Michael himself drew—Steve, he believes, is clearly incorrect and has misled people in a number of fields. So far as I have been able to discover, virtually nobody has reacted to Michael's article by seeing it as a refutation of Steve—with two major exceptions: it is used in this way in Herrnstein and Murray (1994) and is much ballyhooed by J. Philippe Rushton (indeed, an internet search for citations of Michael led me quickly to various sites that feature Rushton's highly controversial claims about race, and to virtually nothing else). Sesardic seems much concerned to assign to Michael a heroic role that Michael himself does not claim and that remarkably few others seem to envisage for him. Pending further measurement of the skulls and further analysis of the data, it seems best to let this grubby affair rest in a footnote.

4. I am grateful to an astute member of the audience at the *Philosophy of Science Association* meeting for raising this point.

Chapter 11

1. Unpublished lecture notes, Science B-16, Harvard University, 1985.
2. William Wordsworth and S. T. Coleridge, *Lyrical Ballads* (London: J. and A. Arch, 1798). No. 4 (Victoria College Library, Toronto).
3. Unpublished lecture notes, Science B-16, Harvard University, 1985.
4. Doug Brown, "Stephen Jay Gould. From Brachiopods to Baseball." www.powells.com/authors/gould.html
5. In a stimulating collaborative endeavor, Steve and his teaching fellows always wrote exam questions together.
6. "The Validation of Continental Drift," the essay in which he made his prediction, appeared in *Ever Since Darwin*, not *The Panda's Thumb*. Since Steve wrote so much and cared little about keeping track of his publication record, this mistake is not surprising.

Index